Distributed Control of Robotic Networks

Princeton Series in Applied Mathematics

Editors: Ingrid Daubechies (Princeton University); Weinan E (Princeton University); Jan Karel Lenstra (Eindhoven University); Endre Suüli (University of Oxford)

The Princeton Series in Applied Mathematics publishes high quality advanced texts and monographs in all areas of applied mathematics. Books include those of a theoretical and general nature as well as those dealing with the mathematics of specific applications areas and real-world situations.

Chaotic Transitions in Deterministic and Stochastic Dynamical Systems Applications of Melnikov Processes in Engineering, Physics, and Neuroscience, by Emil Simiu

Self-Regularity A New Paradigm for Primal-Dual Interior Point Algorithms, by Jiming Peng, Cornelis Roos, and Tamas Terlaky

Selfsimilar Processes, by Paul Embrechts and Makoto Maejima

Analytic Theory of Global Bifurcation An Introduction, by Boris Buffoni and John Toland Entropy, by Andreas Greven, Gerhard Keller, and Gerald Warnecke, eds.

Auxiliary Signal Design for Failure Detection, by Stephen L. Campbell and Ramine Nikoukhah

Max Plus at Work Modeling and Analysis of Synchronized Systems: A Course on Max-Plus Algebra and Its Applications, by Bernd Heidergott, Geert Jan Olsder, and Jacob van der Woude

Optimization: Insights and Applications, by Jan Brinkhuis and Vladimir Tikhomirov

Thermodynamics: A Dynamical Systems Approach, by Wassim M. Haddadd, VijaySekhar Chellaboina, and Sergey G. Nersesov

Impulsive and Hybrid Dynamical Systems: Stability, Dissipativity, and Control, by Wassim M. Haddad, VijaySekhar Chellaboina, and Sergey G. Nersesov

Genomic Signal Processing, by Ilya Shmulevich and Edward Dougherty

Positive Definite Matrices, by Rajendra Bhatia

The Traveling Salesman Problem: A Computational Study, by David L. Applegate, Robert E. Bixby, Vasek Chvatal, and William J. Cook

Wave Scattering by Time-Dependent Perturbations: An Introduction, by G. F. Roach Algebraic Curves over a Finite Field*, by J.W.P. Hirschfeld, G. Korchmros, and F. Torres

Distributed Control of Robotic Networks

A Mathematical Approach to Motion Coordination Algorithms

Francesco Bullo

Jorge Cortés

Sonia Martínez

PRINCETON UNIVERSITY PRESS

PRINCETON AND OXFORD

Copyright © 2009 by Princeton University Press
Requests for permission to reproduce material from this work should be sent to Permissions, Princeton University Press

Published by Princeton University Press, 41 William Street, Princeton, New Jersey 08540
In the United Kingdom: Princeton University Press, 6 Oxford Street, Woodstock, Oxfordshire OX20 1TW

Library of Congress Cataloging-in-Publication Data

Bullo, Francesco.
Distributed control of robotic networks: A mathematical approach to motion coordination algorithms / Francesco Bullo, Jorge Cortés, Sonia Martínez.
    p. cm. – (Princeton series in applied mathematics.)
Includes bibliographical references and index.
ISBN 978-0-691-14195-4 (hardcover : alk. paper) 1. Robotics. 2. Computer algorithms. 3. Robots–Control systems. I. Cortés, Jorge, 1974- II. Martínez, Sonia, 1974- III. Title.
   TJ211.B82 2009
   629.8'9246–dc22

2009006697

British Library Cataloging-in-Publication Data is available

The publisher thanks the authors for providing the camera-ready copy from which this book was produced.

Printed on acid-free paper. ∞

press.princeton.edu

Printed in the United States of America

10 9 8 7 6 5 4 3 2 1

To Lily, to Sonia, and to Olimpia and Leonardo

Contents

Preface

OBJECTIVES

Recent years have witnessed a thriving research activity on cooperative control and motion coordination. This interest is motivated by the growing possibilities enabled by robotic networks in the monitoring of natural phenomena and the enhancement of human capabilities in hazardous and unknown environments.

Our first objective in this book is to present a coherent introduction to basic distributed algorithms for robotic networks. This emerging discipline sits at the intersection of different areas such as distributed algorithms, parallel processing, control, and estimation. Our second objective is to provide a self-contained, broad exposition of the notions and tools from these areas that are relevant in cooperative control problems. These concepts include graph-theoretic notions (connectivity, adjacency, and Laplacian matrices), distributed algorithms from computer science (leader election, basic tree computations) and from parallel processing (averaging algorithms, convergence rates), and geometric models and optimization (Voronoi partitions, proximity graphs). Our third objective is to put forth a model for robotic networks that helps to rigorously formalize coordination algorithms running on them. We illustrate how computational geometry plays an important role in modeling the interconnection topology of robotic networks. We draw on classical notions from distributed algorithms to provide complexity measures that characterize the execution of coordination algorithms. Such measures allow us to quantify the algorithm performance and implementation costs. Our fourth and last objective is to present various algorithms for coordination tasks such as connectivity maintenance, rendezvous, and deployment. We especially emphasize the analysis of the correctness and of the complexity of the proposed algorithms. The technical treatment combines control-theoretic tools such as Lyapunov functions and invariance principles with techniques from computer science and parallel processing, such as induction and message counting.

THE INTENDED AUDIENCE

The intended audience for this book consists of first- and second-year graduate students in control and robotics from Computer Science, Electrical Engineering, Mechanical Engineering, and Aerospace Engineering. A familiarity with basic concepts from analysis, linear algebra, dynamical systems, and control theory is assumed. The writing style is mathematical: we have aimed at being precise in the introduction of the notions, the statement of the results, and the formal description of the algorithms. This mathematical style is complemented by numerous examples, exercises, intuitive explanations and motivating discussions for the introduction of novel concepts.

Researchers in the fields of control theory and robotics who are not aware of the literature on distributed algorithms will also benefit from the book. The book uses notions with a clear computer-science flavor such as synchronous networks, complexity measures, basic tree computations, and linear distributed iterations, and integrates them into the study of robotic networks. Likewise, researchers in the fields of distributed algorithms and automata theory who are not aware of robotic networks and distributed control will also find the book useful. The numerous connections that can be drawn between the classical study of distributed algorithms and the present book provide a friendly roadmap with which to step into the field of controlled coordination of robotic networks.

AN OUTLINE OF THE BOOK

Chapter 1 presents a broad introduction to distributed algorithms on synchronous networks. We start by presenting basic matrix notions and a primer on graph theory that gives special emphasis to linear algebraic aspects such as adjacency and Laplacian matrices. After this, we introduce the notion of synchronous networks, and we present time, communication, and space complexity notions. We examine these notions in basic algorithms such as broadcast, tree computation, and leader election. The chapter ends with a thorough treatment of linear iterations and averaging algorithms.

Chapter 2 presents basic geometric notions that are relevant in motion coordination. Robotic networks have a spatial component which is not always present in synchronous networks as studied in computer science. Geometric objects such as polytopes, Voronoi partitions, and geometric centers play an important role in modeling the interaction of robotic networks with the physical environment. Proximity graphs allow us to rigorously formalize the interconnection topology of a network of robotic agents, and characterize the spatially distributed character of coordination algorithms. This notion is a natural translation of the notion of distributed algorithms treated in the previous chapter. The chapter concludes with a detailed discussion on concepts from geometric optimization and multicenter functions.

Chapter 3 introduces a model for a group of robots that synchronously communicate/sense locally, process information, and move. We describe the physical components of the robotic network and we introduce a formal notion of a motion coordination algorithm as a control and communication law. Generalizing the notions introduced in Chapter 1, we introduce the notions of task and of time, communication, and space complexity. We illustrate these concepts by means of a simple and insightful example of a group of robots moving on a circle.

Chapter 4 analyzes in detail two coordination tasks: connectivity maintenance and rendezvous. The objective of "connectivity maintenance" is to establish local rules that allow agents to move without losing the connectivity of the overall networks. The objective of "rendezvous" is to establish local rules that allow agents to agree on a common spatial location. We present coordination algorithms that achieve these tasks, making use of the geometric concepts introduced in the previous chapters. Furthermore, we provide results on the correctness and complexity of these algorithms.

Chapter 5 considers deployment problems. The objective of "deployment" is to establish local rules that allow agents to achieve optimal network configurations in an environment of interest. Here, optimality is defined using the multicenter functions from geometric optimization introduced in Chapter 2. We present coordination algorithms that achieve these tasks, characterizing their correctness and complexity.

Chapter 6 has a dual purpose. First, we introduce an event-driven control and communication law, in which computation and communication actions are triggered by asynchronous events, rather than taking place on a periodic schedule. Second, we consider a boundary tracking problem, and propose an "estimation and balancing" algorithm that allows a robotic network to monitor a moving boundary efficiently.

The reader will note that, as the discussion progresses, the selection of topics emphasizes problems in which we have been directly involved. There are exciting topics that have been considered in the literature and are not presented here in depth, albeit we briefly discuss a number of them throughout the exposition. In this, our first effort, we decided to tackle the problems that we knew better, postponing the rest for the future. We hope the reader will appreciate the result and share, while reading it, some of the fun we had in writing it.

HOW TO USE THIS BOOK AS A TEXT

Our experience and opinion is that this text can be used for a quarter- or semester-long course on "Distributed Control" or on "Robotic Networks." Such a course could be taught in an Engineering or a Computer Science department. We taught such a course at our respective institutions over a

10 weeks, 3 hours a week, period, skipping some material and some proofs (e.g., skipping some algebraic graph theory in Chapter 1, some of the multi-center functions and the nonconvex geometry treatment in Chapter 2, and the relative-sensing model in Chapter 3). With proofs and more complete treatment, we estimate that the material might require 45 hours of lecture time.

Finally, a complete electronic version of the book with supplementary material, such as slides and software, is freely available on the internet at:

http://coordinationbook.info

At this website, we plan to maintain an up-to-date version of the manuscript that incorporates corrections and minor improvements. The official book website at Princeton University Press is:

http://press.princeton.edu/titles/9101.html

ACKNOWLEDGMENTS

We would like to thank Emilio Frazzoli, Anurag Ganguli and Sara Susca for their contributions to joint work that is contained in this text. Specifically, Chapter 3 contains joint work with Emilio Frazzoli on robotic networks, Chapter 4 contains joint work with Anurag Ganguli on nonconvex rendezvous, and Chapter 6 contains joint work with Sara Susca on boundary estimation.

We would like to thank all our collaborators who have helped shape our perspective on cooperative control over the past few years, including Giuseppe Notarstefano and Ketan Savla. We thank the students and colleagues who followed courses based on early versions of this text and gave constructive feedback, including Ruggero Carli, Paolo Frasca, Rishi Graham, Karl J. Obermeyer, Michael Schuresko, and Stephen L. Smith. The first author thanks Andrew D. Lewis for his positive influence on this text. The first author is grateful for support from the National Science Foundation and the Department of Defense. The second and third authors are grateful for support from the National Science Foundation. We would like to thank the anonymous reviewers and editors who helped us improve the book. Finally, we thank our families for their tremendous patience, love, and ongoing support.

Santa Barbara, California	*Francesco Bullo*
San Diego, California	*Jorge Cortés*
San Diego, California	*Sonia Martínez*
June 29, 2006 — March 31, 2009	

Chapter One

An introduction to distributed algorithms

Graph theory, distributed algorithms, and linear distributed algorithms are a fascinating scientific subject. In this chapter we provide a broad introduction to distributed algorithms by reviewing some preliminary graphical concepts and by studying some simple algorithms. We begin the chapter with one section introducing some basic notation and another section stating a few useful facts from matrix theory, dynamical systems, and convergence theorems based on invariance principles. In the third section of the chapter, we provide a primer on graph theory with a particular emphasis on algebraic aspects, such as the properties of adjacency and Laplacian matrices associated to a weighted digraph. In the next section of the chapter, we introduce the notion of synchronous network and of distributed algorithm. We state various complexity notions and study them in simple example problems such as the broadcast problem, the tree computation problem, and the leader election problem. In the fifth section of the chapter, we discuss linear distributed algorithms. We focus on linear algorithms defined by sequences of stochastic matrices and review the results on their convergence properties. We end the chapter with three sections on, respectively, bibliographical notes, proofs of the results presented in the chapter, and exercises.

1.1 ELEMENTARY CONCEPTS AND NOTATION

1.1.1 Sets and maps

We assume that the reader is familiar with basic notions from topology, such as the notions of open, closed, bounded, and compact sets. In this section, we just introduce some basic notation. We let $x \in S$ denote a point x belonging to a set S. If S is finite, we let $|S|$ denote the number of its elements. For a set S, we let $\mathbb{P}(S)$ and $\mathbb{F}(S)$ denote the collection of subsets of S and the collection of finite subsets of S, respectively. The empty set is denoted by $\emptyset$. The interior and the boundary of a set S are denoted by $\operatorname{int}(S)$ and ∂S, respectively. If R is a subset of or equal to S, then we write $R \subset S$. If R is a strict subset of S, then we write $R \subsetneq S$. We describe subsets of S defined by specific conditions via the notation

$$\{x \in S \mid \text{condition(s) on } x\}.$$

Given two sets S_1 and S_2, we let $S_1 \cup S_2$, $S_1 \cap S_2$, and $S_1 \times S_2$ denote the union, intersection, and Cartesian product of S_1 and S_2, respectively. Given a collection of sets $\{S_a\}_{a \in A}$ indexed by a set A, we interchangeably denote their Cartesian product by $\prod_{a \in A} S_a$ or by $\prod\{S_a \mid a \in A\}$. We adopt analogous notations for union and intersection. We denote by S^n the Cartesian product of n copies of the same S. The *diagonal set* $\text{diag}(S^n)$ of S^n is given by $\text{diag}(S^n) = \{(s, \ldots, s) \in S^n \mid s \in S\}$. The set $S_1 \setminus S_2$ contains all points in S_1 that do not belong to S_2.

We let $\mathbb{N}$ and $\mathbb{Z}_{\geq 0}$ denote the set of natural numbers and of non-negative integers, respectively. We let $\mathbb{R}$, $\mathbb{R}_{>0}$, $\mathbb{R}_{\geq 0}$, and $\mathbb{C}$ denote the set of real numbers, strictly positive real numbers, non-negative real numbers, and complex numbers, respectively. The sets $\mathbb{R}^d$, $\mathbb{C}^d$, and $\mathbb{S}^d \subset \mathbb{R}^{d+1}$ are the d-dimensional Euclidean space, the d-dimensional complex space, and the d-dimensional sphere, respectively. The tangent space of $\mathbb{R}^d$, denoted by $T\mathbb{R}^d$, is the set of all vectors tangent to $\mathbb{R}^d$. Note that $T\mathbb{R}^d$ can be identified with $\mathbb{R}^d \times \mathbb{R}^d$ by mapping a vector v tangent to $\mathbb{R}^d$ at $x \in \mathbb{R}^d$ to the pair (x, v). Likewise, $T\mathbb{S}^d$ is the set of all vectors tangent to $\mathbb{S}^d$, and can be identified with $\mathbb{S}^d \times \mathbb{R}^d$. The Euclidean space $\mathbb{R}^d$ contains the vectors $\mathbf{0}_d = (0, \ldots, 0)$, $\mathbf{1}_d = (1, \ldots, 1)$, and the standard basis $e_1 = (1, 0, \ldots, 0), \ldots, e_d = (0, \ldots, 0, 1)$. Given $a < b$, we let $[a, b]$ and $]a, b[$ denote the closed interval and the open interval between a and b, respectively.

Given two sets S and T, we let $f : S \to T$ denote a map from S to T, that is, a unique way of associating an element of T to an element of S. The *image* of the map $f : S \to T$ is the set $\text{image}(f) = \{f(s) \in T \mid s \in S\}$. Given the map $f : S \to T$ and a set $S_1 \subset S$, we let $f(S_1) = \{f(s) \mid s \in S_1\}$ denote the image of the set S_1 under the map f. Given $f : S \to T$ and $g : U \to S$, we let $f \circ g : U \to T$, defined by $f \circ g(u) = f(g(u))$, denote the composition of f and g. The map $\text{id}_S : S \to S$ is the identity map on S. Given $f : S \to \mathbb{R}$, the *support* of f is the set of elements s such that $f(s) \neq 0$. Given a subset $R \subsetneq S$, the indicator map $1_R : S \to \mathbb{R}$ associated with R is given by $1_R(q) = 1$ if $q \in R$, and $1_R(q) = 0$ if $q \notin R$. Given two sets S and T, a *set-valued map*, denoted by $h : S \rightrightarrows T$, associates to an element of S a subset of T. Given a map $f : S \to T$, the *inverse map* $f^{-1} : T \rightrightarrows S$ is defined by

$$f^{-1}(t) = \{s \in S \mid f(s) = t\}.$$

If f is a real-valued function, that is, a function of the form $f : S \to \mathbb{R}$, then $f^{-1}(x) \subset S$, for any $x \in \mathbb{R}$, is a *level set* of f. In what follows, we require the reader to be familiar with some basic smoothness notions for functions. Specifically, we will use the notions of locally and globally Lipschitz functions, differentiable, piecewise differentiable and continuously differentiable functions, and functions that are multiple times differentiable.

Finally, we introduce the so-called *Bachmann–Landau symbols*. For $f, g : \mathbb{N} \to \mathbb{R}_{\geq 0}$, we say that $f \in O(g)$ (resp., $f \in \Omega(g)$) if there exist $n_0 \in \mathbb{N}$ and $K \in \mathbb{R}_{>0}$ (resp., $k \in \mathbb{R}_{>0}$) such that $f(n) \leq K g(n)$ for all $n \geq n_0$ (resp., $f(n) \geq k g(n)$ for all $n \geq n_0$). If $f \in O(g)$ and $f \in \Omega(g)$, then we use the notation $f \in \Theta(g)$.

1.1.2 Distance functions

A function $\mathrm{dist} : S \times S \to \mathbb{R}_{\geq 0}$ defines a *distance* on a set S if it satisfies: (i) $\mathrm{dist}(x, y) = 0$ if and only if $x = y$; (ii) $\mathrm{dist}(x, y) = \mathrm{dist}(y, x)$, for all $x, y \in S$; and (iii) $\mathrm{dist}(x, y) \leq \mathrm{dist}(x, z) + \mathrm{dist}(z, y)$, for all $x, y, z \in S$. The pair (S, dist) is usually called a *metric space*.

Some relevant examples of distance functions include the following:

L^p**-distance on** $\mathbb{R}^d$**.** For $p \in [1, +\infty[$, consider the L^p-norm on $\mathbb{R}^d$ defined by $\|x\|_p = (\sum_{i=1}^{d} |x_i|^p)^{1/p}$. For $p = +\infty$, consider the L^∞-norm on $\mathbb{R}^d$ defined by $\|x\|_\infty = \max_{i \in \{1,\dots,d\}} |x_i|$. Any of these norms defines naturally a L^p-*distance* in $\mathbb{R}^d$ by $\mathrm{dist}_p(x, y) = \|y - x\|_p$. In particular, the most widely used is the Euclidean distance, corresponding to $p = 2$. Unless otherwise noted, it is always understood that $\mathbb{R}^d$ is endowed with this notion of distance. We will also use the L^1- and the L^∞-distances. Finally, it is convenient to define the norm $\|z\|_{\mathbb{C}}$ of a complex number $z \in \mathbb{C}$ to be the Euclidean norm of z regarded as a vector in $\mathbb{R}^2$.

Geodesic distance on $\mathbb{S}^d$**.** Another example is the notion of *geodesic distance* on $\mathbb{S}^d$. This is defined as follows. For $x, y \in \mathbb{S}^d$, $\mathrm{dist}_g(x, y)$ is the length of the shortest curve in $\mathbb{S}^d$ connecting x and y. We will use this notion of distance in dimensions $d = 1$ and $d = 2$. On the unit circle $\mathbb{S}^1$, by convention, let us define positions as angles measured counterclockwise from the positive horizontal axis. Then, the geodesic distance can be expressed as

$$\mathrm{dist}_g(x, y) = \min\{\mathrm{dist}_c(x, y), \mathrm{dist}_{cc}(x, y)\}, \quad x, y \in \mathbb{S}^1,$$

where $\mathrm{dist}_c(x, y) = (x - y) \mod 2\pi$ is the clockwise distance and $\mathrm{dist}_{cc}(x, y) = (y - x) \mod 2\pi$ is the counterclockwise distance. Here the clockwise distance between two angles is the path length from an angle to the other traveling clockwise, and $x \mod 2\pi$ is the remainder

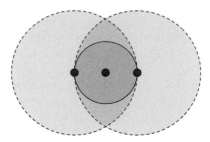

Figure 1.1 Open balls (dashed lines), a closed ball (solid line), and an open lune for the Euclidean distance on the plane.

of the division of x by 2π. On the sphere $\mathbb{S}^2$, the geodesic distance can be computed as follows. Given $x, y \in \mathbb{S}^2$, one considers the great circle determined by x and y. Then, the geodesic distance between x and y is exactly the length of the shortest arc in the great circle connecting x and y.

Cartesian product distance on $\mathbb{R}^{d_1} \times \mathbb{S}^{d_2}$. Consider $\mathbb{R}^{d_1}$ endowed with an L^p-distance, $p \in [1, +\infty]$, and $\mathbb{S}^{d_2}$ endowed with the geodesic distance. Given $(x_1, y_1), (x_2, y_2) \in \mathbb{R}^{d_1} \times \mathbb{S}^{d_2}$, their *Cartesian product distance* is given by $\mathrm{dist}_p(x_1, x_2) + \mathrm{dist}_g(y_1, y_2)$. Unless otherwise noted, it is understood that $\mathbb{R}^{d_1} \times \mathbb{S}^{d_2}$ is endowed with the Cartesian product distance $(\mathrm{dist}_2, \mathrm{dist}_g)$.

Given a metric space (S, dist), the *open* and *closed balls* of center $x \in S$ and radius $\varepsilon \in \mathbb{R}_{>0}$ are defined by, respectively,

$$B(x, \varepsilon) = \{y \in S \mid \mathrm{dist}(x, y) < \varepsilon\},$$
$$\overline{B}(x, \varepsilon) = \{y \in S \mid \mathrm{dist}(x, y) \le \varepsilon\}.$$

Consider a point $x \in X$ and a set $S \subset X$. A *neighborhood* of a point $x \in X$ is a subset of X that contains an open ball centered at x. A *neighborhood* of a set $Y \subset X$ is a subset of X that, for each point $y \in Y$, contains an open ball centered at y. The *open lune* associated to $x, y \in S$ is $B(x, \mathrm{dist}(x, y)) \cap B(y, \mathrm{dist}(x, y))$. These notions are illustrated in Figure 1.1 for the plane equipped with the Euclidean distance.

Given a metric space (S, dist), the distance between a point $x \in S$ and a set $W \subset S$ is the infimum of all distances between x and each of the points in W. Formally, we set

$$\mathrm{dist}(x, W) = \inf\{\mathrm{dist}(x, y) \mid y \in W\}.$$

The projection of a point $x \in S$ onto a set $W \subset S$ is the set-valued map

$\text{proj}_W : S \rightrightarrows W$ defined by

$$\text{proj}_W(x) = \{y \in W \mid \text{dist}(x, y) = \text{dist}(x, W)\}.$$

If W is a closed set, then $\text{proj}_W(x) \neq \emptyset$ for any $x \in S$. The *diameter* of a set is the maximum distance between any two points in the set; formally, we set $\text{diam}(S) = \sup\{\text{dist}(x, y) \mid x, y \in S\}$. With a slight abuse of notation, we often use $\text{diam}(P)$ to denote $\text{diam}(\{p_1, \ldots, p_n\})$ for $P = (p_1, \ldots, p_n)$.

1.1.3 Curves

A curve is the image of a continuous map $\gamma : [a, b] \to \mathbb{R}^d$. The map γ is called a *parameterization* of the curve. We usually identify a parameterization with the curve it defines. Without loss of generality, any curve can be given a parametrization with $a = 0$ and $b = 1$. A curve connects the two points p and q if $\gamma(0) = p$ and $\gamma(1) = q$. A curve $\gamma : [0, 1] \to \mathbb{R}^d$ is *not self-intersecting* if γ is injective on $(0, 1)$. A curve is *closed* if $\gamma(0) = \gamma(1)$.

A set $S \subset \mathbb{R}^d$ is *path connected* if any two points in S can be joined by a curve. A set $S \subset X$ is *simply connected* if it is path connected and any not self-intersecting closed curve can be continuously deformed to a point in the set; that is, for any injective continuous map $\gamma : [0, 1] \to S$ that satisfies $\gamma(0) = \gamma(1)$, there exist $p \in S$ and a continuous map $H : [0, 1] \times [0, 1] \to S$ such that $H(t, 0) = \gamma(t)$ and $H(t, 1) = p$ for all $t \in [0, 1]$. Informally, a simply connected set is a set that consists of a single piece and does not have any holes.

Next, consider a piecewise continuously differentiable curve $\gamma : [0, 1] \to \mathbb{R}^d$; the *length* of γ is

$$\text{length}(\gamma) = \int_0^1 \|\dot{\gamma}(s)\|_2 ds,$$

and its *arc-length parameter* is

$$s_{\text{arc}}(s) = \int_0^s \|\dot{\gamma}(t)\|_2 dt.$$

Note that as the parameter t varies in $[0, 1]$, the arc-length parameter $s_{\text{arc}}(t)$ varies in $[0, \text{length}(\gamma)]$. The *arc-length parameterization* of the curve is the map $\gamma_{\text{arc}} : [0, \text{length}(\gamma)] \to \mathbb{R}^d$ defined by the equation $\gamma_{\text{arc}}(s_{\text{arc}}(s)) = \gamma(s)$. With a slight abuse of notation, we will often drop the subindex arc and denote the arc-length parameterization by γ too.

For closed, not self-intersecting curves in the plane, we introduce the notion of signed and absolute curvatures as follows. Let $\gamma : [0, \text{length}(\gamma)] \to$

$\mathbb{R}^2$ be the counterclockwise arc-length parameterization of a curve. Assume γ is closed, not self-intersecting and twice continuously differentiable. Define the *tangent vector* $\gamma' : [0, \text{length}(\gamma)] \to \mathbb{R}^2$ by $\gamma'(s) = \frac{d\gamma}{ds}$. Note that the tangent vector has unit length, that is, $\|\gamma'(s)\|_2 = 1$ for all s. Additionally, define the *outward normal vector* $\text{n}_{\text{out}} : [0, \text{length}(\gamma)] \to \mathbb{R}^2$ to be the unit-length vector that is point-wise orthogonal to the tangent vector and directed outside the set enclosed by the closed curve γ. With these notations, the *signed curvature* $\kappa_{\text{signed}} : [0, \text{length}(\gamma)] \to \mathbb{R}$ is defined by requiring that it satisfies

$$\gamma''(s) = -\kappa_{\text{signed}}(s)\,\text{n}_{\text{out}}(s), \quad \text{and} \quad \text{n}'_{\text{out}}(s) = \kappa_{\text{signed}}(s)\gamma'(s).$$

If the set enclosed by the closed curve γ is strictly convex, then the signed curvature of γ is strictly positive. In general, the *(absolute) curvature* $\kappa_{\text{abs}} : [0, \text{length}(\gamma)] \to \mathbb{R}_{\geq 0}$ and the *radius of curvature* $\rho : [0, \text{length}(\gamma)] \to \mathbb{R}_{\geq 0}$ of the curve γ are defined by, respectively,

$$\kappa_{\text{abs}}(s) = |\kappa_{\text{signed}}(s)|, \quad \text{and} \quad \rho(s) = |\kappa_{\text{signed}}(s)|^{-1}.$$

1.2 MATRIX THEORY

Here, we present basic notions and results about matrix theory, following the treatments in Horn and Johnson (1985) and Meyer (2001). We let $\mathbb{R}^{n \times m}$ and $\mathbb{C}^{n \times m}$ denote the set of $n \times m$ real and complex matrices. Given a real matrix A and a complex matrix U, we let A^T and U^* denote the transpose of A and the conjugate transpose matrix of U, respectively. We let I_n denote the $n \times n$ identity matrix. For a square matrix A, we write $A > 0$, resp. $A \geq 0$, if A is symmetric positive definite, resp. symmetric positive semidefinite. For a real matrix A, we let $\text{kernel}(A)$ and $\text{rank}(A)$ denote the kernel and rank of A, respectively. Given a vector v, we let $\text{diag}(v)$ denote the square matrix whose diagonal elements are equal to the component v and whose off-diagonal elements are zero.

1.2.1 Matrix sets

A matrix $A \in \mathbb{R}^{n \times n}$ with entries a_{ij}, $i, j \in \{1, \dots, n\}$, is

   (i) *Orthogonal* if $AA^T = I_n$, and is *special orthogonal* if it is orthogonal with $\det(A) = +1$. The set of orthogonal matrices is a group.[1]

---

[1] A set G with a binary operation, denoted by $G \times G \ni (a, b) \mapsto a \star b \in G$, is a *group* if: (i) $a \star (b \star c) = (a \star b) \star c$ for all $a, b, c \in G$ (associativity property); (ii) there exists $e \in G$ such that $a \star e = e \star a = a$ for all $a \in G$ (existence of an identity element); and (iii) there exists $a^{-1} \in G$ such that $a \star a^{-1} = a^{-1} \star a = e$ for all $a \in G$ (existence of inverse elements).

(ii) *Nonnegative* (resp., *positive*) if all its entries are nonnegative (resp., positive).

(iii) *Row-stochastic* (or *stochastic* for brevity) if it is nonnegative and $\sum_{j=1}^{n} a_{ij} = 1$, for all $i \in \{1, \ldots, n\}$; in other words, A is row-stochastic if

$$A\mathbf{1}_n = \mathbf{1}_n.$$

(iv) *Column-stochastic* if it is nonnegative and $\sum_{i=1}^{n} a_{ij} = 1$, for all $j \in \{1, \ldots, n\}$.

(v) *Doubly stochastic* if A is row-stochastic and column-stochastic.

(vi) *Normal* if $A^T A = A A^T$.

(vii) A *permutation matrix* if A has precisely one entry equal to one in each row, one entry equal to one in each column, and all other entries equal to zero. The set of permutation matrices is a group.

The scalars $\mu_1, \ldots, \mu_k$ are *convex combination coefficients* if $\mu_i \geq 0$, for $i \in \{1, \ldots, k\}$, and $\sum_{i=1}^{k} \mu_i = 1$. (Each row of a row-stochastic matrix contains convex combination coefficients.) A *convex combination* of vectors is a linear combination of the vectors with convex combination coefficients. A subset U of a vector space V is *convex* if the convex combination of any two elements of U takes value in U. For example, the set of stochastic matrices and the set of doubly stochastic matrices are convex.

Theorem 1.1 (Birkhoff–von Neumann). *A square matrix is doubly stochastic if and only if it is a convex combination of permutation matrices.*

Next, we review two families of relevant matrices with useful properties. *Toeplitz matrices* are square matrices with equal entries along each diagonal parallel to the main diagonal. In other words, a Toeplitz matrix is a matrix of the form

$$
\begin{bmatrix}
t_0 & t_1 & \ddots & \ddots & \ddots & t_{n-2} & t_{n-1} \\
t_{-1} & t_0 & t_1 & \ddots & \ddots & \ddots & t_{n-2} \\
\ddots & t_{-1} & t_0 & t_1 & \ddots & \ddots & \ddots \\
\ddots & \ddots & t_{-1} & t_0 & t_1 & \ddots & \ddots \\
\ddots & \ddots & \ddots & t_{-1} & t_0 & t_1 & \ddots \\
t_{-n+2} & \ddots & \ddots & \ddots & t_{-1} & t_0 & t_1 \\
t_{-n+1} & t_{-n+2} & \ddots & \ddots & \ddots & t_{-1} & t_0
\end{bmatrix}.
$$

An $n \times n$ Toeplitz matrix is determined by its first row and column, and hence by $2n - 1$ scalars.

Circulant matrices are square Toeplitz matrices where each two subsequent row vectors v_i and v_{i+1} have the following two properties: the last entry of v_i is the first entry of v_{i+1} and the first $(n-1)$ entries of v_i are the second $(n-1)$ entries of v_{i+1}. In other words, a circulant matrix is a matrix of the form

$$\begin{bmatrix} c_0 & c_1 & \ddots & \ddots & \ddots & c_{n-2} & c_{n-1} \\ c_{n-1} & c_0 & c_1 & \ddots & \ddots & \ddots & c_{n-2} \\ \ddots & c_{n-1} & c_0 & c_1 & \ddots & \ddots & \ddots \\ \ddots & \ddots & c_{n-1} & c_0 & c_1 & \ddots & \ddots \\ \ddots & \ddots & \ddots & c_{n-1} & c_0 & c_1 & \ddots \\ c_2 & \ddots & \ddots & \ddots & c_{n-1} & c_0 & c_1 \\ c_1 & c_2 & \ddots & \ddots & \ddots & c_{n-1} & c_0 \end{bmatrix},$$

and, therefore, it is determined by its first row.

1.2.2 Eigenvalues, singular values, and induced norms

We require the reader to be familiar with the notion of eigenvalue and of simple eigenvalue, that is, an eigenvalue with algebraic and geometric multiplicity[2] equal to 1. The set of eigenvalues of a matrix $A \in \mathbb{R}^{n \times n}$ is called its *spectrum* and is denoted by $\mathrm{spec}(A) \subset \mathbb{C}$. The *singular values* of the matrix $A \in \mathbb{R}^{n \times n}$ are the positive square roots of the eigenvalues of $A^T A$.

We begin with a well-known property of the spectrum of a matrix.

Theorem 1.2 (Geršgorin disks). *Let A be an $n \times n$ matrix. Then*

$$\mathrm{spec}(A) \subset \bigcup_{i \in \{1,\dots,n\}} \left\{ z \in \mathbb{C} \mid \|z - a_{ii}\|_{\mathbb{C}} \le \sum_{j=1, j \ne i}^{n} |a_{ij}| \right\}.$$

Next, we review a few facts about normal matrices, their eigenvectors and their singular values.

---

[2]The algebraic multiplicity of an eigenvalue is the multiplicity of the corresponding root of the characteristic equation. The geometric multiplicity of an eigenvalue is the number of linearly independent eigenvectors corresponding to the eigenvalue. The algebraic multiplicity is greater than or equal to the geometric multiplicity.

Lemma 1.3 (Normal matrices). *For a matrix $A \in \mathbb{R}^{n \times n}$, the following statements are equivalent:*

(i) A is normal;

(ii) A has a complete orthonormal set of eigenvectors; and

(iii) A is unitarily similar to a diagonal matrix, that is, there exists a unitary[3] matrix U such that U^*AU is diagonal.

Lemma 1.4 (Singular values of a normal matrix). *If a normal matrix has eigenvalues $\{\lambda_1, \ldots, \lambda_n\}$, then its singular values are $\{|\lambda_1|, \ldots, |\lambda_n|\}$.*

It is well known that real symmetric matrices are normal, are diagonalizable by orthogonal matrices, and have real eigenvalues. Additionally, circulant matrices are normal.

We conclude by defining the notion of induced norm of a matrix. For $p \in \mathbb{N} \cup \{\infty\}$, the *p-induced norm* of $A \in \mathbb{R}^{n \times n}$ is

$$\|A\|_p = \max\{\|Ax\|_p \mid \|x\|_p = 1\}.$$

One can see that

$$\|A\|_1 = \max_{j \in \{1,\ldots,n\}} \sum_{i=1}^n |a_{ij}|, \quad \|A\|_\infty = \max_{i \in \{1,\ldots,n\}} \sum_{j=1}^n |a_{ij}|,$$

$$\|A\|_2 = \max\{\sigma \mid \sigma \text{ is a singular value of } A\}.$$

1.2.3 Spectral radius and convergent matrices

The *spectral radius* of a matrix $A \in \mathbb{R}^{n \times n}$ is

$$\rho(A) = \max\{\|\lambda\|_{\mathbb{C}} \mid \lambda \in \text{spec}(A)\}.$$

In other words, $\rho(A)$ is the radius of the smallest disk centered at the origin that contains the spectrum of A.

Lemma 1.5 (Induced norms and spectral radius). *For any square matrix A and in any norm $p \in \mathbb{N} \cup \{\infty\}$, $\rho(A) \le \|A\|_p$.*

We will often deal with matrices with an eigenvalue equal to 1 and all other eigenvalues strictly inside the unit disk. Accordingly, we generalize the notion of spectral radius as follows. For a square matrix A with $\rho(A) = 1$, we define the *essential spectral radius*

$$\rho_{\text{ess}}(A) = \max\{\|\lambda\|_{\mathbb{C}} \mid \lambda \in \text{spec}(A) \setminus \{1\}\}. \tag{1.2.1}$$

---

[3]A complex matrix $U \in \mathbb{C}^{n \times n}$ is *unitary* if $U^{-1} = U^*$.

Next, we will consider matrices with useful convergence properties.

Definition 1.6 (Convergent and semi-convergent matrices). A matrix $A \in \mathbb{R}^{n \times n}$ is

(i) *semi-convergent* if $\lim_{\ell \to +\infty} A^{\ell}$ exists; and

(ii) *convergent* if it is semi-convergent and $\lim_{\ell \to +\infty} A^{\ell} = 0$.   •

These two notions are characterized as follows.

Lemma 1.7 (Convergent and semi-convergent matrices). *The square matrix A is convergent if and only if $\rho(A) < 1$. Furthermore, A is semi-convergent if and only if the following three properties hold:*

(i) $\rho(A) \le 1$;

(ii) $\rho_{\mathrm{ess}}(A) < 1$, that is, 1 is an eigenvalue and 1 is the only eigenvalue on the unit circle; and

(iii) the eigenvalue 1 is semisimple, *that is, it has equal algebraic and geometric multiplicity (possibly larger than one).*

In other words, A is semi-convergent if and only if there exists a nonsingular matrix T such that

$$A = T \begin{bmatrix} I_k & 0 \\ 0 & B \end{bmatrix} T^{-1},$$

where $B \in \mathbb{R}^{(n-k) \times (n-k)}$ is convergent, that is, $\rho(B) < 1$. With this notation, we have $\rho_{\mathrm{ess}}(A) = \rho(B)$ and the algebraic and geometric multiplicity of the eigenvalue 1 is k.

1.2.4 Perron–Frobenius theory

Positive and nonnegative matrices have useful spectral properties. In what follows, the first theorem amounts to the original Perron's Theorem for positive matrices and the following theorems are the extension due to Frobenius for certain nonnegative matrices. We refer to (Horn and Johnson, 1985, Chapter 8) for a detailed treatment.

Theorem 1.8 (Perron-Frobenius for positive matrices). *If the square matrix A is positive, then*

(i) $\rho(A) > 0$;

(ii) $\rho(A)$ *is an eigenvalue, it is simple, and* $\rho(A)$ *is strictly larger than the magnitude of any other eigenvalue; and*

(iii) $\rho(A)$ *has an eigenvector with positive components.*

Requiring the matrix to be strictly positive is a key assumption that limits the applicability of this theorem. It turns out that it is possible to obtain the same results of the theorem under weaker assumptions.

Definition 1.9 (Irreducible matrix). A nonnegative matrix $A \in \mathbb{R}^{n \times n}$ is *irreducible* if, for any nontrivial partition $J \cup K$ of the index set $\{1, \dots, n\}$, there exist $j \in J$ and $k \in K$ such that $a_{jk} \neq 0$.

Remark 1.10 (Properties of irreducible matrices). An equivalent definition of irreducibility is given as follows. A matrix $A \in \mathbb{R}^{n \times n}$ is *irreducible* if it is not reducible, and is *reducible* if either:

(i) $n = 1$ and $A = 0$; or

(ii) there exists a permutation matrix $P \in \mathbb{R}^{n \times n}$ and a number $r \in \{1, \dots, n-1\}$ such that $P^T A P$ is block upper triangular with diagonal blocks of dimensions $r \times r$ and $(n-r) \times (n-r)$.

It is an immediate consequence that the property of irreducibility depends upon only the patterns of zeros and nonzero elements of the matrix.   •

We can now weaken the assumption in Theorem 1.8 and obtain a comparable, but weaker, result for irreducible matrices.

Theorem 1.11 (Perron–Frobenius for irreducible matrices). *If the nonnegative square matrix A is irreducible, then*

(i) $\rho(A) > 0$;

(ii) $\rho(A)$ *is an eigenvalue, and it is simple; and*

(iii) $\rho(A)$ *has an eigenvector with positive components.*

In general, the spectral radius of a nonnegative irreducible matrix does not need to be the only eigenvalue of maximum magnitude. For example, the matrix $\begin{bmatrix} 0 & 1 \\ 1 & 0 \end{bmatrix}$ has eigenvalues $\{1, -1\}$. In other words, irreducible matrices do indeed have weaker spectral properties than positive matrices. Therefore, it remains unclear which nonnegative matrices have the same properties as those stated for positive matrices in Theorem 1.8.

Definition 1.12 (Primitive matrix). A nonnegative square matrix A is *primitive* if there exists $k \in \mathbb{N}$ such that A^k is positive.   •

It is easy to see that if a nonnegative square matrix is primitive, then it is irreducible. In later sections we will provide a graph-theoretical characterization of primitive matrices; for now, we are finally in a position to sharpen the results of Theorem 1.11.

Theorem 1.13 (Perron–Frobenius for primitive matrices). *If the nonnegative square matrix A is primitive, then*

(i) $\rho(A) > 0$;

(ii) $\rho(A)$ *is an eigenvalue, it is simple, and $\rho(A)$ is strictly larger than the magnitude of any other eigenvalue; and*

(iii) $\rho(A)$ *has an eigenvector with positive components.*

We conclude this section by noting the following convergence property that is an immediate corollary to Lemma 1.7 and to Theorem 1.13.

Corollary 1.14. *If the nonnegative square matrix A is primitive, then the matrix $\rho(A)^{-1}A$ is semi-convergent.*

1.3 DYNAMICAL SYSTEMS AND STABILITY THEORY

In this section, we introduce some basic concepts about dynamical and control systems; see, for example Sontag (1998) and Khalil (2002). We discuss stability and attractivity notions as well as the invariance principle. We conclude with a treatment of set-valued systems and time-dependent systems.

1.3.1 State machines and dynamical systems

Here, we introduce three classes of dynamical and control systems: (i) state machines or discrete-time discrete-space dynamical systems; (ii) discrete-time continuous-space control systems; and (iii) continuous-time continuous-space control systems.

We begin with our specific definition of state machine. A *(deterministic, finite) state machine* is a tuple (X, U, X_0, f), where X is a finite set called the *state space*, U is a finite set called the *input space*, $X_0 \subset X$ is the set of *allowable initial states*, and $f : X \times U \to X$ is the *evolution map*. Given an input sequence $u : \mathbb{Z}_{\geq 0} \to U$, the state machine evolution $x : \mathbb{Z}_{\geq 0} \to X$ starting from $x(0) \in X_0$ is given by

$$x(\ell + 1) = f(x(\ell), u(\ell)), \quad \ell \in \mathbb{Z}_{\geq 0}.$$

We will often refer to a state machine as a *processor*. Note that, in a state machine, both the state and the input spaces are finite or *discrete*. Often times, we will find it useful to consider systems that evolve in continuous space and that are time dependent. Let us then provide two additional definitions in the following paragraphs.

A *(time-dependent) discrete-time continuous-space control system* is a tuple (X, U, X_0, f), where X is a d-dimensional space chosen among $\mathbb{R}^d$, $\mathbb{S}^d$, and the Cartesian products $\mathbb{R}^{d_1} \times \mathbb{S}^{d_2}$, for some $d_1 + d_2 = d$, U is a compact subset of $\mathbb{R}^m$ containing $\mathbf{0}_m$, $X_0 \subset X$, and $f : \mathbb{Z}_{\geq 0} \times X \times U \to X$ is a continuous map. As before, the individual objects X, U, X_0, and f are termed the *state space*, *input space*, *allowable initial states*, and *evolution map*, respectively. Given an input sequence $u : \mathbb{Z}_{\geq 0} \to U$, the evolution $x : \mathbb{Z}_{\geq 0} \to X$ of the dynamical system starting from $x(0) \in X_0$ is given by

$$x(\ell + 1) = f(\ell, x(\ell), u(\ell)), \quad \ell \in \mathbb{Z}_{\geq 0}.$$

A *(time-dependent) continuous-time continuous-space control system* is a tuple (X, U, X_0, f), where X is a d-dimensional space chosen among $\mathbb{R}^d$, $\mathbb{S}^d$, and the Cartesian products $\mathbb{R}^{d_1} \times \mathbb{S}^{d_2}$, for some $d_1 + d_2 = d$, U is a compact subset of $\mathbb{R}^m$ containing $\mathbf{0}_m$, $X_0 \subset X$, and $f : \mathbb{R}_{\geq 0} \times X \times U \to TX$ is a continuously differentiable map. The individual objects X, U, X_0, and f are termed the *state space*, *input space*, *allowable initial states*, and *control vector field*, respectively. Given an input function $u : \mathbb{R}_{\geq 0} \to U$, the evolution $x : \mathbb{R}_{\geq 0} \to X$ of the dynamical system starting from $x(0) \in X_0$ is given by

$$\dot{x}(t) = f(t, x(t), u(t)), \quad t \in \mathbb{R}_{\geq 0}.$$

We often consider the case when the control vector field can be written as $f(t, x, u) = f_0(t, x) + \sum_{a=1}^{m} f_a(t, x) u_a$, for some continuously differentiable maps $f_0, f_1, \ldots, f_m : \mathbb{R}_{\geq 0} \times X \to TX$. Each of these individual maps is called a *(time-dependent) vector field*, and f is said to be a *control-affine vector field*. The control vector field f is *driftless* if $f(t, x, \mathbf{0}_m) = 0$ for all $x \in X$ and $t \in \mathbb{R}_{\geq 0}$.

Finally, the term *dynamical system* denotes a control system that is not subject to any external control action; this terminology is applicable both in discrete and continuous time. Furthermore, we will sometimes neglect to define a specific set of allowable initial states; in this case we mean that any point in the state space is allowable as initial condition.

1.3.2 Stability and attractivity notions

In this section, we consider a continuous-space dynamical system (X, f). We first consider the discrete-time case and later we briefly present the analogous continuous-time case. We study dynamical systems that are *time-invariant*. In discrete time, a time-invariant system is simply described by an evolution map of the form $f : X \to X$.

Definition 1.15 (Equilibrium point). A point $x_* \in X$ is an *equilibrium point* for the time-invariant dynamical system (X, f) if the constant curve $x : \mathbb{Z}_{\geq 0} \to X$, defined by $x(\ell) = x_*$ for all $\ell \in \mathbb{Z}_{\geq 0}$, is an evolution of the system.   ●

It can immediately be seen that a point x_* is an equilibrium point if and only if $f(x_*) = x_*$. We denote the set of equilibrium points of the dynamical system by $\mathrm{Equil}(X, f)$.

Definition 1.16 (Trajectories and sets). Let (X, f) be a time-invariant dynamical system and let W be a subset of X. Then:

(i) The set W is *positively invariant* for (X, f) if each evolution with initial condition in W remains in W for all subsequent times.

(ii) A trajectory $x : \mathbb{Z}_{\geq 0} \to X$ *approaches* a set $W \subset X$ if, for every neighborhood Y of W, there exists a time $\ell_0 > 0$ such that $x(\ell)$ takes values in Y for all subsequent times $\ell \geq \ell_0$. In such a case, we write $x(\ell) \to W$ as $\ell \to +\infty$.   ●

In formal terms, W is positively invariant if $x(0) \in W$ implies $x(\ell) \in W$ for all $\ell \in \mathbb{Z}_{\geq 0}$, where $x : \mathbb{Z}_{\geq 0} \to X$ is the evolution of (X, f) starting from $x(0)$.

Definition 1.17 (Stability and attractivity). For a time-invariant dynamical system (X, f), a set S is:

(i) *stable* if, for any neighborhood Y of S, there exists a neighborhood W of S such that every evolution of (X, f) with initial condition in W remains in Y for all subsequent times;

(ii) *unstable* if it is not stable;

(iii) *locally attractive* if there exists a neighborhood Y of S such that every evolution with initial condition in Y approaches the set S; and

(iv) *locally asymptotically stable* if it is stable and locally attractive.

Additionally, the set S is *globally attractive* if every evolution of the dynamical system approaches it and it is *globally asymptotically stable* if it is stable and globally attractive.   •

Remark 1.18 (Continuous-time dynamical systems). It is straightforward to extend the previous definitions to the setting of continuous-time continuous-space dynamical systems. These notions are illustrated in Figure 1.2.   •

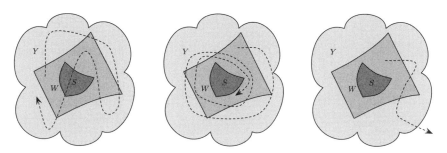

Figure 1.2 Illustrations of stability, asymptotic stability, and instability.

1.3.3 Invariance principles

Before discussing various versions of the invariance principle, we begin with a useful notion. Given a discrete-time time-invariant continuous-space dynamical system (X, f) and a set $W \subset X$, a function $V : X \to \mathbb{R}$ is *non-increasing along f in W* if $V(f(x)) \leq V(x)$ for all $x \in W$. (Such functions are often referred to as *Lyapunov functions*.) In other words, if a function V is non-increasing along f, then the composite function $\ell \mapsto V(y(\ell))$ is non-increasing for each evolution y of the dynamical system (X, f). The following theorem exploits this fact to establish useful properties of the evolutions of (X, f).

Theorem 1.19 (LaSalle Invariance Principle for discrete-time dynamical systems). *Let (X, f) be a discrete-time time-independent dynamical system. Assume that:*

  (i) *there exists a closed set $W \subset X$ that is positively invariant for (X, f);*

 (ii) *there exists a function $V : X \to \mathbb{R}$ that is non-increasing along f on W;*

(iii) *all evolutions of (X, f) with initial conditions in W are bounded; and*

 (iv) *f and V are continuous on W.*

Then each evolution with initial condition in W approaches a set of the form $V^{-1}(c) \cap S$, where c is a real constant and S is the largest positively invariant set contained in $\{w \in W \mid V(f(w)) = V(w)\}$.

We refer to Section 1.8.1 for a discussion about the proof of this result. Next, we present the continuous-time version of the invariance principle. In other words, we now assume that (X, f) is a continuous-time time-invariant continuous-space dynamical system.

We begin by revisiting the notion of non-increasing function. Given a continuously differentiable function $V : X \to \mathbb{R}$, the *Lie derivative* of V along f, denoted by $\mathcal{L}_f V : X \to \mathbb{R}$, is defined by

$$\mathcal{L}_f V(x) = \frac{\mathrm{d}}{\mathrm{d}t} V(\gamma(t))\Big|_{t=0},$$

where the trajectory $\gamma :]-\varepsilon, \varepsilon[\to X$ satisfies $\dot{\gamma}(t) = f(\gamma(t))$ and $\gamma(0) = x$. If $X = \mathbb{R}^d$, then we can write x in components $(x_1, \ldots, x_d)$ and we can give the following explicit formula for the Lie derivative:

$$\mathcal{L}_f V(x) = \sum_{i=1}^{d} \frac{\partial V}{\partial x_i}(x) f_i(x).$$

Similar formulas can be obtained for more general state spaces. Note that, given a set $W \subset X$, a function $V : X \to \mathbb{R}$ is non-increasing along f in W if $\mathcal{L}_f V(x) \leq 0$ for all $x \in W$.

Finally, we state the invariance principle for continuous-time systems.

Theorem 1.20 (LaSalle Invariance Principle for continuous-time dynamical systems). *Let (X, f) be a continuous-time time-independent dynamical system. Assume that:*

(i) there exists a closed set $W \subset X$ that is positively invariant for (X, f);

(ii) there exists a function $V : X \to \mathbb{R}$ that is non-increasing along f on W;

(iii) all evolutions of (X, f) with initial conditions in W are bounded; and

(iv) f and V are continuously differentiable[4] on W.

Then, each evolution with initial condition in W approaches a set of the form $V^{-1}(c) \cap S$, where c is a real constant and S is the largest positively invariant set contained in $\{w \in W \mid \mathcal{L}_f V(w) = 0\}$.

---

[4]It suffices that f be locally Lipschitz and V be continuously differentiable; see Cortés (2008b).

1.3.4 Notions and results for set-valued systems

Next, we focus on a more sophisticated version of the LaSalle Invariance Principle for more general dynamical systems, that is, dynamical systems described by set-valued maps that allow for non-deterministic evolutions. To do so, we need to present numerous notions, including set-valued dynamical systems, closedness properties, and weak positive invariance.

Specifically, a *discrete-time continuous-space set-valued dynamical system* (in short, *set-valued dynamical system*) is determined by a tuple (X, X_0, T), where X is a d-dimensional space chosen among $\mathbb{R}^d$, $\mathbb{S}^d$, and the Cartesian products $\mathbb{R}^{d_1} \times \mathbb{S}^{d_2}$, for some $d_1 + d_2 = d$, $X_0 \subset X$, and $T : X \rightrightarrows X$ is a set-valued map. We assume that T assigns to each point $x \in X$ a nonempty set $T(x) \subset X$. The individual objects X, X_0, and T are termed the *state space*, *allowable initial states*, and *evolution map*, respectively. An *evolution* of the dynamical system (X, X_0, T) is any trajectory $x : \mathbb{Z}_{\geq 0} \to X$ satisfying

$$x(\ell + 1) \in T(x(\ell)), \quad \ell \in \mathbb{Z}_{\geq 0}.$$

Figure 1.3 illustrates this notion. In particular, a (time-invariant) discrete-time continuous-space dynamical system (X, X_0, f) can be seen as a discrete-time continuous-space set-valued dynamical system (X, X_0, T), where the evolution set-valued map is just the singleton-valued map $x \mapsto T(x) = \{f(x)\}$.

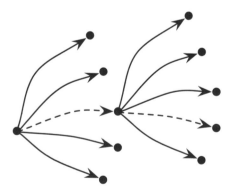

Figure 1.3 A discrete-time continuous-space set-valued dynamical system. A sample evolution is shown dashed.

Next, we introduce a notion of continuity for set-valued maps. The evolution map T is said to be *closed at* $x \in X$ if, for any sequences $\{x_k \mid k \in \mathbb{Z}_{\geq 0}\}$

and $\{y_k \mid k \in \mathbb{Z}_{\geq 0}\}$ such that

$$\lim_{k \to +\infty} x_k = x, \quad \lim_{k \to +\infty} y_k = y, \quad \text{and} \quad y_k \in T(x_k),$$

it holds that $y \in T(x)$. The evolution set-valued map T is *closed at* $W \subset X$ if for any $x \in W$, T is closed at x. Note that a continuous map $f : X \to X$ is closed when viewed as a singleton-valued map.

(i) A set $C \subset X$ is *weakly positively invariant with respect to T* if, for any $x \in C$, there exists $y \in C$ such that $y \in T(x)$.

(ii) A set $C \subset X$ is *strongly positively invariant with respect to T* if $T(x) \subset C$ for any $x \in C$.

A point x_0 is said to be a *fixed point of T* if $x_0 \in T(x_0)$. A continuous function $V : X \to \mathbb{R}$ is *non-increasing along T in $W \subset X$* if $V(y) \leq V(x)$ for all $x \in W$ and $y \in T(x)$.

We finally state and prove a general version of the invariance principle, whose proof is presented in Section 1.8.1.

Theorem 1.21 (LaSalle Invariance Principle for set-valued discrete-time dynamical systems). *Let (X, X_0, T) be a discrete-time set-valued dynamical system. Assume that:*

(i) *there exists a closed set $W \subset X$ that is strongly positively invariant for (X, X_0, T);*

(ii) *there exists a function $V : X \to \mathbb{R}$ that is non-increasing along T on W;*

(iii) *all evolutions of (X, X_0, T) with initial conditions in W are bounded; and*

(iv) *T is nonempty and closed at W and V is continuous on W.*

Then, each evolutions with initial condition in W approaches a set of the form $V^{-1}(c) \cap S$, where c is a real constant and S is the largest weakly positively invariant set contained in $\{w \in W \mid \exists w' \in T(w) \text{ such that } V(w') = V(w)\}$.

1.3.5 Notions and results for time-dependent systems

In this final subsection, we consider time-dependent discrete-time dynamical systems and discuss *uniform* stability and convergence notions. We begin with some uniform boundedness, stability, and attractivity definitions.

In what follows, given a time-dependent discrete-time dynamical system (X, X_0, f), an *evolution with initial condition in W at time $\ell_0 \in \mathbb{Z}_{\geq 0}$* is a trajectory $x : [\ell_0, +\infty[\to X$ of the dynamical system (X, X_0, f) defined by the initial condition $x(\ell_0) = x_0$, for some $x_0 \in W$. In other words, for time-dependent systems we will often consider trajectories that begin at time ℓ_0 not necessarily equal to zero.

Definition 1.22 (Uniformly bounded evolutions). A time-dependent discrete-time dynamical system (X, X_0, f) has *uniformly bounded evolutions* if, given any bounded set Y, there exists a bounded set W such that every evolution with initial condition in Y at any time $\ell_0 \in \mathbb{Z}_{\geq 0}$ remains in W for all subsequent times $\ell \geq \ell_0$. ●

Definition 1.23 (Uniform stability and attractivity notions). For a time-dependent discrete-time dynamical system (X, X_0, f), the set S is:

(i) *uniformly stable* if, for any neighborhood Y of S, there exists a neighborhood W of S such that every evolution with initial condition in W at any time $\ell_0 \in \mathbb{Z}_{\geq 0}$ remains in Y for all subsequent times $\ell \geq \ell_0$;

(ii) *uniformly locally attractive* if there exists a neighborhood Y of S such that every evolution with initial condition in Y at any time ℓ_0 approaches the set S in the following *time-uniform* manner:

> for all $\ell_0 \in \mathbb{Z}_{\geq 0}$, for all $x_0 \in Y$, and for all neighborhoods W of S, there exists a single $\tau_0 \in \mathbb{Z}_{\geq 0}$ such that the evolution $x : [\ell_0, +\infty[\to X$ defined by $x(\ell_0) = x_0$ takes value in W for all times $\ell \geq \ell_0 + \tau_0$; and

(iii) *uniformly locally asymptotically stable* if it is uniformly stable and uniformly locally attractive.

Additionally, the set S is *uniformly globally attractive* if every evolution of the dynamical system approaches the set in a time-uniform manner, and it is *uniformly globally asymptotically stable* if it is uniformly stable and uniformly globally attractive. ●

With the same notation in the definition, the set S is *(non-uniformly) locally attractive* if for all $\ell_0 \in \mathbb{Z}_{\geq 0}$, $x_0 \in Y$, and neighborhoods W of S, the evolution $x : [\ell_0, +\infty[\to X$ defined by $x(\ell_0) = x_0$, takes value in W for all times $\ell \geq \ell_0 + \tau_0(\ell_0)$, for some $\tau_0(\ell_0) \in \mathbb{Z}_{\geq 0}$.

To establish uniform stability and attractivity results we will overapproximate the evolution of the time-dependent dynamical system by considering the larger set of evolutions of an appropriate set-valued dynamical system. Given a time-dependent evolution map $f : \mathbb{Z}_{\geq 0} \times X \to X$, define a set-valued *overapproximation map* $T_f : X \rightrightarrows X$ by

$$T_f(x) = \{f(\ell, x) \mid \ell \in \mathbb{Z}_{\geq 0}\}.$$

With this notion we can state a useful result, whose proof is left to the reader as an exercise.

Lemma 1.24 (Overapproximation Lemma). *Consider a discrete-time time-dependent dynamical system* (X, X_0, f):

   (i) *If* $x : [\ell_0, +\infty[\rightarrow X$ *is an evolution of the dynamical system* (X, f), *then* $y : \mathbb{Z}_{\geq 0} \rightarrow X$ *defined by* $y(\ell) = x(\ell + \ell_0)$ *is an evolution of the set-valued overapproximation system* (X, T_f).

   (ii) *If the set* S *is locally attractive for the set-valued overapproximation system* (X, T_f), *then it is uniformly locally attractive for* (X, f).

   In other words, every evolution of the time-dependent dynamical system from any initial time is an evolution of the set-valued overapproximation system and, therefore, the set of trajectories of the set-valued overapproximation system contains the set of trajectories of the original time-dependent system. Uniform attractivity is a consequence of attractivity for the time-invariant set-valued overapproximation.

1.4 GRAPH THEORY

Here we present basic definitions about graph theory, following the treatments in the literature; see, for example Biggs (1994), Godsil and Royle (2001), and Diestel (2005).

   A *directed graph*—in short, *digraph*—of order n is a pair $G = (V, E)$, where V is a set with n elements called *vertices* (or *nodes*) and E is a set of ordered pair of vertices called *edges*. In other words, $E \subseteq V \times V$. We call V and E the *vertex set* and *edge set*, respectively. When convenient, we let $V(G)$ and $E(G)$ denote the vertices and edges of G, respectively. For $u, v \in V$, the ordered pair (u, v) denotes an edge *from u to v*.

   An *undirected graph*—in short, *graph*—consists of a vertex set V and of a set E of unordered pairs of vertices. For $u, v \in V$ and $u \neq v$, the set $\{u, v\}$ denotes an unordered edge. A digraph is *undirected* if $(v, u) \in E$ anytime $(u, v) \in E$. It is possible and convenient to identify an undirected digraph with the corresponding graph; vice versa, the *directed version* of a graph (V, E) is the digraph (V', E') with the property that $(u, v) \in E'$ if and only if $\{u, v\} \in E$. In what follows, our convention is to allow self-loops in both graphs and digraphs.

A digraph (V', E') is a *subgraph* of a digraph (V, E) if $V' \subset V$ and $E' \subset E$; additionally, a digraph (V', E') is a *spanning subgraph* if it is a subgraph and $V' = V$. The subgraph of (V, E) *induced by* $V' \subset V$ is the digraph (V', E'), where E' contains all edges in E between two vertices in V'. For two digraphs $G = (V, E)$ and $G' = (V', E')$, the *intersection* and *union* of G and G' are defined by

$$G \cap G' = (V \cap V', E \cap E'),$$
$$G \cup G' = (V \cup V', E \cup E').$$

Analogous definitions may be given for graphs.

In a digraph G with an edge $(u, v) \in E$, u is called an *in-neighbor* of v, and v is called an *out-neighbor* of u. We let $\mathcal{N}_G^{\text{in}}(v)$ (resp., $\mathcal{N}_G^{\text{out}}(v)$) denote the set of in-neighbors, (resp. the set of out-neighbors) of v in the digraph G. We will drop the subscript when the graph G is clear from the context. The *in-degree* and *out-degree* of v are the cardinality of $\mathcal{N}^{\text{in}}(v)$ and $\mathcal{N}^{\text{out}}(v)$, respectively. A digraph is *topologically balanced* if each vertex has the same in- and out-degrees (even if distinct vertices have distinct degrees). Likewise, in an undirected graph G, the vertices u and v are *neighbors* if $\{u, v\}$ is an undirected edge. We let $\mathcal{N}_G(v)$ denote the set of neighbors of v in the undirected graph G. As in the directed case, we will drop the subscript when the graph G is clear from the context. The *degree* of v is the cardinality of $\mathcal{N}(v)$.

Remark 1.25 (Additional notions). For a digraph $G = (V, E)$, the *reverse digraph* $\text{rev}(G)$ has vertex set V and edge set $\text{rev}(E)$ composed of all edges in E with reversed direction. A digraph $G = (V, E)$ is *complete* if $E = V \times V$. A *clique* (V', E') of a digraph (V, E) is a subgraph of (V, E) which is complete, that is, such that $E' = V' \times V'$. Note that a clique is fully determined by its set of vertices, and hence there is no loss of precision in denoting it by V'. A *maximal clique V' of an edge* of a digraph is a clique of the digraph with the following two properties: it contains the edge, and any other subgraph of the digraph that strictly contains $(V', V' \times V')$ is not a clique.   •

1.4.1 Connectivity notions

Let us now review some basic connectivity notions for digraphs and graphs. We begin with the setting of undirected graphs because of its simplicity.

A *path* in a graph is an ordered sequence of vertices such that any pair of consecutive vertices in the sequence is an edge of the graph. A graph is *connected* if there exists a path between any two vertices. If a graph is not connected, then it is composed of multiple *connected components*, that is, multiple connected subgraphs. A path is *simple* if no vertices appear more than once in it, except possibly for initial and final vertex. A *cycle* is a simple path that starts and ends at the same vertex. A graph is *acyclic* if it contains no cycles. A connected acyclic graph is a *tree*. A *forest* is a graph that can be written as the disjoint union of trees. Trees have interesting properties: for example, $G = (V, E)$ is a tree if and only if G is connected and $|E| = |V| - 1$. Alternatively, $G = (V, E)$ is a tree if and only if G is acyclic and $|E| = |V| - 1$. Figure 1.4 illustrates these notions.

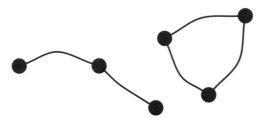

Figure 1.4 An illustration of connectivity notions on a graph. The graph has two connected components. The leftmost connected component is a tree, while the rightmost connected component is a cycle.

Next, we generalize these notions to the case of digraphs. A *directed path* in a digraph is an ordered sequence of vertices such that any ordered pair of vertices appearing consecutively in the sequence is an edge of the digraph. A *cycle* in a digraph is a directed path that starts and ends at the same vertex and that contains no repeated vertex except for the initial and the final vertex. A digraph is *acyclic* if it contains no cycles. In an acyclic graph, every vertex of in-degree 0 is named a *source*, and every vertex of out-degree 0 is named a *sink*. Every acyclic digraph has at least one source and at least one sink. Figure 1.5 illustrates these notions.

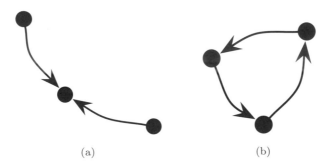

(a)   (b)

Figure 1.5 Illustrations of connectivity notions on a digraph: (a) shows an acyclic digraph with one sink and two sources; (b) shows a directed path which is also a cycle.

The set of cycles of a directed graph is finite. A directed graph is *aperiodic* if there exists no $k > 1$ that divides the length of every cycle of the graph.

In other words, a digraph is aperiodic if the greatest common divisor of the lengths of its cycles is one. A digraph is *periodic* if it is not aperiodic. Figure 1.6 shows examples of a periodic and an aperiodic digraph.

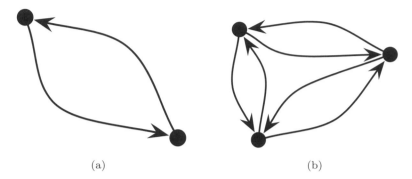

(a)   (b)

Figure 1.6 (a) A periodic digraph. (b) An aperiodic digraph with cycles of length 2 and 3.

A vertex of a digraph is *globally reachable* if it can be reached from any other vertex by traversing a directed path. A digraph is *strongly connected* if every vertex is globally reachable. The decomposition of a digraph into its strongly connected components and the notion of condensation digraph are discussed in Exercise E1.13.

A *directed tree* (sometimes called a rooted tree) is an acyclic digraph with the following property: there exists a vertex, called the *root*, such that any other vertex of the digraph can be reached by one and only one directed path starting at the root. In a directed tree, every in-neighbor of a vertex is called a *parent* and every out-neighbor is called a *child*. Two vertices with the same parent are called *siblings*. A *successor* of a vertex u is any other node that can be reached with a directed path starting at u. A *predecessor* of a vertex v is any other node such that a directed path exists starting at it and reaching v. A *directed spanning tree*, or simply a *spanning tree*, of a digraph is a spanning subgraph that is a directed tree. Clearly, a digraph contains a spanning tree if and only if the reverse digraph contains a globally reachable vertex. A *(directed) chain* is a directed tree with exactly one source and one sink. A *(directed) ring digraph* is the cycle obtained by adding to the edge set of a chain a new edge from its sink to its source. Figure 1.7 illustrates some of these notions.

The proof of the following result is given in Section 1.8.2.

Lemma 1.26 (Connectivity in topologically balanced digraphs). *Let G be a digraph. The following statements hold:*

(i) if G is strongly connected, then it contains a globally reachable vertex

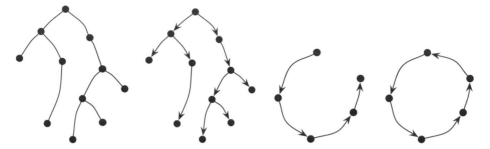

Figure 1.7 From left to right, tree, directed tree, chain, and ring digraphs.

and a spanning tree; and

(ii) if G is topologically balanced and contains either a globally reach-able vertex or a spanning tree, then G is strongly connected and is Eulerian.[5]

Given a digraph $G = (V, E)$, an *in-neighbor* of a nonempty set of nodes U is a node $v \in V \setminus U$ for which there exists an edge $(v, u) \in E$ for some $u \in U$.

Lemma 1.27 (Disjoint subsets and spanning trees). *Given a digraph G with at least two nodes, the following two properties are equivalent:*

(i) G has a spanning tree; and

(ii) for any pair of nonempty disjoint subsets U_1, $U_2 \subset V$, either U_1 has an in-neighbor or U_2 has an in-neighbor.

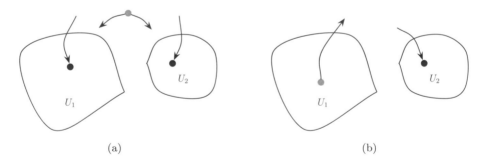

(a)   (b)

Figure 1.8 An illustration of Lemma 1.27. The root of the spanning tree is plotted in gray. In (a), the root is outside the sets U_1 and U_2. Because these sets are non-empty, there exists a directed path from the root to a vertex in each one of these sets. Therefore, both U_1 and U_2 have in-neighbors. In (b), the root is contained in U_1. Because U_2 is non-empty, there exists a directed path from the root to a vertex in U_2, and, therefore, U_2 has in-neighbors. The case when the root belongs to U_2 is treated analogously.

---

[5]A graph is Eulerian if it has a cycle that visits all the graph edges exactly once.

We will postpone the proof to Section 1.8.2. The result is illustrated in Figure 1.8. We can also state the result in terms of global reachability: G has a globally reachable node if and only if, for any pair of nonempty disjoint subsets U_1, $U_2 \subset V$, either U_1 has an out-neighbor or U_2 has an out-neighbor. We let the reader give a proper definition of the out-neighbor of a set.

1.4.2 Weighted digraphs

A *weighted digraph* is a triplet $G = (V, E, A)$, where the pair (V, E) is a digraph with nodes $V = \{v_1, \ldots, v_n\}$, and where the nonnegative matrix $A \in \mathbb{R}_{\geq 0}^{n \times n}$ is a *weighted adjacency matrix* with the following property: for $i, j \in \{1, \ldots, n\}$, the entry $a_{ij} > 0$ if (v_i, v_j) is an edge of G, and $a_{ij} = 0$ otherwise. In other words, the scalars a_{ij}, for all $(v_i, v_j) \in E$, are a set of weights for the edges of G. Note that the edge set is uniquely determined by the weighted adjacency matrix and it can therefore be omitted. When convenient, we denote the adjacency matrix of a weighted digraph G by $A(G)$. Figure 1.9 shows an example of a weighted digraph.

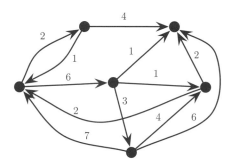

Figure 1.9 A weighted digraph with natural weights.

A digraph $G = (V, E)$ can be naturally thought of as a weighted digraph by defining the weighted adjacency matrix $A \in \{0, 1\}^{n \times n}$ as

$$a_{ij} = \begin{cases} 1, & \text{if } (v_i, v_j) \in E, \\ 0, & \text{otherwise,} \end{cases} \tag{1.4.1}$$

where $V = \{v_1, \ldots, v_n\}$. The adjacency matrix of a graph is the adjacency matrix of the directed version of the graph. Reciprocally, given a weighted digraph $G = (V, E, A)$, we refer to the digraph (V, E) as the *unweighted version* of G and to its associated adjacency matrix as the *unweighted adjacency matrix*. A weighted digraph is *undirected* if $a_{ij} = a_{ji}$ for all $i, j \in \{1, \ldots, n\}$. Clearly, G is undirected if and only if $A(G)$ is symmetric.

Numerous concepts introduced for digraphs remain equally valid for the case of weighted digraphs, including the connectivity notions and the definitions of in- and out-neighbors.

Finally, we generalize the notions of in- and out-degree to weighted digraphs. In a weighted digraph $G = (V, E, A)$ with $V = \{v_1, \ldots, v_n\}$, the *weighted out-degree* and the *weighted in-degree* of vertex v_i are defined by, respectively,

$$d_{\text{out}}(v_i) = \sum_{j=1}^{n} a_{ij}, \quad \text{and} \quad d_{\text{in}}(v_i) = \sum_{j=1}^{n} a_{ji}.$$

The weighted digraph G is *weight-balanced* if $d_{\text{out}}(v_i) = d_{\text{in}}(v_i)$ for all $v_i \in V$. The *weighted out-degree matrix* $D_{\text{out}}(G)$ and the *weighted in-degree matrix* $D_{\text{in}}(G)$ are the diagonal matrices defined by

$$D_{\text{out}}(G) = \text{diag}(A\mathbf{1}_n), \quad \text{and} \quad D_{\text{in}}(G) = \text{diag}(A^T\mathbf{1}_n).$$

That is, $(D_{\text{out}}(G))_{ii} = d_{\text{out}}(v_i)$ and $(D_{\text{in}}(G))_{ii} = d_{\text{in}}(v_i)$, respectively.

1.4.3 Distances on digraphs and weighted digraphs

We first present a few definitions for unweighted digraphs. Given a digraph G, the *(topological) length* of a directed path is the number of the edges composing it. Given two vertices u and v in the digraph G, the *distance* from u to v, denoted $\text{dist}_G(u, v)$, is the smallest length of any directed path from u to v, or $+\infty$ if there is no directed path from u to v. That is,

$$\text{dist}_G(u, v) = \min\left(\{\text{length}(p) \mid p \text{ is a directed path from } u \text{ to } v\} \cup \{+\infty\}\right).$$

Given a vertex v of a digraph G, the *radius* of v in G is the maximum of all the distances from v to any other vertex in G. That is,

$$\text{radius}(v, G) = \max\{\text{dist}_G(v, u) \mid u \in V(G)\}.$$

If T is a directed tree and v is its root, then the *depth* of T is $\text{radius}(v, T)$. Finally, the *diameter* of the digraph G is

$$\text{diam}(G) = \max\{\text{dist}_G(u, v) \mid u, v \in V(G)\}.$$

These definitions lead to the following simple results:

(i) $\text{radius}(v, G) \leq \text{diam}(G)$ for all vertices v of G;

(ii) G contains a spanning tree rooted at v if and only if $\text{radius}(v, G) < +\infty$; and

(iii) G is strongly connected if and only if $\text{diam}(G) < +\infty$.

The definitions of path length, distance between vertices, radius of a vertex, and diameter of a digraph can be easily applied to undirected graphs.

Next, we consider weighted digraphs. Given two vertices u and v in the weighted digraph G, the *weighted distance* from u to v, denoted $\text{wdist}_G(u, v)$, is the smallest weight of any directed path from u to v, or $+\infty$ if there is no directed path from u to v. That is,

$$\text{wdist}_G(u, v) = \min\left(\{\text{weight}(p) \mid p \text{ is a directed path from } u \text{ to } v\} \cup \{+\infty\}\right).$$

Here, the weight of a subgraph of a weighted digraph is the sum of the weights of all the edges of the subgraph. Note that when a digraph is thought of as a weighted digraph (with the unweighted adjacency matrix (1.4.1)), the notions of weight and weighted distance correspond to the usual notions of length and distance, respectively. We leave it the reader to provide the definitions of weighted radius, weighted depth, and weighted diameter.

1.4.4 Graph algorithms

In this section, we present a few algorithms defined on graphs. We present only high-level descriptions and we refer to Cormen et al. (2001) for a comprehensive discussion including a detailed treatment of computationally efficient data structures and algorithmic implementations.

1.4.4.1 Breadth-first spanning tree

Let v be a vertex of a digraph G with $\text{radius}(v, G) < +\infty$. A *breadth-first spanning (BFS) tree* of G with respect to v, denoted T_{BFS}, is a spanning directed tree rooted at v that contains a shortest path from v to every other vertex of G. (Here, a shortest path is one with the shortest topological length.) Let us provide the BFS ALGORITHM that, given a digraph G of order n and a vertex v with $\text{radius}(v, G) < +\infty$, computes a BFS tree T_{BFS} rooted at v:

> *[Informal description]* Initialize a subgraph to contain only the root v. Repeat $\text{radius}(v, G)$ times the following instructions: attach to the subgraph all out-neighbors of the subgraph as well as a single connecting edge for each out-neighbor. The final subgraph is the desired directed tree.

The algorithm is formally stated as follows:

```
function BFS(G, v)
```
1: $(V_1, E_1) := (\{v\}, \emptyset)$
2: **for** $k = 2$ to radius(v, G) **do**
3:    find all vertices $w_1, \ldots, w_m$ not in V_{k-1} that are out-neighbors of
      some vertex in V_{k-1} and, for $j \in \{1, \ldots, m\}$, let e_j be an edge
      connecting a vertex in V_{k-1} to w_j
4:    $V_k := V_{k-1} \cup \{w_1, \ldots, w_m\}$
5:    $E_k := E_{k-1} \cup \{e_1, \ldots, e_m\}$
6: **return** (V_n, E_n)

Note that the output of this algorithm is not necessarily unique, since the choice of edges at step 3: in the algorithm is not unique. Figure 1.10 shows an execution of the BFS ALGORITHM.

Figure 1.10 Execution of the BFS ALGORITHM. In the leftmost frame, vertex v is colored
in gray. The other frames correspond to incremental additions of vertices and
edges as specified by the function BFS. The output of the algorithm is a
BFS tree of the digraph. The BFS tree is represented in the last frame with
vertices and edges colored in gray.

Some properties of the BFS ALGORITHM are characterized as follows.

Lemma 1.28 (BFS tree). *For a digraph G with a vertex v, any digraph T computed by the BFS ALGORITHM, $T \in \mathrm{BFS}(G, v)$, has the following properties:*

(i) *T is a directed tree with root v;*

(ii) *T contains a shortest path from v to any other vertex reachable from v inside G, that is, if there is a path in G from v to w, then $w \in T$ and $\mathrm{dist}_G(v, w) = \mathrm{dist}_T(v, w)$; and*

(iii) *if G contains a spanning tree rooted at v, then T is spanning too and therefore, T is a BFS tree of G.*

We leave the proof to the reader. The key property of the algorithm is that (V_k, E_k), $k \in \{1, \ldots, n\}$, is a sequence of directed trees with the property that $(V_k, E_k) \subset (V_{k+1}, E_{k+1})$, for $k \in \{1, \ldots, n-1\}$.

1.4.4.2 The depth-first spanning tree

Next, we define the DFS ALGORITHM that, given a digraph G and a vertex v with radius$(v, G) < +\infty$, computes what we term a *depth-first spanning (DFS) tree* T_{DFS} rooted at v:

> *[Informal description]* Visit all nodes of the graph recording the traveled edges to form the desired tree. Visit the nodes in the following recursive way: (1) as long as a node has an unvisited child, visit it; (2) when the node has no more unvisited children, then return to its parent (and recursively attempt to visit its unvisited children).

The algorithm is formally stated as a recursive procedure, as follows:

function DFS(G, v)
1: $(V_{\mathrm{visited}}, E_{\mathrm{visited}}) := (\{v\}, \emptyset)$
2: DFS-VISIT(G, v)
3: **return** $(V_{\mathrm{visited}}, E_{\mathrm{visited}})$

function DFS-VISIT(G, w)
1: **for** u out-neighbor of w **do**
2:     **if** u does not belong to V_{visited} **then**
3:         $V_{\mathrm{visited}} := V_{\mathrm{visited}} \cup \{u\}$
4:         $E_{\mathrm{visited}} := E_{\mathrm{visited}} \cup \{(w, u)\}$
5:         DFS-VISIT(G, u)

Note that the output of this algorithm is not necessarily unique, since the order in which the vertices are chosen in step 1: of DFS-VISIT is not unique. Any digraph T computed by the DFS ALGORITHM, $T \in \mathrm{DFS}(G, v)$, is a directed spanning tree with root v. Figure 1.11 shows an execution of the algorithm.

Some properties of the DFS ALGORITHM are characterized as follows.

Lemma 1.29 (DFS tree). *For a digraph G with a vertex v, any digraph T computed by the DFS ALGORITHM, $T \in \mathrm{DFS}(G, v)$, has the following properties:*

(i) T is a directed tree with root v; and

(ii) if G contains a spanning tree rooted at v, then T is spanning too.

Note that both BFS and DFS trees are uniquely defined once a lexicographic order is introduced for the children of a node.

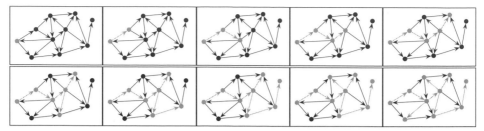

Figure 1.11 Execution of the DFS ALGORITHM. In the top leftmost frame, vertex v is
           colored in gray. The other frames correspond to incremental additions of
           vertices and edges as specified by the function DFS. The output of the
           algorithm is a DFS tree of the digraph. The DFS tree is represented in the
           last frame with vertices and edges in gray.

1.4.4.3 The shortest-paths tree in weighted digraphs via the Dijkstra algorithm

Finally, we focus on weighted digraphs and on the notion of weighted path
length. Given a weighted digraph G of order n with weighted adjacency
matrix A and a vertex v with radius$(v, G) < +\infty$, a *shortest-paths tree* of G
with respect to v, denoted $T_{\text{shortest-paths}}$, is a spanning directed tree rooted
at v that contains a (weighted) shortest path from v to every other vertex
of G. This tree differs from the BFS tree defined above because here the
path length is measured using the digraph weights.

We now provide the DIJKSTRA ALGORITHM that, given a digraph G of
order n and a vertex v with radius$(v, G) < +\infty$, computes a shortest-paths
tree $T_{\text{shortest-paths}}$ rooted at v:

> *[Informal description]* Incrementally construct a tree that con-
> tains only shortest paths. In each round, add to the tree (1) the
> node that is closest to the source and is not yet in the tree, and
> (2) the edge corresponding to the shortest path. The weighted
> distance to the source (required to perform step (1)) is computed
> via an array of distance estimates that is updated as follows:
> when a node is added to the tree, the distance estimates of all
> its out-neighbors are updated.

The algorithm is formally stated as follows:

function DIJKSTRA$\big((V, E, A), v\big)$
1: $T_{\text{shortest-paths}} := \emptyset$
   % Initialize estimated distances and estimated parent nodes
2: **for** $u \in V$ **do**

3:   $\texttt{dist}(u) := \begin{cases} 0, & u = v, \\ +\infty, & \text{otherwise.} \end{cases}$

4:   $\texttt{parent}(u) := u$

     % Main loop to grow the tree and update estimates

5: **while** ($T_{\text{shortest-paths}}$ does not contain all vertices) **do**

6:     find vertex u outside $T_{\text{shortest-paths}}$ with smallest $\texttt{dist}(u)$

7:     add to $T_{\text{shortest-paths}}$ the vertex u

8:     if $u \neq v$, add to $T_{\text{shortest-paths}}$ the edge $(\texttt{parent}(u), u)$

9:     **for** each node w that is an out-neighbor of u in (V, E, A) **do**

10:       **if** $\texttt{dist}(w) > \texttt{dist}(u) + a_{uw}$ **then**

11:          $\texttt{dist}(w) := \texttt{dist}(u) + a_{uw}$

12:          $\texttt{parent}(w) := u$

13: **return** $T_{\text{shortest-paths}}$

Note that the output of this algorithm is not necessarily unique, since the choice of vertex at step 6: in the algorithm is not unique. Figure 1.12 shows an execution of the the DIJKSTRA ALGORITHM.

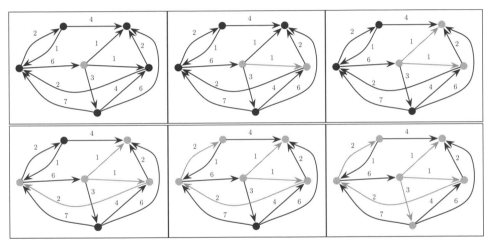

Figure 1.12 Execution of the DIJKSTRA ALGORITHM on the weighted digraph plotted in Figure 1.9. In the top leftmost frame, vertex v is colored in gray. The other frames correspond to incremental additions of vertices and edges as specified by the function DIJKSTRA. The output of the algorithm is a shortest-paths tree of the digraph rooted at v. This tree is represented in the last frame with vertices and edges colored in gray.

The following properties of the DIJKSTRA ALGORITHM mirror those of the BFS ALGORITHM in Lemma 1.28.

Lemma 1.30 (Dijkstra algorithm). *For a weighted digraph G with a vertex v, any digraph T computed by the* DIJKSTRA ALGORITHM, $T \in$ DIJKSTRA(G, v), *has the following properties:*

(i) T is a directed tree with root v;

(ii) T contains a shortest path from v to any other vertex reachable from v inside G, that is, if there is a path in G from v to w, then $w \in T$ and $\mathrm{wdist}_G(v, w) = \mathrm{wdist}_T(v, w)$; and

(iii) if G contains a spanning tree rooted at v, then T is spanning too, and therefore, T is a shortest-paths tree of G.

1.4.4.4 On combinatorial optimization problems

We conclude this section on graph algorithms with a brief mention of classic optimization problems defined on graphs. Standard references on combinatorial optimization include Vazirani (2001) and Korte and Vygen (2005). Given a weighted directed graph G, classical combinatorial optimization problems include the following:

Minimum-weight spanning tree. A *minimum-weight spanning tree* of G, denoted MST, is a spanning tree with the minimum possible weight. In order for the MST to exist, G must contain a spanning tree. If all the weights of the individual edges are different, then the MST is unique.

Traveling salesperson problem. A *traveling salesperson tour* of G, denoted TSP, is a cycle that passes through all the nodes of the digraph and has the minimum possible weight. In order for the TSP to exist, G must contain a cycle through all nodes.

Multicenter optimization problems. Given a weighted digraph G with vertices $V = \{v_1, \ldots, v_n\}$ and a set $U = \{u_1, \ldots, u_k\} \subset V$, the weighted distance from $v \in V$ to the set U is the smallest weighted distance from v to any vertex in $\{u_1, \ldots, u_k\}$. We now consider the cost functions $\mathcal{H}_{\max}, \mathcal{H}_{\Sigma} : V^k \to \mathbb{R}$ defined by

$$\mathcal{H}_{\max}(u_1, \ldots, u_k) = \max_{i \in \{1, \ldots, n\}} \min_{h \in \{1, \ldots, k\}} \mathrm{wdist}_G(v_i, u_h),$$

$$\mathcal{H}_{\Sigma}(u_1, \ldots, u_k) = \sum_{i=1}^{n} \min_{h \in \{1, \ldots, k\}} \mathrm{wdist}_G(v_i, u_h).$$

The *k-center problem* and the *k-median problem* consist of finding a set of vertices $\{u_1, \ldots, u_k\}$ that minimizes the *k-center function* $\mathcal{H}_{\max}$ and the *k-median function* $\mathcal{H}_{\Sigma}$, respectively. We refer to Vazirani (2001) for a discussion of the *k*-center and *k*-median problems (as well as the more general *uncapacited facility location* problem) over complete undirected graphs with edge costs satisfying the triangle inequality.

The Euclidean versions of these combinatorial optimization problems refer to the situation where one considers a weighted complete digraph whose vertex set is a point set in $\mathbb{R}^d$, $d \in \mathbb{N}$, and whose weight map assigns to each edge the Euclidean distance between the two nodes connected by the edge.

1.4.5 Algebraic graph theory

Algebraic graph theory (Biggs, 1994; Godsil and Royle, 2001) is the study of matrices defined by digraphs: in this section, we expose two topics. First, we study the equivalence between properties of graphs and of their associated adjacency matrices. We also specify how to associate a digraph to a nonnegative matrix. Second, we introduce and characterize the Laplacian matrix of a weighted digraph.

We begin by studying adjacency matrices. Note that the adjacency matrix of a weighted digraph is nonnegative and, in general, not stochastic. The following lemma expands on this point.

Lemma 1.31 (Weight-balanced digraphs and doubly stochastic adjacency matrices). *Let G be a weighted digraph of order n with weighted adjacency matrix A and weighted out-degree matrix D_{out}. Define the matrix*

$$
F = \begin{cases} D_{\text{out}}^{-1}A, & \text{if each out-degree is strictly positive,} \\ (I_n + D_{\text{out}})^{-1}(I_n + A), & \text{otherwise.} \end{cases}
$$

Then

(i) F is row-stochastic; and

(ii) F is doubly stochastic if G is weight-balanced and the weighted degree is constant for all vertices.

Proof. Consider first the case when each vertex has an outgoing edge so that D_{out} is invertible. We first note that $\operatorname{diag}(v)^{-1}v = \mathbf{1}_n$, for each $v \in (\mathbb{R}\backslash\{0\})^n$. Therefore

$$
\left(D_{\text{out}}^{-1}A\right)\mathbf{1}_n = \operatorname{diag}(A\mathbf{1}_n)^{-1}\left(A\mathbf{1}_n\right) = \mathbf{1}_n,
$$

which proves (i). Furthermore, if $D_{\text{out}} = D_{\text{in}} = dI_n$ for some $d \in \mathbb{R}_{>0}$, then

$$
\left(D_{\text{out}}^{-1}A\right)^T\mathbf{1}_n = \frac{1}{d}\left(A^T\mathbf{1}_n\right) = D_{\text{in}}^{-1}\left(A^T\mathbf{1}_n\right) = \operatorname{diag}(A^T\mathbf{1}_n)^{-1}\left(A^T\mathbf{1}_n\right) = \mathbf{1}_n,
$$

which proves (ii). Finally, if (V, E, A) does not have outgoing edges at each vertex, then apply the statement to the weighted digraph $(V, E \cup \{(i,i) \mid i \in \{1,\ldots,n\}\}, A + I_n)$. ∎

The next result characterizes the relationship between the adjacency matrix and directed paths in the digraph.

Lemma 1.32 (Directed paths and powers of the adjacency matrix).
Let G be a weighted digraph of order n with weighted adjacency matrix A, with unweighted adjacency matrix $A_{0,1} \in \{0,1\}^{n \times n}$, and possibly with self-loops. For all $i, j, k \in \{1, \ldots, n\}$

(i) *the (i,j) entry of $A_{0,1}^k$ equals the number of directed paths of length k (including paths with self-loops) from node i to node j; and*

(ii) *the (i,j) entry of A^k is positive if and only if there exists a directed path of length k (including paths with self-loops) from node i to node j.*

Proof. The second statement is a direct consequence of the first. The first statement is proved by induction. The statement is clearly true for $k = 1$. Next, we assume the statement is true for $k \geq 1$ and we prove it for $k+1$. By assumption, the entry $(A^k)_{ij}$ equals the number of directed paths from i to j of length k. Note that each path from i to j of length $k+1$ identifies (1) a unique node ℓ such that (i, ℓ) is an edge of G and (2) a unique path from ℓ to j of length k. We write $A^{k+1} = AA^k$ in components as

$$(A^{k+1})_{ij} = \sum_{\ell=1}^{n} A_{i\ell} (A^k)_{\ell j}.$$

Therefore, it is true that the entry $(A^{k+1})_{ij}$ equals the number of directed paths from i to j of length $k+1$. This concludes the induction argument. ∎

The following proposition characterizes in detail the relationship between various connectivity properties of the digraph and algebraic properties of the adjacency matrix. The result is illustrated in Figure 1.13 and its proof is postponed until Section 1.8.3.

Proposition 1.33 (Connectivity properties of the digraph and positive powers of the adjacency matrix). *Let G be a weighted digraph of order n with weighted adjacency matrix A. The following statements are equivalent:*

(i) *G is strongly connected;*

(ii) *A is irreducible; and*

(iii) *$\sum_{k=0}^{n-1} A^k$ is positive.*

For any $j \in \{1, \ldots, n\}$, the following two statements are equivalent:

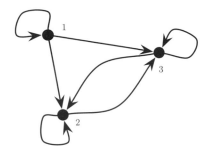

Figure 1.13 An illustration of Proposition 1.33. Even though vertices 2 and 3 are globally reachable, the digraph is not strongly connected because vertex 1 has no in-neighbor other than itself. Therefore, the associated adjacency matrix $A = (a_{ij})$ with $(a_{1j}) = \mathbf{1}_3$, $(a_{2j}) = (a_{3j}) = (0,1,1)$, is reducible.

(iv) the jth node of G is globally reachable; and

(v) the jth column of $\sum_{k=0}^{n-1} A^k$ has positive entries.

Stronger statements can be given for digraphs with self-loops.

Proposition 1.34 (Connectivity properties of the digraph and positive powers of the adjacency matrix: cont'd). *Let G be a weighted digraph of order n with weighted adjacency matrix A and with self-loops at each node. The following statements are equivalent:*

(iv) G is strongly connected; and

(v) A^{n-1} is positive.

For any $j \in \{1, \ldots, n\}$, the following two statements are equivalent:

(iv) the jth node of G is globally reachable; and

(v) the jth column of A^{n-1} has positive entries.

Next, we characterize the relationship between irreducible aperiodic digraphs and primitive matrices (recall Definition 1.12). We will postpone the proof to Section 1.8.3.

Proposition 1.35 (Strongly connected and aperiodic digraph and primitive adjacency matrix). *Let G be a weighted digraph of order n with weighted adjacency matrix A. The following two statements are equivalent:*

(i) G is strongly connected and aperiodic; and

(ii) A is primitive, that is, there exists $k \in \mathbb{N}$ such that A^k is positive.

This concludes our study of adjacency matrices associated to weighted digraphs. Next, we emphasize how all results obtained so far have analogs that hold when the original object is a nonnegative matrix, instead of a weighted digraph.

Remark 1.36 (From a nonnegative matrix to its associated digraphs). Given a nonnegative $n \times n$ matrix A, its *associated weighted digraph* is the weighted digraph with nodes $\{1, \ldots, n\}$, and weighted adjacency matrix A. The unweighted version of this weighted digraph is called the *associated digraph*. The following statements are analogs of the previous lemmas:

(i) if A is stochastic, then its associated digraph has weighted out-degree matrix equal to I_n;

(ii) if A is doubly stochastic, then its associated weighted digraph is weight-balanced and, additionally, both in-degree and out-degree matrices are equal to I_n; and

(iii) A is irreducible if and only if its associated weighted digraph is strongly connected.   •

So far, we have analyzed in detail the properties of adjacency matrices. We conclude this section by studying a second relevant matrix associated to a digraph, called the Laplacian matrix. The *Laplacian matrix* of the weighted digraph G is

$$L(G) = D_{\text{out}}(G) - A(G).$$

Some immediate consequences of this definition are the following:

(i) $L(G)\mathbf{1}_n = \mathbf{0}_n$, that is, 0 is an eigenvalue of $L(G)$ with eigenvector $\mathbf{1}_n$;

(ii) G is undirected if and only if $L(G)$ is symmetric; and

(iii) $L(G)$ equals the Laplacian matrix of the digraph obtained by adding to or removing from G any self-loop with arbitrary weight.

Further properties are established as follows.

Theorem 1.37 (Properties of the Laplacian matrix). *Let G be a weighted digraph of order n. The following statements hold:*

(i) *all eigenvalues of $L(G)$ have nonnegative real part (thus, if G is undirected, then $L(G)$ is symmetric positive semidefinite);*

(ii) *if G is strongly connected, then $\text{rank}(L(G)) = n - 1$, that is, 0 is a simple eigenvalue of $L(G)$;*

(iii) G contains a globally reachable vertex if and only if $\operatorname{rank}(L(G)) = n - 1;$

(iv) the following three statements are equivalent:

    (a) G is weight-balanced;

    (b) $\mathbf{1}_n^T L(G) = \mathbf{0}_n^T;$ *and*

    (c) $L(G) + L(G)^T$ *is positive semidefinite.*

1.5 DISTRIBUTED ALGORITHMS ON SYNCHRONOUS NETWORKS

Here, we introduce a synchronous network as a group of processors with the ability to exchange messages and perform local computations. What we present is a basic classic model studied extensively in the distributed algorithms literature. Our treatment is directly adopted with minor variations, from the texts by Lynch (1997) and Peleg (2000).

1.5.1 Physical components and computational models

Loosely speaking, a synchronous network is a group of processors, or nodes, that possess a local state, exchange messages along the edges of a digraph, and compute an update to their local state based on the received messages. Each processor alternates the two tasks of exchanging messages with its neighboring processors and of performing a computation step. Let us begin by describing what constitutes a network.

Definition 1.38 (Network). The physical component of a *synchronous network* $\mathcal{S}$ is a digraph (I, E_{cmm}), where:

   (i) $I = \{1, \ldots, n\}$ is called the *set of unique identifiers (UIDs)*; and

   (ii) E_{cmm} is a set of directed edges over the vertices $\{1, \ldots, n\}$, called the communication links.   •

In general, the set of unique identifiers does not need to be n consecutive natural numbers, but we adopt this convention for simplicity. The set E_{cmm} models the topology of the communication service among the nodes: for $i, j \in \{1, \ldots, n\}$, processor i can send a message to processor j if the directed edge (i, j) is present in E_{cmm}. Note that, unlike the standard treatments in Lynch (1997) and Peleg (2000), we do not assume the digraph to be strongly connected; the required connectivity assumption will be specified on a case-by-case basis.

Next, we discuss the state and the algorithms that each processor possesses and executes, respectively. By convention, we let the superscript $[i]$ denote any quantity associated with the node i.

Definition 1.39 (Distributed algorithm). A *distributed algorithm* $\mathcal{DA}$ for a network $\mathcal{S}$ consists of the sets

(i) $\mathbb{A}$, a set containing the `null` element, called the *communication alphabet*—elements of $\mathbb{A}$ are called *messages*;

(ii) $W^{[i]}$, $i \in I$, called the *processor state sets*; and

(iii) $W_0^{[i]} \subseteq W^{[i]}$, $i \in I$, sets of *allowable initial values*;

and of the maps

(i) $\mathrm{msg}^{[i]} : W^{[i]} \times I \to \mathbb{A}$, $i \in I$, called *message-generation functions*; and

(ii) $\mathrm{stf}^{[i]} : W^{[i]} \times \mathbb{A}^n \to W^{[i]}$, $i \in I$, called *state-transition functions*.

If $W^{[i]} = W$, $\mathrm{msg}^{[i]} = \mathrm{msg}$, and $\mathrm{stf}^{[i]} = \mathrm{stf}$ for all $i \in I$, then $\mathcal{DA}$ is said to be *uniform* and is described by a tuple $(\mathbb{A}, W, \{W_0^{[i]}\}_{i \in I}, \mathrm{msg}, \mathrm{stf})$. •

Now, with all elements in place, we can explain in more detail how a synchronous network executes a distributed algorithm (see Figure 1.14). The *state* of processor i is a variable $w^{[i]} \in W^{[i]}$, initially set equal to an

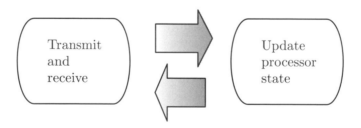

Figure 1.14 The execution of a distributed algorithm by a synchronous network.

allowable value in $W_0^{[i]}$. At each time instant $\ell \in \mathbb{Z}_{\geq 0}$, processor i sends to each of its out-neighbors j in the communication digraph (I, E_{cmm}) a message (possibly the `null` message) computed by applying the message-generation function $\mathrm{msg}^{[i]}$ to the current values of its state $w^{[i]}$ and to the identity j. Subsequently, but still at time instant $\ell \in \mathbb{Z}_{\geq 0}$, processor i updates the value of its state $w^{[i]}$ by applying the state-transition function $\mathrm{stf}^{[i]}$ to the current value of $w^{[i]}$ and to the messages it receives from its in-neighbors. At each round, the first step is transmission and the second one is computation. These notions are formalized in the following definition.

Definition 1.40 (Network evolution). Let $\mathcal{DA}$ be a distributed algorithm for the network $\mathcal{S}$. The *evolution* of $(\mathcal{S}, \mathcal{DA})$ from initial conditions $w_0^{[i]} \in W_0^{[i]}$, $i \in I$, is the collection of trajectories $w^{[i]} : \mathbb{Z}_{\geq 0} \to W^{[i]}$, $i \in I$, satisfying

$$w^{[i]}(\ell) = \mathrm{stf}^{[i]}(w^{[i]}(\ell - 1), y^{[i]}(\ell)),$$

where $w^{[i]}(-1) = w_0^{[i]}$, $i \in I$, and where the trajectory $y^{[i]} : \mathbb{Z}_{\geq 0} \to \mathbb{A}^n$ (describing the messages received by processor i) has components $y_j^{[i]}(\ell)$, for $j \in I$, given by

$$y_j^{[i]}(\ell) = \begin{cases} \mathrm{msg}^{[j]}(w^{[j]}(\ell - 1), i), & \text{if } (j, i) \in E_{\mathrm{cmm}}, \\ \texttt{null}, & \text{otherwise.} \end{cases}$$

Let $\ell \mapsto w(\ell) = (w^{[1]}(\ell), \dots, w^{[n]}(\ell))$ denote the collection of trajectories. •

We conclude this section with two sets of remarks. We first discuss some aspects of our communication model that have a large impact on the subsequent development. We then collect a few general comments about control structures and failure modes relevant in the study of distributed algorithms on networks.

Remarks 1.41 (Aspects of the communication model).

(i) The network $\mathcal{S}$ and the algorithm $\mathcal{DA}$ are referred to as *synchronous* because the communications between all processors takes place at the same time for all processors.

(ii) Communication is modeled as a so-called "point-to-point" service: a processor can specify different messages for different out-neighbors and knows the processor identity corresponding to any incoming message.

(iii) Information is exchanged between processors as messages, that is, elements of the alphabet $\mathbb{A}$; the message $\texttt{null}$ indicates no communication. Messages might encode logical expressions such as $\texttt{true}$ and $\texttt{false}$, or finite-resolution quantized representations of integer and real numbers.

(iv) In some uniform algorithms, the messages between processors are the processors' states. In such cases, the corresponding communication alphabet is $\mathbb{A} = W \cup \{\texttt{null}\}$ and the message-generation function $\mathrm{msg}_{\mathrm{std}}(w, j) = w$ is referred to as the *standard message-generation function*. •

Remarks 1.42 (Advanced topics: Control structures and failures).

(i) Processors in a network have only partial information about the network topology. In general, each processor only knows its own UID, and the UID of its in- and out-neighbors. Sometimes we will assume that the processor knows the network diameter. In some cases (Peleg, 2000), actively running networks might depend upon "control structures," that is, structures that are computed at initial time and are exploited in subsequent algorithms. For example, routing tables might be computed for routing problems, "leader" processors might be elected, and tree structures might be computed and represented in a distributed manner for various tasks; for example, coloring or maximal independent set problems. We present some sample algorithms to compute these structures below.

(ii) A key issue in the study of distributed algorithms is the possible occurrence of failures. A network might experience intermittent or permanent communication failures: along given edges, a `null` message or an arbitrary message might be delivered instead of the intended value. Alternatively, a network might experience various types of processor failures: a processor might transmit only `null` messages (i.e., the msg function always returns `null`), a processor might quit updating its state (i.e., the stf function neglects incoming messages and returns the current state value), or a processor might implement arbitrarily modified msg and stf functions. The latter situation, in which completely arbitrary and possibly malicious behavior is adopted by faulty nodes, is referred to as a Byzantine failure in the distributed algorithms literature.   •

1.5.2 Complexity notions

Here, we begin our analysis of the performance of distributed algorithms. We introduce a notion of algorithm completion and, in turn, we introduce the classic notions of time, space, and communication complexity.

Definition 1.43 (Algorithm completion). We say that an algorithm *terminates* when only `null` messages are transmitted and all processors' states become constants.   •

Remarks 1.44 (Alternative termination notions).

(i) In the interest of simplicity, we have defined evolutions to be unbounded in time and we do not explicitly require algorithms to actually have termination conditions, that is, to be able to detect when termination takes place.

(ii) It is also possible to define the termination time as the first instant when a given problem or task is achieved, independently of the fact that the algorithm might continue to transmit data subsequently. •

Definition 1.45 (Time complexity). The *(worst-case) time complexity* of a distributed algorithm $\mathcal{DA}$ on a network $\mathcal{S}$, denoted $\mathrm{TC}(\mathcal{DA})$, is the maximum number of rounds required by the execution of $\mathcal{DA}$ on $\mathcal{S}$ among all allowable initial states until termination.   •

Next, it is of interest to quantify the memory and communication requirements of distributed algorithms. From an information theory viewpoint (Gallager, 1968), the information content of a memory variable or of a message is properly measured in bits. On the other hand, it is convenient to use the alternative notions of "basic memory unit" and "basic message." It is customary (Peleg, 2000) to assume that a "basic memory unit" or a "basic message" contains $\log(n)$ bits; so that, for example, the information content of a robot identifier $i \in \{1, \ldots, n\}$ is $\log(n)$ bits or, equivalently, one "basic memory unit." Note that elements of the processor state set W or of the alphabet set $\mathbb{A}$ might amount to multiple basic memory units or basic messages; the `null` message has zero cost. Unless specified otherwise, the following definitions and examples are stated in terms of basic memory units and basic messages.

Definition 1.46 (Space complexity). The *(worst-case) space complexity* of a distributed algorithm $\mathcal{DA}$ on a network $\mathcal{S}$, denoted by $\mathrm{SC}(\mathcal{DA})$, is the maximum number of basic memory units required by a processor executing $\mathcal{DA}$ on $\mathcal{S}$ among all processors and among all allowable initial states until termination.   •

Remark 1.47 (Space complexity conventions). By convention, each processor knows its identity, that is, it requires $\log(n)$ bits to represent its unique identifier in a set with n distinct elements. We do not count this cost in the space complexity of an algorithm.   •

Next, we introduce a notion of communication complexity.

Definition 1.48 (Communication complexity). The *(worst-case) communication complexity* of a distributed algorithm $\mathcal{DA}$ on a network $\mathcal{S}$, denoted by $\mathrm{CC}(\mathcal{DA})$, is the maximum number of basic messages transmitted over the entire network during the execution of $\mathcal{DA}$ among all allowable initial states until termination.   •

We conclude this section by discussing ways of quantifying time, space and communication complexity. The idea, borrowed from combinatorial optimization, is to adopt asymptotic "order of magnitude" measures. Formally, complexity bounds will be expressed with respect to the Bachmann–Landau symbols O, Ω and Θ defined in Section 1.1. Let us be more specific:

(i) we will say that an algorithm has time complexity *of order* $\Omega(f(n))$ *over some network* if, for all n, there exists a network of order n and initial processor values such that the time complexity of the algorithm is greater than a constant factor times $f(n)$;

(ii) we will say that an algorithm has time complexity *of order* $O(f(n))$ *over arbitrary networks* if, for all n, for all networks of order n and for all initial processor values, the time complexity of the algorithm is lower than a constant factor times $f(n)$; and

(iii) we will say that an algorithm has time complexity *of order* $\Theta(f(n))$ if its time complexity is of order $\Omega(f(n))$ over some network and $O(f(n))$ over arbitrary networks at the same time.

Similar conventions will be used for space and communication complexity.

In many cases, the complexity of an algorithm will typically depend upon the number of vertices of the network. It is therefore useful to present a few simple facts about these functions now. Over arbitrary digraphs $\mathcal{S} = (I, E_{\mathrm{cmm}})$ of order n, we have

$$\mathrm{diam}(\mathcal{S}) \in \Theta(n), \quad |E_{\mathrm{cmm}}(\mathcal{S})| \in \Theta(n^2) \quad \text{and} \quad \mathrm{radius}(v, \mathcal{S}) \in \Theta(\mathrm{diam}(\mathcal{S})),$$

where v is any vertex of $\mathcal{S}$.

Remark 1.49 (Additional complexity notions). Numerous variations of the proposed complexity notions are possible and may be of interest.

Global lower bounds. In the definition of lower bound, consider the logic quantifier describing the role of the network. The lower bound statement is "existential" rather than "global," in the sense that the bound does not hold for all graphs. As discussed in Peleg (2000), it is possible to define also "global" lower bounds, that is, lower bounds over all graphs, or lower bounds over specified classes of graphs.

Average complexity notions. The proposed complexity notions focus on the worst-case situation. It is possible to define *expected* or *average* complexity notions, where one is interested in characterizing, for example, the average number of rounds required or the average number of basic messages transmitted over the entire network during the algorithm execution among all allowable initial states until termination.

Problem complexity. It is possible to define complexity notions for problems, rather than algorithms, by considering, for example, the worst-case optimal performance among all algorithms that solve the given problem, or over classes of algorithms or classes of graphs. •

1.5.3 Broadcast and BFS tree computation

In the following, we consider some basic algorithmic problems such as the simple one-to-all communication task—that is, broadcasting—and the establishment of some "control structures" (see Remarks 1.42), such as the construction of a BFS spanning tree and the election of a leader.

Problem 1.50 (Broadcast). Assume that a processor, called the *source*, has a message, called the *token*. Transmit the token to all others processors in the network. •

Note that existence of a spanning tree rooted at the source is a necessary requirement for the broadcast problem to be solvable. We begin by establishing some analysis results for the broadcast problem.

Lemma 1.51 (Complexity lower bounds for the broadcast problem). *Let $\mathcal{S}$ be a network containing a spanning tree rooted at v. The broadcast problem for $\mathcal{S}$ from the source v has communication complexity lower bounded by $n - 1$ and time complexity lower bounded by* radius$(v, \mathcal{S})$.

In what follows, we shall solve the broadcast problem while simultaneously also considering the following problem.

Problem 1.52 (BFS tree computation). Let $\mathcal{S}$ be a network containing a spanning tree rooted at v. Compute a distributed representation for a BFS tree rooted at v. •

We add two remarks on the BFS tree computation problem:

(i) By a distributed representation of a directed tree with bounded memory at each node, we mean the following: each child vertex knows the identity of its parent and the root vertex knows that it has no parents. A more informative structure would require each parent to know the identity of its children; this is easy to achieve on undirected digraphs.

(ii) The BFS tree computation has the same lower bounds as the broadcast problem.

An elegant and classic solution to the broadcast and BFS tree computation problems is given by the FLOODING ALGORITHM. This algorithm implements the same "breadth-first search" mechanism of the (centralized) BFS ALGORITHM characterized in Lemma 1.28:

> *[Informal description]* The source broadcasts the token to its out-neighbors. In each communication round, each node determines whether it has received a non-**null** message from one of its in-neighbors. When a non-**null** message is received—that is, the token is received—the node performs two actions. First, the node stores the token in the variable **data** (this solves the Broadcast problem). Second, the node stores the identity of one of the transmitting in-neighbors in the variable **parent** (this solves the BFS tree computation problem). Specifically, if the message is received simultaneously from multiple in-neighbors, then the node stores the smallest among the identities of the transmitting in-neighbors. In the subsequent communication round, the node broadcasts the token to its out-neighbors.

To formally describe the algorithm, we assume that the node with the message to be broadcast is $v = 1$. Also, we assume that the token is a letter of the Greek alphabet $\{\alpha, \ldots, \omega\}$:

Synchronous Network: $\mathcal{S} = (\{1, \ldots, n\}, E_{\text{cmm}})$

Distributed Algorithm: FLOODING

Alphabet: $\mathbb{A} = \{\alpha, \ldots, \omega\} \cup \texttt{null}$

Processor State: $w = (\texttt{parent}, \texttt{data}, \texttt{snd-flag})$, where

parent	$\in \{0, \ldots, n\}$,	initially:	$\texttt{parent}^{[1]} = 1$,
			$\texttt{parent}^{[j]} = 0$ for all $j \neq 1$
data	$\in \mathbb{A}$,	initially:	$\texttt{data}^{[1]} = \mu$,
			$\texttt{data}^{[j]} = \texttt{null}$ for all $j \neq 1$
snd-flag	$\in \{\texttt{false}, \texttt{true}\}$,	initially:	$\texttt{snd-flag}^{[1]} = \texttt{true}$,
			$\texttt{snd-flag}^{[j]} = \texttt{false}$ for $j \neq 1$

function $\text{msg}(w, i)$
  1: **if** (parent $\neq i$) AND (snd-flag $=$ true) **then**
  2:    **return data**
  3: **else**
  4:    **return null**

function $\text{stf}(w, y)$

1: **case**
2:   (data = null) AND (y contains only null messages):
      % The node has not yet received the token
3:       new-parent := null
4:       new-data := null
5:       new-snd-flag := false
6:   (data = null) AND (y contains a non-null message):
      % The node has just received the token
7:       new-parent := smallest UID among transmitting in-neighbors
8:       new-data := a non-null message
9:       new-snd-flag := true
10:   (data ≠ null):
      % If the node already has the token, then do not re-broadcast it
11:       new-parent := parent
12:       new-data := data
13:       new-snd-flag := false
14: **return** (new-parent, new-data, new-snd-flag)

An execution of the FLOODING ALGORITHM is shown in Figure 1.15.

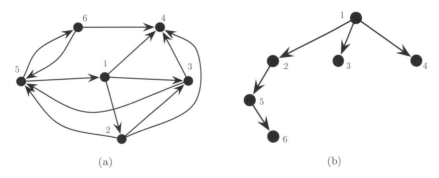

(a)	(b)

Figure 1.15 An example execution of the FLOODING ALGORITHM. The source is vertex 1: (a) shows the network and (b) shows the BFS tree that results from the execution.

This algorithm can analyzed by induction: one can show that, for $d \in \{1, \ldots, \text{radius}(v, \mathcal{S})\}$, every node at a distance d from the root receives a non-null message at round d. A summary of the results is given as follows.

Lemma 1.53 (Complexity upper bounds for the flooding algorithm). *For a network $\mathcal{S}$ containing a spanning tree rooted at v, the* FLOODING ALGORITHM *has communication complexity in $\Theta(|E_{\text{cmm}}|)$, time complexity in $\Theta(\text{radius}(v, \mathcal{S}))$, and space complexity in $\Theta(1)$.*

We conclude the section with a final remark.

Remark 1.54 (Termination condition for the flooding algorithm).
As presented, the FLOODING ALGORITHM does not include a termination
condition, that is, the processors do not have a mechanism to detect when
the broadcast and tree computation are complete. If an upper bound on the
graph diameter is known, then it is easy to design a termination condition
based on this information; we do this in the next subsection. If no *a priori*
knowledge is available, then one can design more sophisticated algorithms
for networks with stronger connectivity properties. We refer to Lynch (1997)
and Peleg (2000) for a complete discussion about this.   •

1.5.4 Leader election

Next, we formulate another interesting problem for a network.

Problem 1.55 (Leader election). Assume that all processors of a network have a state variable, say `leader`, initially set to `unknwn`. We say that
a leader is elected when one and only one processor has the state variable
set to `true` and all others have it set to `false`. Elect a leader.   •

  This task that is a bit more global in nature. We display here a solution
that requires individual processors to know the diameter of the network,
denoted by $\mathrm{diam}(\mathcal{S})$, or an upper bound on it:

> *[Informal description]* In each communication round, each agent
> sends to its out-neighbors the maximum UID it has received up
> to that time. This is repeated for $\mathrm{diam}(\mathcal{S})$ rounds. At the last
> round, each agent compares the maximum received UID with its
> own, and declares itself a leader if they coincide, or a non-leader
> otherwise.

The algorithm is called the FLOODMAX ALGORITHM: the maximum UID in
the network is transmitted to other agents in an incremental fashion. At
the first communication round, agents that are neighbors of the agent with
the maximum UID receive the message from it. At the next communication
round, the neighbors of these agents receive the message with the maximum
UID. This process goes on for $\mathrm{diam}(\mathcal{S})$ rounds, to ensure that every agent
receives the maximum UID. Note that there are networks for which all agents
receive the message with the maximum UID in fewer communication rounds
than $\mathrm{diam}(\mathcal{S})$. The algorithm is formally stated as follows:

Synchronous Network: $\mathcal{S} = (\{1, \ldots, n\}, E_{\mathrm{cmm}})$

Distributed Algorithm: FLOODMAX

Alphabet: $\mathbb{A} = \{1, \ldots, n\} \cup \{\texttt{null}\}$

Processor State: $w = (\texttt{my-id}, \texttt{max-id}, \texttt{leader}, \texttt{round})$, where

$\texttt{my-id} \in \{1, \ldots, n\}$,   initially: $\texttt{my-id}^{[i]} = i$ for all i

$\texttt{max-id} \in \{1, \ldots, n\}$,   initially: $\texttt{max-id}^{[i]} = i$ for all i

$\texttt{leader} \in \{\texttt{false}, \texttt{true}, \texttt{unknwn}\}$, initially: $\texttt{leader}^{[i]} = \texttt{unknwn}$ for all i

$\texttt{round} \in \{0, 1, \ldots, \operatorname{diam}(\mathcal{S})\}$,   initially: $\texttt{round}^{[i]} = 0$ for all i

function $\operatorname{msg}(w, i)$
1: **if** $\operatorname{round} < \operatorname{diam}(\mathcal{S})$ **then**
2:     **return** max-id
3: **else**
4:     **return** null

function $\operatorname{stf}(w, y)$
1: $\texttt{new-id}:= \max\{\texttt{max-id}, \text{largest identifier in } y\}$
2: **case**
3:     $\operatorname{round} < \operatorname{diam}(\mathcal{S})$:   new-lead := unknwn
4:     $\operatorname{round} = \operatorname{diam}(\mathcal{S})$ AND max-id = my-id: new-lead := true
5:     $\operatorname{round} = \operatorname{diam}(\mathcal{S})$ AND max-id > my-id: new-lead := false
6: **return** $(\texttt{my-id}, \texttt{new-id}, \texttt{new-lead}, \texttt{round} + 1)$

Figure 1.16 shows an execution of the FLOODMAX ALGORITHM. Some properties of this algorithm are characterized in the following lemma. A complete analysis of this algorithm, including modifications to improve the communication complexity, is discussed in Lynch (1997, Section 4.1).

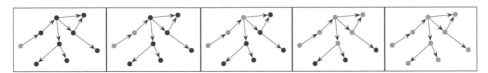

Figure 1.16 Execution of the FLOODMAX ALGORITHM. The diameter of the network is 4. In the leftmost frame, the agent with the maximum UID is colored in gray. After 4 communication rounds, its message has been received by all agents.

Lemma 1.56 (Complexity upper bounds for the floodmax algorithm). *For a network $\mathcal{S}$ containing a spanning tree, the* FLOODMAX AL-GORITHM *has communication complexity in* $O(\operatorname{diam}(\mathcal{S})|E_{\mathrm{cmm}}|)$, *time complexity equal to* $\operatorname{diam}(\mathcal{S})$, *and space complexity in* $\Theta(1)$.

A simplification of the FLOODMAX ALGORITHM leads to the Le Lann–Chang–Roberts algorithm (or LCR ALGORITHM in short) for leader election in rings, see (Lynch, 1997, Chapter 3.3), which we describe next. The

LCR ALGORITHM runs on a ring digraph and does not require the agents to know the diameter of the network. We provide an informal and a formal description of the algorithm.

> *[Informal description]* In each communication round, each agent sends to its neighbors the maximum UID it has received up to that time. (Agents do not record the number of communication rounds.) When the agent with the maximum UID receives its own UID from an in-neighbor, it declares itself the leader.

Synchronous Network: ring digraph

Distributed Algorithm: LCR

Alphabet: $\mathbb{A} = \{1, \dots, n\} \cup \{\texttt{null}\}$

Processor State: $w = (\texttt{my-id}, \texttt{max-id}, \texttt{leader}, \texttt{snd-flag})$, where

$\texttt{my-id} \in \{1, \dots, n\}$,	initially: $\texttt{my-id}^{[i]} = i$ for all i
$\texttt{max-id} \in \{1, \dots, n\}$,	initially: $\texttt{max-id}^{[i]} = i$ for all i
$\texttt{leader} \in \{\texttt{true}, \texttt{false}, \texttt{unknwn}\}$,	initially: $\texttt{leader}^{[i]} = \texttt{unknwn}$ for all i
$\texttt{snd-flag} \in \{\texttt{true}, \texttt{false}\}$,	initially: $\texttt{snd-flag}^{[i]} = \texttt{true}$ for all i

function $\mathrm{msg}(w, i)$
  1: **if** $\texttt{snd-flag} = \texttt{true}$ **then**
  2:     **return** $\texttt{max-id}$
  3: **else**
  4:     **return** $\texttt{null}$

function $\mathrm{stf}(w, y)$
  1: **case**
  2:     (y contains only $\texttt{null}$ msgs) OR (largest identifier in $y <$ $\texttt{my-id}$):
  3:         $\texttt{new-id} := \texttt{max-id}$
  4:         $\texttt{new-lead} := \texttt{leader}$
  5:         $\texttt{new-snd-flag} := \texttt{false}$
  6:     (largest identifier in $y = \texttt{my-id}$):
  7:         $\texttt{new-id} := \texttt{max-id}$
  8:         $\texttt{new-lead} := \texttt{true}$
  9:         $\texttt{new-snd-flag} := \texttt{false}$
 10:     (largest identifier in $y > \texttt{my-id}$):
 11:         $\texttt{new-id} :=$ largest identifier in y
 12:         $\texttt{new-lead} := \texttt{false}$
 13:         $\texttt{new-snd-flag} := \texttt{true}$
 14: **return** $(\texttt{my-id}, \texttt{new-id}, \texttt{new-lead}, \texttt{new-snd-flag})$

Figure 1.17 shows an execution of the LCR ALGORITHM. The properties of the LCR ALGORITHM can be characterized as follows.

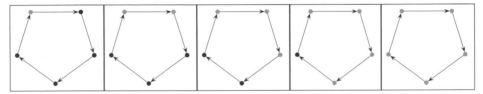

Figure 1.17 Execution of the LCR ALGORITHM. In the leftmost frame, the agent with the maximum UID is colored in gray. After 5 communication rounds, this agent receives its own UID from its in-neighbor and declares itself the leader.

Lemma 1.57 (Complexity upper bounds for the LCR algorithm). *For a ring network S of order n, the* LCR ALGORITHM *has communication complexity in* $\Theta(n^2)$, *time complexity equal to* n, *and space complexity in* $\Theta(1)$.

1.5.5 Shortest-paths tree computation

Finally, we consider the shortest-paths tree problem in a weighted digraph: in Section 1.4.4 we presented the DIJKSTRA ALGORITHM to solve this problem in a centralized setting; we present here the BELLMAN-FORD ALGORITHM for the distributed setting. We consider a synchronous network associated to a weighted digraph, that is, we assume that a strictly positive weight is associated to each communication edge. We aim to compute a tree containing shortest paths from a source, say node 1, to all other nodes. As for the computation of a BFS tree, we aim to obtain a distributed representation of a directed tree with bounded memory at each node:

> *[Informal description]* Each agent maintains in its memory an estimate `dist` of its weighted distance from the source, and an estimate `parent` of the in-neighbor corresponding to the (weighted) shortest path from the source. The `dist` estimate is initialized to 0 for the source and to $+\infty$ for all other nodes. In each communication round, each agent performs the following tasks: (1) it transmits its `dist` value estimate to its out-neighbors, (2) it computes the smallest quantity among "the `dist` value received from an in-neighbor summed with the edge weight corresponding to that in-neighbor," and (3) if the agent's estimate `dist` is larger than this quantity, then the agent updates its `dist` and its estimate `parent`.

The algorithm is formally stated as follows:

Synchronous Network with Weights: $\mathcal{S} = (\{1, \ldots, n\}, E_{\mathrm{cmm}}, A)$

Distributed Algorithm: DISTRIBUTED BELLMAN-FORD

Alphabet: $\mathbb{A} = \mathbb{R}_{>0} \cup \mathtt{null} \cup \{+\infty\}$

Processor State: $w = (\mathtt{parent}, \mathtt{dist})$, where

> $\mathtt{parent} \in \{1, \ldots, n\}$,   initially: $\mathtt{parent}^{[j]} = j$ for all j
>
> $\mathtt{dist} \quad \in \mathbb{A}$,   initially: $\mathtt{data}^{[1]} = 0$,
>
> $\mathtt{data}^{[j]} = +\infty$ for all $j \neq 1$

function $\mathrm{msg}(w, i)$

1: **if** round $< n$ **then**
2:      **return dist**
3: **else**
4:      **return null**

function $\mathrm{stf}(w, y)$

1:   $i :=$ processor UID
2:   $k := \mathrm{arginf}\{y_j + a_{ji} \mid$ for all $y_j \neq \mathtt{null}\}$
3:   **if** $(\mathtt{dist} < k)$ **then**
4:      **return** $(\mathtt{parent}, \mathtt{dist})$
5:   **else**
6:      **return** $(k, y_k + a_{ki})$

In other words, if we let $d_i \in \mathbb{R}_{\geq 0} \cup \{+\infty\}$ denote the $\mathtt{dist}$ variable for each processor i, then the BELLMAN-FORD ALGORITHM is equivalent to the following discrete-time dynamical system:

$$d_i(\ell + 1) = \inf \left\{ d_i(\ell), \inf\{d_j(\ell) + a_{ji} \mid (j, i) \in E_{\mathrm{cmm}}\} \right\},$$

with initial conditions $d(0) = (1, +\infty, \ldots, +\infty)$. (Recall that E_{cmm} is the edge set and that the weights a_{ij} are strictly positive for all $(i, j) \in E_{\mathrm{cmm}}$.)

The following formal statements may be made about the evolution of this algorithm. If there exists a directed spanning tree rooted at vertex 1, then all variables d_i will take a final value in time equal to their topological distance from vertex 1. After k communication rounds, the estimated distance at node i equals the shortest path of topological length at most k from the source to node i. Therefore, after $n - 1$ communication rounds, all possible distinct topological paths connecting source to node i have been investigated.

The complexity properties of the DISTRIBUTED BELLMAN-FORD ALGO-RITHM are described as follows.

Lemma 1.58 (Complexity upper bounds for the distributed Bell-man-Ford algorithm). *For a network $\mathcal{S}$ of order n containing a spanning tree rooted at v, the* DISTRIBUTED BELLMAN-FORD ALGORITHM *has communication complexity in $\Theta(n|E_{\mathrm{cmm}}|)$, time complexity equal to $n - 1$, and space complexity in $\Theta(1)$.*

Figure 1.18 shows an execution of the DISTRIBUTED BELLMAN-FORD ALGORITHM in a weighted digraph with four nodes and six edges.

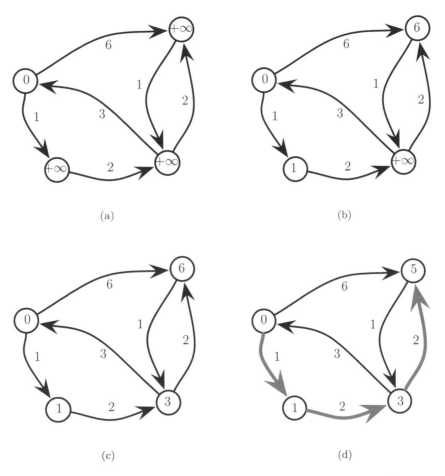

Figure 1.18 Execution of the DISTRIBUTED BELLMAN-FORD ALGORITHM. (a) The processor state initialization. The vertex 1 is the only one whose variable dist is 0. After three iterations, as guaranteed by Lemma 1.58, (d) depicts the resulting shortest-paths tree of the digraph rooted at vertex 1. This tree is represented in the last frame, with edges colored in gray.

1.6 LINEAR DISTRIBUTED ALGORITHMS

Computing a linear combination of the initial states of the processors is one of the most basic computation that we might be interested in implementing on a synchronous network. More accurately, linear distributed algorithms on synchronous networks are discrete-time linear dynamical systems whose evolution map is linear and has a sparsity structure related to the network. These algorithms represent an important class of iterative algorithms that find applications in optimization, in the solution of systems of equations, and in distributed decision making; see, for instance Bertsekas and Tsitsiklis (1997). In this section, we present some relevant results on distributed linear algorithms.

1.6.1 Linear iterations on synchronous networks

Given a synchronous network $\mathcal{S} = (\{1,\dots,n\}, E_{\mathrm{cmm}})$, assign a scalar $f_{ji} \neq 0$ to each directed edge $(i,j) \in E_{\mathrm{cmm}}$. Given such scalars f_{ji} for $(i,j) \in E_{\mathrm{cmm}}$, the LINEAR COMBINATION ALGORITHM over $\mathcal{S}$ is defined as follows:

Distributed Algorithm: LINEAR COMBINATION

Alphabet: $\mathbb{A} = \mathbb{R} \cup \mathtt{null}$

Processor state: $w \in \mathbb{R}$

function $\mathrm{msg}(w, i) = \mathrm{msg}_{\mathrm{std}}(w, i)$

function $\mathrm{stf}(w, y)$
  1: $i :=$ processor UID
  2: **return** $f_{ii}w + \sum_{j \in \mathcal{N}^{\mathrm{in}}(i)} f_{ij} y_j$

We assume that each processor $i \in \{1,\dots,n\}$ knows the scalars f_{ij}, for $j \in \mathcal{N}^{\mathrm{in}}(i) \cup \{i\}$, so that it can evaluate the state-transition function. Also, we assume that real numbers may be transmitted through a communication channel, that is, we neglect quantization issues in the message-generation function.

In the language of Section 1.3, one can regard the LINEAR COMBINATION ALGORITHM over $\mathcal{S}$ as the discrete-time continuous-space dynamical system (X, X_0, f), with $X = X_0 = \mathbb{R}^n$ and an evolution map defined by $f(w) = F \cdot w$, where we define a matrix $F \in \mathbb{R}^{n \times n}$ with vanishing entries except for f_{ji}, for $(i,j) \in E_{\mathrm{cmm}}$. Note that, if $A(\mathcal{S})$ denotes the adjacency matrix of the digraph $\mathcal{S}$, then the entries of F vanish precisely when the entries of

$A(\mathcal{S})^T$ vanish. With this notation, the evolution $w : \mathbb{Z}_{\geq 0} \to \mathbb{R}^n$ with initial condition $w_0 \in \mathbb{R}^n$ is given by

$$w(0) = w_0, \quad w(\ell + 1) = F \cdot w(\ell), \quad \ell \in \mathbb{Z}_{\geq 0}. \tag{1.6.1}$$

Conversely, any linear algorithm of the form (1.6.1) can easily be cast as a LINEAR COMBINATION ALGORITHM over a suitable synchronous network. We do this bookkeeping carefully, in order to be consistent with the notion of associated weighted digraph from Remark 1.36. Given $F \in \mathbb{R}^{n \times n}$, let $\mathcal{S}_F$ be the synchronous network with node set $\{1, \dots, n\}$ and with edge set $E_{\mathrm{cmm}}(F)$, defined by any of the equivalent statements:

(i) $(i, j) \in E_{\mathrm{cmm}}(F)$ if and only if $f_{ji} \neq 0$; or

(ii) $\mathcal{S}_F$ is the reversed and unweighted version of the digraph associated to F.

1.6.2 Averaging algorithms

Here, we study linear combination algorithms over time-dependent weighted directed graphs; we restrict our analysis to nonnegative weights.

Definition 1.59 (Averaging algorithms). The *averaging algorithm* associated to a sequence of stochastic matrices $\{F(\ell) \mid \ell \in \mathbb{Z}_{\geq 0}\} \subset \mathbb{R}^{n \times n}$ is the discrete-time dynamical system

$$w(\ell + 1) = F(\ell) \cdot w(\ell), \quad \ell \in \mathbb{Z}_{\geq 0}. \tag{1.6.2}$$

•

In the literature, such algorithms are often referred to as agreement algorithms, or as consensus algorithms.

There are useful ways to compute a stochastic matrix, and therefore, a time-independent averaging algorithm, from a weighted digraph; see Exercise E1.15.

Definition 1.60 (Adjacency- and Laplacian-based averaging). Let G be a weighted digraph with node set $\{1, \dots, n\}$, weighted adjacency matrix A, weighted out-degree matrix D_{out}, and weighted Laplacian L. Then

(i) the *adjacency-based averaging algorithm* is defined by the stochastic matrix $(I_n + D_{\mathrm{out}})^{-1}(I_n + A)$ and reads in components

$$w_i(\ell + 1) = \frac{1}{1 + d_{\mathrm{out}}(i)}\Big(w_i(\ell) + \sum_{j=1}^{n} a_{ij} w_j(\ell)\Big); \tag{1.6.3}$$

(ii) given a positive scalar ε upper bounded by $\min\{1/d_{\mathrm{out}}(i) \mid i \in \{1,\ldots,n\}\}$, the *Laplacian-based averaging algorithm* is defined by the stochastic matrix $I_n - \varepsilon L(G)$ and reads in components

$$w_i(\ell+1) = \Big(1 - \varepsilon \sum_{j=1, j\neq i}^{n} a_{ij}\Big) w_i(\ell) + \varepsilon \sum_{j=1, j\neq i}^{n} a_{ij} w_j(\ell). \quad (1.6.4)$$

These notions are immediately extended to sequences of stochastic matrices arising from sequences of weighted digraphs. •

Adjacency-based averaging algorithms arising from unweighted undirected graphs without self-loops are also known as *equal-neighbor averaging rule* or the *Vicsek's model* (see Vicsek et al., 1995). Specifically, if G is an unweighted graph with vertices $\{1,\ldots,n\}$ and without self-loops, then the equal-neighbor averaging rule is

$$w_i(\ell+1) = \mathrm{avrg}\Big(\{w_i(\ell)\} \cup \{w_j(\ell) \mid j \in \mathcal{N}_G(i)\}\Big), \quad (1.6.5)$$

where we adopt the shorthand $\mathrm{avrg}(\{x_1,\ldots,x_k\}) = (x_1 + \cdots + x_k)/k$.

Remark 1.61 (Sensing versus communication interpretation of directed edges). In the definition of averaging algorithms arising from digraphs, the digraph edges play the role of "sensing edges," not that of "communication edges." In other words, a nonzero entry a_{ij}, corresponding to the digraph edge (i, j), implies that the ith component of the state is updated with the jth component of the state. It is as if node i could sense the state of node j, rather than node i transmitting to node j its own state. •

Next, we present the main stability and convergence results for averaging algorithms associated to a sequence of stochastic matrices. We start by discussing equilibrium points and their stability. Recall that $\mathbf{1}_n$ is an eigenvector of any stochastic matrix with eigenvalue 1 and that the diagonal set $\mathrm{diag}(\mathbb{R}^n)$ is the vector subspace generated by $\mathbf{1}_n$. Therefore, any point in $\mathrm{diag}(\mathbb{R}^n)$ is an equilibrium for any averaging algorithm. We refer to the points of the $\mathrm{diag}(\mathbb{R}^n)$ as *agreement configurations*, since all the components of an element in $\mathrm{diag}(\mathbb{R}^n)$ are equal to the same value. We will informally say that an algorithm *achieves agreement* if it steers the network state toward the set of agreement configurations.

Lemma 1.62 (Stability of agreement configurations). *Any averaging algorithm in $\mathbb{R}^n$ is uniformly stable and uniformly bounded with respect to $\mathrm{diag}(\mathbb{R}^n)$.*

Regarding convergence results, we need to introduce a useful property of collections of stochastic matrices. Given $\alpha \in \,]0,1]$, the set of *non-degenerate matrices with respect to* α consists of all stochastic matrices F with entries f_{ij}, for $i,j \in \{1,\dots,n\}$, satisfying

$$f_{ii} \in [\alpha,1], \quad \text{and} \quad f_{ij} \in \{0\} \cup [\alpha,1] \text{ for } j \neq i.$$

Additionally, the sequence of stochastic matrices $\{F(\ell) \mid \ell \in \mathbb{Z}_{\geq 0}\}$ is *non-degenerate* if there exists $\alpha \in \,]0,1]$ such that $F(\ell)$ is non-degenerate with respect to α for all $\ell \in \mathbb{Z}_{\geq 0}$. We now state the main convergence result and postpone its proof to Section 1.8.5.

Theorem 1.63 (Convergence for time-dependent stochastic matrices). *Let $\{F(\ell) \mid \ell \in \mathbb{Z}_{\geq 0}\} \subset \mathbb{R}^{n \times n}$ be a non-degenerate sequence of stochastic matrices. For $\ell \in \mathbb{Z}_{\geq 0}$, let $G(\ell)$ be the unweighted digraph associated to $F(\ell)$, according to Remark 1.36. The following statements are equivalent:*

(i) the set $\mathrm{diag}(\mathbb{R}^n)$ is uniformly globally attractive for the averaging algorithm associated to $\{F(\ell) \mid \ell \in \mathbb{Z}_{\geq 0}\}$; and

(ii) there exists a duration $\delta \in \mathbb{N}$ such that, for all $\ell \in \mathbb{Z}_{\geq 0}$, the digraph

$$G(\ell+1) \cup \cdots \cup G(\ell+\delta)$$

contains a globally reachable vertex.

We collect a few observations about this result.

Remarks 1.64 (Discussion of Theorem 1.63).

(i) The statement in Theorem 1.63(i) means that each solution to the time-dependent linear dynamical system (1.6.2) converges uniformly and asymptotically to the vector subspace generated by $\mathbf{1}_n$.

(ii) The necessary and sufficient condition in Theorem 1.63(ii) amounts to the existence of a uniformly bounded time duration δ with the property that a weak connectivity assumption holds over each collection of δ consecutive digraphs. We refer to Blondel et al. (2005) for a counterexample showing that if the duration in Theorem 1.63 is not uniformly bounded, then there exist algorithms that do not converge.

(iii) According to Definition 1.23, uniform convergence is a property of all solutions to system (1.6.2) starting at *any arbitrary time*, and not only at time equal to zero. If we restrict our attention to solutions that only start at time zero, then Theorem 1.63 should be modified as follows: the statement in Theorem 1.63(i) implies, but is not implied by, the statement in Theorem 1.63(ii).

(iv) The theorem applies only to sequences of non-degenerate matrices. Indeed, there exist sequences of degenerate stochastic matrices whose associated averaging algorithms converge. Furthermore, one does not even need to consider sequences, because it is possible to define converging algorithms by just considering a single stochastic matrix. Precisely when the stochastic matrix is primitive, we already know that the associated averaging algorithm will converge (see Theorem 1.13). Examples of degenerate primitive stochastic matrices (with converging associated averaging algorithms) are given in Exercise E1.23. We discuss time-invariant averaging algorithms in Proposition 1.68 below.   •

Theorem 1.63 gives a general result about non-degenerate stochastic matrices that are not necessarily symmetric. The following theorem presents a convergence result for the case of symmetric matrices (i.e., undirected digraphs) under connectivity requirements that are weaker (i.e., the duration does not need to be uniformly bounded) than those expressed in statement (ii) of Theorem 1.63.

Theorem 1.65 (Convergence for time-dependent stochastic symmetric matrices). *Let $\{F(\ell) \mid \ell \in \mathbb{Z}_{\geq 0}\} \subset \mathbb{R}^{n \times n}$ be a non-degenerate sequence of symmetric, stochastic matrices. For $\ell \in \mathbb{Z}_{\geq 0}$, let $G(\ell)$ be the unweighted graph associated to $F(\ell)$, according to Remark 1.36. The following statements are equivalent:*

(i) the set $\mathrm{diag}(\mathbb{R}^n)$ is globally attractive for the averaging algorithm associated to $\{F(\ell) \mid \ell \in \mathbb{Z}_{\geq 0}\}$; and

(ii) for all $\ell \in \mathbb{Z}_{\geq 0}$, the graph

$$\bigcup_{\tau \geq \ell} G(\tau)$$

   is connected.

Let us particularize our discussion here on adjacency- and Laplacian-based averaging algorithms.

Corollary 1.66 (Convergence of adjacency- and Laplacian-based averaging algorithms). *Let $\{G(\ell) \mid \ell \in \mathbb{Z}_{\geq 0}\} \subset \mathbb{R}^{n \times n}$ be a sequence of weighted digraphs. The following statements are equivalent:*

(i) there exists $\delta \in \mathbb{N}$ such that, for all $\ell \in \mathbb{Z}_{\geq 0}$, the digraph

$$G(\ell + 1) \cup \cdots \cup G(\ell + \delta)$$

   contains a globally reachable vertex;

(ii) the set $\mathrm{diag}(\mathbb{R}^n)$ is uniformly globally attractive for the adjacency-

based averaging algorithm (1.6.3) *associated to* $\{G(\ell) \mid \ell \in \mathbb{Z}_{\geq 0}\}$; *and*

(iii) the set $\mathrm{diag}(\mathbb{R}^n)$ *is uniformly globally attractive for the Laplacian-based averaging algorithm* (1.6.4) *(defined with* $\varepsilon < 1/n$*) associated to* $\{G(\ell) \mid \ell \in \mathbb{Z}_{\geq 0}\}$.

Finally, we refine the results presented thus far by discussing some further aspects.

Proposition 1.67 (Convergence to a point in the invariant set). *Under the assumptions in Theorem 1.63 and assuming that* $\mathrm{diag}(\mathbb{R}^n)$ *is uniformly globally attractive for the averaging algorithm, each individual evolution converges to a specific point of* $\mathrm{diag}(\mathbb{R}^n)$.

In general, the final value upon which all w_i, $i \in \{1, \ldots, n\}$, agree in the limit is unknown. This final value depends on the initial condition and the specific sequence of matrices defining the time-dependent linear algorithm. In some cases, however, one can compute the final value by restricting the class of allowable matrices. We consider two settings: time-independent averaging algorithms and doubly stochastic averaging algorithms.

First, we specialize the main convergence result to the case of time-independent averaging algorithms. Note that, given a stochastic matrix F, convergence of the averaging algorithm associated to F for all initial conditions is equivalent to the matrix F being semi-convergent (see Definition 1.6).

Proposition 1.68 (Time-independent averaging algorithm). *Consider the linear dynamical system on* $\mathbb{R}^n$

$$w(\ell + 1) = Fw(\ell), \quad \ell \in \mathbb{Z}_{\geq 0}. \tag{1.6.6}$$

Assume that $F \in \mathbb{R}^{n \times n}$ *is stochastic, let* $G(F)$ *denote its associated weighted digraph, and let* $v \in \mathbb{R}^n$ *be a left eigenvector of* F *with eigenvalue* 1. *Assume either one of the two following properties:*

(i) F is primitive (i.e., $G(F)$ is strongly connected and aperiodic); or

(ii) F has non-zero diagonal terms and a column of F^{n-1} has positive entries (i.e., $G(F)$ has self-loops at each node and has a globally reachable node).

Then every trajectory w of system (1.6.6) *converges to* $(v^T w(0) / v^T \mathbf{1}_n) \mathbf{1}_n$.

Proof. From Theorem 1.63 we know that the dynamical system (1.6.6) converges if property (ii) holds. The same conclusion follows if F satisfies property (i) because of the Perron–Frobenius Theorem 1.13 and Lemma 1.7. To computing the limiting value, note that

$$v^T w(\ell + 1) = v^T F w(\ell) = v^T w(\ell),$$

that is, the quantity $\ell \mapsto v^T w(\ell)$ is constant. Because F is semi-convergent and stochastic, we know that $\lim_{\ell \to +\infty} w(\ell) = \alpha \mathbf{1}_n$ for some α. To conclude, we compute α from the relationship $\alpha(v^T \mathbf{1}_n) = \lim_{\ell \to +\infty} v^T w(\ell) = v^T w(0)$. ∎

Remarks 1.69 (Alternative conditions for time-independent averaging).

(i) The following necessary and sufficient condition generalizes and is weaker than the two sufficient conditions given in Proposition 1.68: every trajectory of system (1.6.6) is asymptotically convergent if and only if all sinks of the condensation digraph of $G(F)$ are aperiodic subgraphs of $G(F)$. We refer the interested reader to Meyer (2001, Chapter 8) for the proof of this statement and for the related notion of ergodic classes of a Markov chain. Also, we refer the interested reader to Exercise E1.13 for the notion of condensation digraph.

(ii) Without introducing any trajectory w, the result of the proposition can be equivalently stated by saying that

$$\lim_{\ell \to +\infty} F^\ell = (v^T \mathbf{1}_n)^{-1} \mathbf{1}_n v^T. \qquad \bullet$$

Second, we focus on the case of doubly stochastic averaging algorithms.

Corollary 1.70 (Average consensus). *Let $\{F(\ell) \mid \ell \in \mathbb{Z}_{\geq 0}\}$ be a sequence of stochastic matrices as in Theorem 1.63. If all matrices $F(\ell)$, $\ell \in \mathbb{Z}_{\geq 0}$, are doubly stochastic, then every trajectory w of the averaging algorithms satisfies*

$$\sum_{i=1}^{n} w_i(\ell) = \sum_{i=1}^{n} w_i(0), \quad \text{for all } \ell,$$

that is, the sum of the initial conditions is a conserved quantity. Therefore, if $\{F(\ell) \mid \ell \in \mathbb{Z}_{\geq 0}\}$ is non-degenerate and satisfies property (ii) in Theorem 1.63, then

$$\lim_{\ell \to +\infty} w_j(\ell) = \frac{1}{n} \sum_{i=1}^{n} w_i(0), \quad j \in \{1, \dots, n\}.$$

Proof. The proof of the first fact is an immediate consequence of

$$\sum_{i=1}^{n} w_i(\ell+1) = \mathbf{1}_n^T w(\ell+1) = \mathbf{1}_n^T F(\ell) w(\ell) = \mathbf{1}_n^T w(\ell) = \sum_{i=1}^{n} w_i(\ell).$$

The second fact is an immediate consequence of the first fact. ∎

In other words, if the matrices are doubly stochastic, then each component of the trajectories will converge to the average of the initial condition. We therefore adopt the following definition: an *average-consensus averaging algorithm* is an averaging algorithm whose sequence of stochastic matrices are all doubly stochastic.

1.6.3 The convergence speed of averaging algorithms

We know that any trajectory of the associated averaging algorithm converges to the diagonal set $\mathrm{diag}(\mathbb{R}^n)$; in what follows we characterize how fast this convergence takes place. We begin with some general definitions for semi-convergent matrices (recall the discussion culminating in Lemma 1.7).

Definition 1.71 (Convergence time and exponential convergence factor). Let $A \in \mathbb{R}^{n \times n}$ be semi-convergent with limit $\lim_{\ell \to +\infty} A^\ell = A^*$.

(i) For $\varepsilon \in]0,1[$, the *ε-convergence time* of A is the smallest time $T_\varepsilon(A) \in \mathbb{Z}_{\geq 0}$ such that, for all $x_0 \in \mathbb{R}^n$ and $\ell \geq T_\varepsilon(A)$,

$$\left\| A^\ell x_0 - A^* x_0 \right\|_2 \leq \varepsilon \| x_0 - A^* x_0 \|_2.$$

(ii) The *exponential convergence factor* of A, denoted by $r_{\exp}(A) \in [0,1[$, is

$$r_{\exp}(A) = \sup_{x_0 \neq A^* x_0} \limsup_{\ell \to +\infty} \left(\frac{\| A^\ell x_0 - A^* x_0 \|_2}{\| x_0 - A^* x_0 \|_2} \right)^{1/\ell}. \qquad \bullet$$

The exponential convergence factor has the following interpretation: If the trajectory $x(\ell) = A^\ell x_0$ maximizing the sup operator has the form $x(\ell) = \rho^\ell(x_0 - x^*) + x^*$, for $\rho < 1$, then it is immediate to see that $r_{\exp}(A) = \rho$.

Lemma 1.72 (Exponential convergence factor of a convergent matrix). *If A is a convergent matrix, then $r_{\exp}(A) = \rho(A)$.*

In what follows, we are interested in studying how the convergence time and the exponential convergence factor of a matrix depend upon ε and upon the dimension of the matrix itself.

Remark 1.73 (Complexity notions). Analogously to the treatment in Section 1.5.2, we introduce some complexity notions. Let $A_n \in \mathbb{R}^{n \times n}$, $n \in \mathbb{N}$, be a sequence of semi-convergent matrices with limit $\lim_{\ell \to +\infty} A_n^\ell = A_n^*$, and let $\varepsilon \in \,]0, 1]$. We say that:

(i) $T_\varepsilon(A_n)$ is *of order* $\Omega(f(n, \varepsilon))$ if, for all n and all ε, there exists an initial condition $x_0 \in \mathbb{R}^n$ such that $\left\| A_n^\ell x_0 - A^* x_0 \right\|_2 > \varepsilon \left\| x_0 - A^* x_0 \right\|_2$ for all times ℓ greater than a constant factor times $f(n, \varepsilon)$;

(ii) $T_\varepsilon(A_n)$ is *of order* $O(f(n, \varepsilon))$ if, for all n and all ε, $T_\varepsilon(A_n)$ is less than or equal to a constant factor times $f(n, \varepsilon)$; and

(iii) $T_\varepsilon(A_n)$ is *of order* $\Theta(f(n, \varepsilon))$ if it is both of order $\Omega(f(n, \varepsilon))$ and of order $O(f(n, \varepsilon))$. $\qquad\qquad\bullet$

Lemma 1.74 (Asymptotic relationship). *Let $A_n \in \mathbb{R}^{n \times n}$, $n \in \mathbb{N}$, be a sequence of semi-convergent matrices and let $\varepsilon \in \,]0, 1]$. In the limit as $\varepsilon \to 0^+$ and as $n \to +\infty$,*

$$T_\varepsilon(A_n) \in O\left(\frac{1}{1 - r_{\exp}(A_n)} \log \varepsilon^{-1} \right).$$

Proof. By the definition of the exponential convergence factor and of lim sup, we know that for all $\eta > 0$, there exists N such that, for all $\ell > N$,

$$\left\| A^\ell x_0 - A^* x_0 \right\|_2 \le (r_{\exp}(A_n) + \eta)^\ell \|x_0 - A^* x_0\|_2.$$

The ε-convergence time is upper bounded by any ℓ such that $(r_{\exp}(A_n) + \eta)^\ell \le \varepsilon$. Selecting $\eta = (1 - r_{\exp}(A_n))/2$, simple manipulations lead to

$$\ell \ge \frac{1}{-\log((r_{\exp}(A_n) + 1)/2)} \log \varepsilon^{-1}.$$

It is also immediate to note that $\frac{2}{1-r} \ge \frac{1}{-\log((r+1)/2)}$, for all $r \in \,]0, 1[$. This establishes the bound in the statement above. $\qquad\blacksquare$

Next, we apply the notion of convergence time and exponential convergence factor to any non-degenerate stochastic matrix whose associated digraph has a globally reachable node.

Lemma 1.75 (Exponential convergence factor of stochastic matrices). *Let F be a stochastic matrix with strictly positive diagonal entries and whose associated digraph has a globally reachable node. Then*

$$r_{\exp}(F) = \rho_{\text{ess}}(F).$$

(From equation (1.2.1), recall that $\rho_{\text{ess}}(F) = \max\{\|\lambda\|_{\mathbb{C}} \mid \lambda \in \text{spec}(F) \setminus \{1\}\}$.)

Proof. If $v \in \mathbb{R}^n$ is a left eigenvector of F, then, as in Proposition 1.68,

$$\lim_{\ell \to +\infty} F^{\ell} = F^* = (v^T \mathbf{1}_n)^{-1} \mathbf{1}_n v^T.$$

Relying upon $v^T F = v^T$ and $F\mathbf{1}_n = \mathbf{1}_n$, straightforward manipulations show that $F^* = F^* F = F F^* = F^* F^*$ and in turn

$$F^{\ell+1} - F^* = (F - F^*)(F^{\ell} - F^*).$$

For any $w_0 \in \mathbb{R}^n$ such that $w_0 \neq F^* w_0$, define the error variable $e(\ell) := F^{\ell} w_0 - F^* w_0$. Note that the error variable evolves according to $e(\ell + 1) = (F - F^*)e(\ell)$ and converges to zero. Additionally, the rate at which $w(\ell) = F^{\ell} w_0$ converges to $F^* w_0$ is the same at which $e(\ell)$ converges to zero, that is,

$$r_{\text{exp}}(F - F^*) = r_{\text{exp}}(F).$$

Therefore,

$$r_{\text{exp}}(F) = r_{\text{exp}}(F - F^*) = \rho(F - F^*) = \rho_{\text{ess}}(F).$$

∎

The following result establishes bounds on convergence factors and convergence times for stochastic matrices arising from the equal-neighbor averaging rule in equation (1.6.5).

Theorem 1.76 (Bounds on the convergence factor and the convergence time). *Let G be an undirected unweighted connected graph of order n and let $\varepsilon \in \,]0, 1]$. Define the stochastic matrix $F = (I_n + D(G))^{-1}(I_n + A(G))$. There exists $\gamma > 0$ (independent of n) such that the exponential convergence factor and convergence time of F satisfy*

$$r_{\text{exp}}(F) \leq 1 - \gamma n^{-3}, \quad and \quad T_{\varepsilon}(F) \in O(n^3 \log \varepsilon^{-1}),$$

as $\varepsilon \to 0^+$ and $n \to +\infty$.

1.6.4 Algorithms defined by tridiagonal Toeplitz and tridiagonal circulant matrices

This section presents a detailed analysis of the convergence rates of linear distributed algorithms defined by tridiagonal Toeplitz matrices and by certain circulant matrices. Let us start by introducing the family of matrices under study. For $n \geq 2$ and $a, b, c \in \mathbb{R}$, define the $n \times n$ matrices $\mathrm{Trid}_n(a, b, c)$ and $\mathrm{Circ}_n(a, b, c)$ by

$$
\mathrm{Trid}_n(a, b, c) = \begin{bmatrix} b & c & 0 & \cdots & 0 \\ a & b & c & \cdots & 0 \\ \vdots & \ddots & \ddots & \ddots & \vdots \\ 0 & \cdots & a & b & c \\ 0 & \cdots & 0 & a & b \end{bmatrix},
$$

and

$$
\mathrm{Circ}_n(a, b, c) = \mathrm{Trid}_n(a, b, c) + \begin{bmatrix} 0 & \cdots & \cdots & 0 & a \\ 0 & \cdots & \cdots & 0 & 0 \\ \vdots & \ddots & \ddots & \ddots & \vdots \\ 0 & 0 & \cdots & 0 & 0 \\ c & 0 & \cdots & 0 & 0 \end{bmatrix}.
$$

We call the matrices Trid_n and Circ_n *tridiagonal Toeplitz* and *tridiagonal circulant*, respectively. The two matrices only differ in their $(1, n)$ and $(n, 1)$ entries. Note our convention that

$$
\mathrm{Circ}_2(a, b, c) = \begin{bmatrix} b & a + c \\ a + c & b \end{bmatrix}.
$$

Note that, for $a = 0$ and $c \neq 0$ (alternatively, $a \neq 0$ and $c = 0$), the synchronous networks defined by $\mathrm{Trid}(a, b, c)$ and $\mathrm{Circ}(a, b, c)$ are, respectively, the chain and the ring digraphs introduced in Section 1.4. If both a and c are non-vanishing, then the synchronous networks are, respectively, the undirected versions of the chain and the ring digraphs.

Now, we characterize the eigenvalues and eigenvectors of Trid_n and Circ_n.

Lemma 1.77 (Eigenvalues and eigenvectors of tridiagonal Toeplitz and tridiagonal circulant matrices). *For $n \geq 2$ and $a, b, c \in \mathbb{R}$, the following statements hold:*

(i) *for $ac \neq 0$, the eigenvalues and eigenvectors of $\mathrm{Trid}_n(a, b, c)$ are, respectively, for $i \in \{1, \ldots, n\}$,*

$$
b + 2c \sqrt{\frac{a}{c}} \cos \left(\frac{i\pi}{n+1} \right) \in \mathbb{C}, \qquad \begin{pmatrix} \left(\frac{a}{c}\right)^{1/2} \sin \left(\frac{i\pi}{n+1} \right) \\ \left(\frac{a}{c}\right)^{2/2} \sin \left(\frac{2i\pi}{n+1} \right) \\ \vdots \\ \left(\frac{a}{c}\right)^{n/2} \sin \left(\frac{ni\pi}{n+1} \right) \end{pmatrix} \in \mathbb{C}^n;
$$

(ii) *the eigenvalues and eigenvectors of $\mathrm{Circ}_n(a, b, c)$ are, respectively,* *for $i \in \{1, \dots, n\}$ and $\omega = \exp(\frac{2\pi\sqrt{-1}}{n})$,*

$$b + (a + c)\cos\left(\frac{i2\pi}{n}\right) + \sqrt{-1}(c - a)\sin\left(\frac{i2\pi}{n}\right) \in \mathbb{C},$$

and $(1, \omega^i, \dots, \omega^{(n-1)i})^T \in \mathbb{C}^n$.

Proof. Both facts are discussed, for example, in Meyer (2001, Example 7.2.5 and Exercise 7.2.20). Fact (ii) requires some straightforward algebraic manipulations. ∎

Figure 1.19 illustrates the location of the eigenvalues of these matrices in the complex plane.

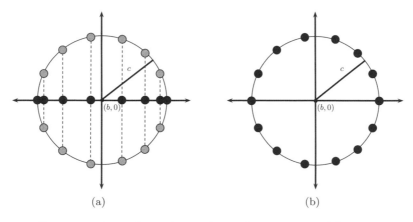

(a)   (b)

Figure 1.19 The eigenvalues of Toeplitz and circulant matrices (cf., Lemma 1.77) are closely related to the roots of unity. Plotted in the complex plane, the black disks correspond in (a) to the eigenvalues of $\mathrm{Trid}_{13}(a, b, c)$, and in (b) to the eigenvalues of $\mathrm{Circ}_{14}(0, b, c)$.

Remarks 1.78 (Inclusion relationships for eigenvalues of tridiagonal Toeplitz and tridiagonal circulant matrices).

(i) The set of eigenvalues of $\mathrm{Trid}_n(a, b, c)$ is contained in the real interval $[b - 2\sqrt{ac}, b + 2\sqrt{ac}]$, if $ac \geq 0$, and in the interval in the complex plane $[b - 2\sqrt{-1}\sqrt{|ac|}, b + 2\sqrt{-1}\sqrt{|ac|}]$, if $ac \leq 0$.

(ii) The set of eigenvalues of $\mathrm{Circ}_n(a, b, c)$ is contained in the ellipse on the complex plane with center b, horizontal axis $2|a+c|$, and vertical axis $2|c - a|$. •

Next, we characterize the convergence rate of linear algorithms defined by tridiagonal Toeplitz and tridiagonal circulant matrices. As in the previous section, we are interested in asymptotic results as the system dimension $n \to +\infty$ and as the accuracy parameter ε goes to 0^+.

Theorem 1.79 (Linear algorithms defined by tridiagonal Toeplitz and tridiagonal circulant matrices). *Let $n \geq 2$, $\varepsilon \in\,]0,1[$, and $a, b, c \in \mathbb{R}$. Let $x : \mathbb{Z}_{\geq 0} \to \mathbb{R}^n$ and $y : \mathbb{Z}_{\geq 0} \to \mathbb{R}^n$ be solutions to*

$$x(\ell + 1) = \mathrm{Trid}_n(a, b, c)\, x(\ell), \qquad y(\ell + 1) = \mathrm{Circ}_n(a, b, c)\, y(\ell),$$

with initial conditions $x(0) = x_0$ and $y(0) = y_0$, respectively. The following statements hold:

(i) if $a = c \neq 0$ and $|b| + 2|a| = 1$, then $\lim_{\ell \to +\infty} x(\ell) = \mathbf{0}_n$ with ε-convergence time in $\Theta\!\left(n^2 \log \varepsilon^{-1}\right)$;

(ii) if $a \neq 0$, $c = 0$ and $0 < |b| < 1$, then $\lim_{\ell \to +\infty} x(\ell) = \mathbf{0}_n$ with ε-convergence time in $O\!\left(n \log n + \log \varepsilon^{-1}\right)$; and

(iii) if $a \geq 0$, $c \geq 0$, $1 > b > 0$ and $a + b + c = 1$, then $\lim_{\ell \to +\infty} y(\ell) = \left(\frac{1}{n}\mathbf{1}_n^T y_0\right)\mathbf{1}_n$ with ε-convergence time in $\Theta\!\left(n^2 \log \varepsilon^{-1}\right)$.

The proof of this result is reported in Section 1.8.6. Next, we extend these results to another interesting set of tridiagonal matrices. For $n \geq 2$ and $a, b \in \mathbb{R}$, define the $n \times n$ matrices $\mathrm{ATrid}_n^+(a, b)$ and $\mathrm{ATrid}_n^-(a, b)$ by

$$\mathrm{ATrid}_n^\pm(a, b) = \mathrm{Trid}_n(a, b, a) \pm \begin{bmatrix} a & 0 & \cdots & \cdots & 0 \\ 0 & 0 & \cdots & \cdots & 0 \\ \vdots & \ddots & \ddots & \ddots & \vdots \\ 0 & \cdots & \cdots & 0 & 0 \\ 0 & \cdots & \cdots & 0 & a \end{bmatrix}.$$

We refer to these matrices as *augmented tridiagonal* matrices. If we define

$$P_+ = \begin{bmatrix} 1 & 1 & 0 & 0 & \cdots & 0 \\ 1 & -1 & 1 & 0 & \cdots & 0 \\ 1 & 0 & -1 & 1 & \cdots & 0 \\ \vdots & & \ddots & \ddots & \ddots & \\ 1 & 0 & \cdots & 0 & -1 & 1 \\ 1 & 0 & \cdots & 0 & 0 & -1 \end{bmatrix},$$

and

$$
P_- = \begin{bmatrix}
1 & 1 & 0 & 0 & \cdots & 0 \\
-1 & 1 & 1 & 0 & \cdots & 0 \\
1 & 0 & 1 & 1 & \cdots & 0 \\
\vdots & & & \ddots & \ddots & \ddots \\
(-1)^{n-2} & 0 & \cdots & 0 & 1 & 1 \\
(-1)^{n-1} & 0 & \cdots & 0 & 0 & 1
\end{bmatrix},
$$

then the following similarity transforms are satisfied:

$$
\text{ATrid}_n^\pm(a,b) = P_\pm \begin{bmatrix} b \pm 2a & 0 \\ 0 & \text{Trid}_{n-1}(a,b,a) \end{bmatrix} P_\pm^{-1}. \tag{1.6.7}
$$

To analyze the convergence properties of the linear algorithms determined by $\text{ATrid}_n^+(a,b)$ and $\text{ATrid}_n^-(a,b)$, we will find it useful to consider the vector

$$
\mathbf{1}_{n-}^T = (1, -1, 1, \ldots, (-1)^{n-2}, (-1)^{n-1})^T \in \mathbb{R}^n.
$$

In the following theorem, we will not assume that the matrices of interest are semi-convergent. We will establish convergence to a trajectory, rather than to a fixed point. For $\varepsilon \in \,]0,1[$, we say that a trajectory $x : \mathbb{Z}_{\geq 0} \to \mathbb{R}^n$ converges to $x_{\text{final}} : \mathbb{Z}_{\geq 0} \to \mathbb{R}^n$ with convergence time $T_\varepsilon \in \mathbb{Z}_{\geq 0}$ if

(i) $\|x(\ell) - x_{\text{final}}(\ell)\|_2 \to 0$ as $\ell \to +\infty$; and

(ii) T_ε is the smallest time such that $\|x(\ell) - x_{\text{final}}(\ell)\|_2 \leq \varepsilon \|x(0) - x_{\text{final}}(0)\|_2$, for all $\ell \geq T_\varepsilon$.

Theorem 1.80 (Linear algorithms defined by augmented tridiagonal matrices). *Let $n \geq 2$, $\varepsilon \in \,]0,1[$, and $a, b \in \mathbb{R}$ with $a \neq 0$ and $|b| + 2|a| = 1$. Let $x : \mathbb{Z}_{\geq 0} \to \mathbb{R}^n$ and $z : \mathbb{Z}_{\geq 0} \to \mathbb{R}^n$ be solutions to*

$$
x(\ell+1) = \text{ATrid}_n^+(a,b)\,x(\ell), \qquad z(\ell+1) = \text{ATrid}_n^-(a,b)\,z(\ell),
$$

with initial conditions $x(0) = x_0$ and $z(0) = z_0$, respectively. The following statements hold:

(i) $\lim_{\ell \to +\infty} \big(x(\ell) - x_{\text{ave}}(\ell)\mathbf{1}_n\big) = \mathbf{0}_n$, *where* $x_{\text{ave}}(\ell) = (\frac{1}{n}\mathbf{1}_n^T x_0)(b+2a)^\ell$, *with ε-convergence time in $\Theta\big(n^2 \log \varepsilon^{-1}\big)$; and*

(ii) $\lim_{\ell \to +\infty} \big(z(\ell) - z_{\text{ave}}(\ell)\mathbf{1}_{n-}\big) = \mathbf{0}_n$, *where* $z_{\text{ave}}(\ell) = (\frac{1}{n}\mathbf{1}_{n-}^T z_0)(b - 2a)^\ell$, *with ε-convergence time in $\Theta\big(n^2 \log \varepsilon^{-1}\big)$.*

The proof of this result is reported in Section 1.8.6.

Remark 1.81 (From Toeplitz to stochastic matrices). A tridiagonal Toeplitz matrix is not stochastic unless its off-diagonal elements are zero. The tridiagonal circulant matrices Circ_n and augmented tridiagonal matrices ATrid_n^+ studied in Theorem 1.79(iii) and Theorem 1.80(i) are slight modifications of tridiagonal Toeplitz matrices and are doubly stochastic. Indeed, the evolutions converge to the average consensus value, as predicted by Corollary 1.70. Note that convergence times obtained for Circ_n and ATrid_n^+ are consistent with the upper bound predicted by Theorem 1.76. •

We conclude this section with some useful bounds.

Lemma 1.82 (Bounds on vector norms). *Assume that $x \in \mathbb{R}^n$, $y \in \mathbb{R}^{n-1}$, and $z \in \mathbb{R}^{n-1}$ jointly satisfy*

$$x = P_+ \begin{bmatrix} 0 \\ y \end{bmatrix}, \qquad x = P_- \begin{bmatrix} 0 \\ z \end{bmatrix}.$$

Then $\frac{1}{2}\|x\|_2 \leq \|y\|_2 \leq (n-1)\|x\|_2$ and $\frac{1}{2}\|x\|_2 \leq \|z\|_2 \leq (n-1)\|x\|_2$.

The proof of this result is based on spelling out the coordinate expressions for x, y, and z, and is left to the reader as Exercise E1.29.

1.7 NOTES

Dynamical systems and stability theory

Our definition of a state machine is very basic; more general definitions of state machines can be found in the literature (see Sipser, 2005), but the one presented in this chapter is sufficient for our purposes.

The literature on dynamical and control systems is vast. The main tool that we use in later chapters is the LaSalle Invariance Principle, obtained by LaSalle (1960) and discussed in LaSalle (1986); see also the earlier works by Barbašin and Krasovskiĭ (1952) and Krasovskiĭ (1963) for related versions. Relevant sample references include modern texts on dynamical systems (Guckenheimer and Holmes, 1990), linear control systems (Chen, 1984), nonlinear control systems (Khalil, 2002), robust control (Dullerud and Paganini, 2000), and discrete-event systems (Cassandras and Lafortune, 2007).

Graph theory

The basic definitions of graph theory are standard in the literature; see, for example, Biggs (1994), Godsil and Royle (2001), and Diestel (2005). The discussion about graph algorithms is taken from Cormen et al. (2001), which also contains detailed discussion on implementation and complexity issues.

Regarding Section 1.4.4.4, standard references on combinatorial optimization include Vazirani (2001) and Korte and Vygen (2005).

In Section 1.4.5, all statements about powers of the adjacency matrix are standard results in algebraic graph theory; see, for example Biggs (1994) and Godsil and Royle (2001). Lemma 1.27 is a recent result from Lin et al. (2005) andMoreau (2005). Proposition 1.35, on the fact that a weighted digraph is aperiodic and irreducible if and only if its adjacency matrix is primitive, is related to standard results in the theory of Markov chains; see, for example Seneta (1981) and Meyn and Tweedie (1999). Our proof adopts the approach in Lin (2005). Laplacian matrices have numerous remarkable properties; two elegant surveys are Mohar (1991) and Merris (1994). Theorem 1.37, characterizing the properties of the Laplacian matrix, contains some recent results. A proof of statement (ii) is given in Olfati-Saber and Murray (2004); in our proof, we follow the approach in Francis (2006). Statement (iii) is proved by Lin et al. (2005) and Francis (2006); the following equivalent version is proved in Ren and Beard (2005): a weighted digraph G contains a spanning tree if and only if $\text{rank}(L(\text{rev}(G))) = n - 1$. Regarding statement (iv), the equivalence between (iv)a and (iv)b is proved by Olfati-Saber and Murray (2004) and the equivalence between (iv)b and (iv)c is proved by Moreau (2005).

Distributed algorithms

Our discussion of distributed algorithms is extremely incomplete. We have only presented a few token ideas and we refer to the textbooks by Lynch (1997) and Peleg (2000) for detailed treatments. Let us mention briefly that many more efficient algorithms are available in the literature—for example, the GHS algorithm (Gallager et al., 1983) for minimum spanning tree computation and consensus algorithms with communication and processors faults; much attention is dedicated to fault tolerance in asynchronous systems with shared memory and in asynchronous network systems.

Linear distributed algorithms

Distributed linear algorithms—and, in particular, averaging iterations that achieve consensus among processors—have a long and rich history. The richness comes from the vivid analogies with physical processes of diffusion, with Markov chain models, and with the sharp theory of positive matrices developed by Perron and Frobenius. What follows is a necessarily incomplete list. An early reference on averaging opinions and achieving consensus is DeGroot (1974). An early reference on the connection between averaging algorithms, the products of stochastic matrices, and ergodicity in inhomogeneous Markov chains is Chatterjee and Seneta (1977) – the history of inhomogeneous Markov chains being a classic topic since the early twentieth century. The stochastic setting was investigated in Cogburn (1984). Load balancing with divisible tasks in parallel computers is discussed in Cybenko (1989). A comprehensive theory of asynchronous parallel processors implementing distributed gradient methods and time-dependent averaging algorithms is developed in the series of works Tsitsiklis (1984), Tsitsiklis et al. (1986), and Bertsekas and Tsitsiklis (1997). Much interest for averaging algorithms arose from the influential work on flocking by Jadbabaie et al. (2003). Sharp conditions for convergence for the time-dependent setting were obtained in Moreau (2005). Finally, proper attention was given to the average consensus problem in Olfati-Saber and Murray (2004).

Regarding Theorem 1.63, characterizing the convergence of averaging algorithms defined by sequences of stochastic matrices, we note that: (1) the PhD thesis Tsitsiklis (1984) established convergence under a strong-connectivity assumption; (2) a sufficient condition was independently rediscovered in Jadbabaie et al. (2003), adopting a result from Wolfowitz (1963); and (3) Moreau (2003, 2005) obtained the necessary and sufficient condition (for uniform convergence in non-degenerate sequences) involving the existence of a uniformly globally reachable node. The work in Moreau (2003, 2005) is an early reference also for Theorem 1.65; additional related results and a historical discussion appeared in Blondel et al. (2005) and Hendrickx (2008). The estimates of the convergence factor given in Theorem 1.76 in Section 1.6.3 were proved by Landau and Odlyzko (1981). Our treatment in Section 1.6.4 follows Martínez et al. (2007a).

Among the numerous recent directions of research on consensus and averaging, we would like to mention the following: continuous-time consensus algorithms (Olfati-Saber and Murray, 2004; Moreau, 2004; Lin et al., 2004; Ren and Beard, 2005; Lin et al., 2005, 2007c), consensus over random networks (Hatano and Mesbahi, 2005; Wu, 2006; Patterson et al., 2007; Picci and Taylor, 2007; Porfiri and Stilwell, 2007; Tahbaz-Salehi and Jadbabaie, 2008; Fagnani and Zampieri, 2009), consensus in finite time (Cortés, 2006; Sundaram and Hadjicostis, 2008), consensus in small-world networks (Olfati-Saber, 2005; Durrett, 2006; Tahbaz-Salehi and Jadbabaie, 2007), consensus algorithms for general functions (Bauso et al., 2006; Cortés, 2008c; Lorenz and Lorenz, 2008; Sundaram and Hadjicostis, 2008), connections with the heat equation and partial difference equation (Ferrari-Trecate et al., 2006), spatially decaying interactions (Cucker and Smale, 2007), convergence in time-delayed and asynchronous settings (Blondel et al., 2005; Angeli and Bliman, 2006; Fang and Antsaklis, 2008), quantized consensus problems (Savkin, 2004; Kashyap et al., 2007; Carli et al., 2009; Zhu and Martínez, 2008b), consensus on manifolds (Scardovi et al., 2007; Sarlette and Sepulchre, 2009; Igarashi et al., 2007), applications to distributed signal processing (Spanos et al., 2005; Xiao et al., 2005; Olfati-Saber et al., 2006; Zhu and Martínez, 2008a), characterization of convergence rates and time complexity (Landau and Odlyzko, 1981; Olshevsky and Tsitsiklis, 2009; Cao et al., 2008; Carli et al., 2008). Numerous interesting results are reported in recent PhD theses (Lin, 2005; Cao, 2007; Lorenz, 2007; Barooah, 2007; Carli, 2008; Hendrickx, 2008; Sarlette, 2009). Finally, we would like to point out two recent surveys (Olfati-Saber et al., 2007; Ren et al., 2007) and the text by (Ren and Beard, 2008).

Synchronization is a fascinating topic related to averaging algorithms. A very early reference is the work by Huygens (1673) on coupled pendula. The synchronization of oscillators in dynamical systems has received increasing attention, and key references include Wiener (1958), Kuramoto (1975), Winfree (1980), Kuramoto (1984), Strogatz (2000), and Nijmeijer (2001); see also the widely accessible Strogatz (2003). Under all-to-all interactions, Mirollo and Strogatz (1990) prove synchronization of a collection of "integrate and fire" biological oscillators. Recent works on the Kuramoto and other synchronized oscillator models include Jadbabaie et al. (2004), Chopra and Spong (2009), Triplett et al. (2006), Papachristodoulou and Jadbabaie (2006), Wang and Slotine (2006).

1.8 PROOFS

This section gathers the proofs of the main results presented in the chapter.

1.8.1 Proof of Theorem 1.21

Here we provide the proof of the LaSalle Invariance Principle for set-valued discrete-time dynamical systems. We remark that Theorem 1.19 is an immediate consequence of Theorem 1.21 and that Theorem 1.20 is proved in a similar way (for details, we refer to (Khalil, 2002)).

Proof of Theorem 1.21. Let γ be any evolution of (X, X_0, T) starting from W. Let $\Omega(\gamma)$ denote the ω-limit set[6] of the sequence $\gamma = \{\gamma(\ell) \mid \ell \in \mathbb{Z}_{\geq 0}\}$; since W is closed, it follows that $\Omega(\gamma) \subset W$. Next, we prove that $\Omega(\gamma)$ is weakly positively invariant. Let $z \in \Omega(\gamma)$. Then there exists a subsequence

---

[6]The ω-limit set of a sequence $\gamma = \{\gamma(\ell) \mid \ell \in \mathbb{Z}_{\geq 0}\}$ is the set of points y for which there exists a subsequence $\{\gamma(\ell_m) \mid m \in \mathbb{Z}_{\geq 0}\}$ of γ such that $\lim\limits_{m \to +\infty} \gamma(\ell_m) = y$.

$\{\gamma(\ell_m) \mid m \in \mathbb{Z}_{\geq 0}\}$ of γ such that $\lim\limits_{m \to +\infty} \gamma(\ell_m) = z$. Consider the sequence $\{\gamma(\ell_m + 1) \mid m \in \mathbb{Z}_{\geq 0}\}$. Since this sequence is bounded, it has a convergent subsequence. For ease of notation, we use the same notation to refer to it, that is, there exists y such that $\lim\limits_{m \to +\infty} \gamma(\ell_m + 1) = y$. By definition, $y \in \Omega(\gamma)$. Moreover, using the fact that T is closed, we deduce that $y \in T(z)$. Therefore, $\Omega(\gamma)$ is weakly positively invariant.

Now, consider the sequence $V \circ \gamma = \{V(\gamma(\ell)) \mid \ell \in \mathbb{Z}_{\geq 0}\}$. Since γ is bounded and V is non-increasing along T on W, the sequence $V \circ \gamma$ is decreasing and bounded from below, and therefore, convergent. Let $c \in \mathbb{R}$ satisfy $\lim\limits_{\ell \to +\infty} V(\gamma(\ell)) = c$. Next, we prove that the value of V on $\Omega(\gamma)$ is constant and equal to c. Take any $z \in \Omega(\gamma)$. Accordingly, there exists a subsequence $\{\gamma(\ell_m) \mid m \in \mathbb{Z}_{\geq 0}\}$ such that $\lim\limits_{m \to +\infty} \gamma(\ell_m) = z$. Since V is continuous, $\lim\limits_{m \to +\infty} V(\gamma(\ell_m)) = V(z)$. From $\lim\limits_{\ell \to +\infty} V(\gamma(\ell)) = c$, we conclude that $V(z) = c$.

Finally, the fact that $\Omega(\gamma)$ is weakly positively invariant and V being constant on $\Omega(\gamma)$ implies that

$$\Omega(\gamma) \subset \{x \in X \mid \exists y \in T(x) \text{ such that } V(y) = V(x)\}.$$

Therefore, we conclude that $\lim\limits_{\ell \to +\infty} \operatorname{dist}(\gamma(\ell), S \cap V^{-1}(c)) = 0$, where S is the largest weakly positively invariant set contained in $\{x \in X \mid \exists y \in T(x) \text{ such that } V(y) = V(x)\}$. ∎

1.8.2 Proofs of Lemmas 1.26 and 1.27

Proof of Lemma 1.26. The first statement is obvious. Regarding the second statement, we prove that a topologically balanced digraph with a globally reachable node is strongly connected, and leave the proof of the other case to the reader. We reason by contradiction. Assume that G is not strongly connected. Let $S \subset V$ be the set of all nodes of G that are globally reachable. By hypothesis, $S \neq \emptyset$. Since G is not strongly connected, we have $S \subsetneq V$. Note that any outgoing edge with origin in a globally reachable node automatically makes the destination a globally reachable node too. This implies that there cannot be any outgoing edges from a node in S to a node in $V \setminus S$. Let $v \in V \setminus S$ such that v has an out-neighbor in S (such a node must exist, since otherwise the nodes in S cannot be globally reachable). Since by hypothesis G is balanced, there must exist an edge of the form $(w, v) \in E$. Clearly, $w \notin S$, since otherwise v would be globally reachable too, which is a contradiction. Therefore, $w \in V \setminus S$. Again, using the fact that G is topologically balanced, there must exist an edge of the form $(z, w) \in E$. As before, $z \in V \setminus S$ (note that $z = v$ is a possibility). Since $V \setminus S$ is finite and so is the number of possible edges between its nodes, applying this argument repeatedly, we find that there exists a vertex whose out-degree is strictly larger than its in-degree, which is a contradiction with the fact that G is topologically balanced. We refer to Cortés (2008c) for the proof that G is Eulerian. ∎

Proof of Lemma 1.27. (i) $\implies$ (ii) Assume that $i \in V$ is the root of the spanning tree and take an arbitrary pair of nonempty, disjoint sub-sets $U_1, U_2 \subset V$. If $i \in U_1$, then there must exist a path from $i \in U_1$ to a node in U_2. Therefore, U_2 must have an in-neighbor. Analogously, if $i \in U_2$, then U_1 must have an in-neighbor. Finally, it is possible that $i \notin U_1 \cup U_2$. In this case, there exist paths from i to both U_1 and U_2, that is, both sets have in-neighbors.

(ii) $\implies$ (i) This is proved by finding a node from which there exists a path to all others. We do this in an algorithmic manner using induction. At each induction step k, except the last one, four sets of nodes are considered, $U_1(k) \subset W_1(k) \subset V$, $U_2(k) \subset W_2(k) \subset V$, with the following properties:

(a) the sets $W_1(k)$ and $W_2(k)$ are disjoint; and

(b) from each node of $U_s(k)$ there exists a path to each other node in $W_s(k) \setminus U_s(k)$, $s \in \{1, 2\}$.

Induction Step k=1: Set $U_1 = W_1 = \{i_1\}$ and $U_2 = W_2 = \{i_2\}$, where i_1, i_2 are two arbitrary different nodes of the graph that satisfy the properties (a) and (b).

Induction Step k > 1: Suppose that for $k - 1$ we found sets $U_1(k - 1) \subset W_1(k - 1)$ and $U_2(k - 1) \subset W(k - 1)$ as in (a) and (b). Since $U_1(k - 1)$ and $U_2(k - 1)$ are disjoint, then there exists either an edge (i_k, j_1) with $j_1 \in U_1(k - 1)$, $i_k \in V \setminus U_1(k - 1)$, or an edge (i_k, j_2) with $j_2 \in U_2(k - 1)$ and $i_k \in V \setminus U_2(k - 1)$. Suppose that an edge (i_k, j_2) exists (the case of a edge (i_k, j_1) can be treated in a similar way). Only four cases are possible.

(A) If $i_k \in W_1(k-1)$ and $W_1(k-1) \cup W_2(k-1) = V$, then we can termi-nate the algorithm and conclude that from any node $h \in U_1(k - 1)$ there exists a path to all other nodes in the graph and thus there is a spanning tree.

(B) If $i_k \in W_1(k-1)$ and $W_1(k-1) \cup W_2(k-1) \neq V$, then set:

$$U_1(k) = U_1(k-1),$$
$$W_1(k) = W_1(k-1) \cup W_2(k-2),$$
$$U_2(k) = W_2(k) = \{h_k\},$$

where h_k is an arbitrary node not belonging to $W_1(k-1) \cup W_2(k-1)$.

(C) If $i_k \notin W_1(k-1) \cup W_2(k-1)$, then set

$$U_1(k) = U_1(k-1),$$
$$W_1(k) = W_1(k-1),$$
$$U_2(k) = \{i_k\},$$
$$W_2(k) = W_2(k-1) \cup \{i_k\}.$$

(D) If $i_k \in W_2(k-1) \setminus U_2(k-1)$ then

$$U_1(k) = U_1(k-1),$$
$$W_1(k) = W_1(k-1),$$
$$U_2(k) = U_2(k-1) \cup \{i_k\},$$
$$W_2(k) = W_2(k-1).$$

The algorithm terminates in a finite number of induction steps because at each step, except when finally case (A) holds true, either the number of nodes in $W_1 \cup W_2$ increases, or the number of nodes in $W_1 \cup W_2$ remains constant and the number of nodes in $U_1 \cup U_2$ increases. ∎

1.8.3 Proofs of Propositions 1.33 and 1.35

Proof of Proposition 1.33. *(ii)* $\implies$ *(i)* We aim to show that there exist directed paths from any node to any other node. Fix $i \in \{1, \dots, n\}$ and let $R_i \subset \{1, \dots, n\}$ be the set of nodes that belong to directed paths originating from node i. Denote the unreachable nodes by $U_i = \{1, \dots, n\} \setminus R_i$. We argue that U_i cannot contain any element, because if it does, then $R_i \cup U_i$ is a nontrivial partition of the index set $\{1, \dots, n\}$ and irreducibility implies the existence of a non-zero entry a_{jk} with $j \in R_i$ and $k \in U_i$. Therefore, $U_i = \emptyset$, and all nodes are reachable from i. The converse statement *(i)* $\implies$ *(ii)* is proved similarly.

(i) $\implies$ *(iii)* If G is strongly connected, then there exists a directed path of length $k \leq n - 1$ connecting any node i to any other node j. Hence, by Lemma 1.32(ii), the entry $(A^k)_{ij}$ is strictly positive. This immediately implies the statement (iii). The converse statement *(iii)* $\implies$ *(i)* is proved similarly. ∎

Next, we present a useful number theory result. This states that relatively co-prime numbers generate all sufficiently large natural numbers.

Lemma 1.83 (Natural number combination). *Let* $a_1, \ldots, a_N \in \mathbb{N}$ *have greatest common divisor* 1. *There exists* $k \in \mathbb{N}$ *such that every number* $m > k$ *can be written as*

$$m = \alpha_1 a_1 + \cdots + \alpha_N a_N,$$

for appropriate numbers $\alpha_1, \ldots, \alpha_N \in \mathbb{N}$.

Proof. Assume that $a_1 \leq \cdots \leq a_N$ without loss of generality. From the generalized Bezout identity we know that, for any numbers $a_1, \ldots, a_N$ with greatest common divisor 1, there exist integers $\gamma_1, \ldots, \gamma_N \in \mathbb{Z}$ such that

$$1 = \gamma_1 a_1 + \cdots + \gamma_N a_N. \qquad (1.8.1)$$

Pick $k = |\gamma_1| a_1^2 + \cdots + |\gamma_N| a_N^2 \in \mathbb{N}$. Every number $m > k$ can be written as

$$m = k + m_{\text{qtnt}} a_1 + m_{\text{rmndr}},$$

for appropriate numbers $m_{\text{qtnt}} \geq 0$ and $1 \leq m_{\text{rmndr}} < a_1$. Using the definition of k and equation (1.8.1), we write

$$m = \left(|\gamma_1| a_1^2 + \cdots + |\gamma_N| a_N^2 \right) + m_{\text{qtnt}} a_1 + m_{\text{rmndr}} (\gamma_1 a_1 + \cdots + \gamma_N a_N)$$
$$= m_{\text{qtnt}} a_1 + (|\gamma_1| a_1 + m_{\text{rmndr}} \gamma_1) a_1 + \cdots + (|\gamma_N| a_N + m_{\text{rmndr}} \gamma_N) a_N.$$

The proof is now completed by noting that each integer number $(|\gamma_1| a_1 + m_{\text{rmndr}} \gamma_1), \ldots, (|\gamma_N| a_N + m_{\text{rmndr}} \gamma_N)$ is strictly positive, because $m_{\text{rmndr}} < a_1 \leq \cdots \leq a_N$. ∎

Proof of Proposition 1.35. $(i) \implies (ii)$ Pick any i. We claim that there exists a number $k(i)$ with the property that, for all $m > k(i)$, we have that $(A^m)_{ii}$ is positive, that is, there exists a directed path from i to i of length m for all m larger than a number $k(i)$. To show this claim, let $\{c_1, \ldots, c_N\}$ be the set of the cycles of G and let $\{\ell_1, \ldots, \ell_N\}$ be their lengths. Because G is aperiodic, Lemma 1.83 implies the existence of a number $h(\ell_1, \ldots, \ell_N)$ such that any number larger than $h(\ell_1, \ldots, \ell_N)$ is a linear combination of $\ell_1, \ldots, \ell_N$ with natural numbers as coefficients. Because G is strongly connected, there exists a path γ of arbitrary length $\Gamma(i)$ that starts at i, contains a vertex of each of the cycles $c_1, \ldots, c_N$, and terminates at i. Now, we claim that $k(i) = \Gamma(i) + h(\ell_1, \ldots, \ell_N)$ has the desired property. Indeed, pick any number $m > k(i)$ and write it as $k = \Gamma(i) + \beta_1 \ell_1 + \cdots + \beta_N \ell_N$ for appropriate numbers $\beta_1, \ldots, \beta_N \in \mathbb{N}$. A directed path from i to i of length m is constructed by attaching to the path γ the following cycles: β_1 times the cycle c_1, β_2 times the cycle c_2, ..., β_N times the cycle c_N. Finally, having proved the existence of $k(i)$ with the desired property, let K be the maximum $k(i)$

over all nodes i, and recall that $\mathrm{diam}(G)$ is the maximum pairwise distance between nodes. Clearly, A^M is positive for all $M > K + \mathrm{diam}(G)$.

(ii) $\implies$ *(i)* From Lemma 1.32 we know that $A^k > 0$ means that there are paths from every node to every other node of length k. Hence, the digraph G is strongly connected. Next, we prove aperiodicity. Because G is strongly connected, each node of G has at least one outgoing edge, that is, for all i, there exists at least one index j such that $a_{ij} > 0$. This fact implies that the matrix $A^{k+1} = AA^k$ is positive via the following simple calculation: $(A^{k+1})_{il} = \sum_{h=1}^{n} a_{ih}(A^k)_{hl} \geq a_{ij}(A^k)_{jl} > 0$. In summary, we have shown that, if A^k is positive for some k, then A^m is positive for all subsequent $m \geq k$. Therefore, there are cycles in G of any length greater than or equal to k, which means that G is aperiodic. ∎

1.8.4 Proof of Theorem 1.37

Proof. We begin with statement (i). Let l_{ij}, for $i, j \in \{1, \ldots, n\}$, be the entries of $L(G)$. Note that $l_{ii} = \sum_{j=1, j \neq i}^{n} a_{ij} \geq 0$ and $l_{ij} = -a_{ij} \leq 0$ for $i \neq j$. By the Geršgorin disks Theorem 1.2, we know that each eigenvalue of $L(G)$ belongs to at least one of the disks

$$\left\{ z \in \mathbb{C} \mid \|z - l_{ii}\|_{\mathbb{C}} \leq \sum_{j=1, j \neq i}^{n} |l_{ij}| \right\} = \left\{ z \in \mathbb{C} \mid \|z - l_{ii}\|_{\mathbb{C}} \leq l_{ii} \right\}.$$

These disks contain the origin $\mathbf{0}_n$ and complex numbers with a positive real part. This concludes the proof of statement (i).

Regarding statement (ii), note that $D_{\mathrm{out}}(G)$ is invertible because G is strongly connected. Define the two matrices $\tilde{A} = D_{\mathrm{out}}(G)^{-1}A(G)$ and $\tilde{L} = D_{\mathrm{out}}(G)^{-1}L(G)$, and note that they satisfy $\tilde{L} = I_n - \tilde{A}$. Since $D_{\mathrm{out}}(G)$ is diagonal, the matrices $A(G)$ and $\tilde{A}$ have the same pattern of zeros and positive entries. This observation and the assumption that G is strongly connected imply that $\tilde{A}$ is nonnegative and irreducible. By the Perron–Frobenius Theorem 1.11, the spectral radius $\rho(\tilde{A})$ is a simple eigenvalue. Furthermore, one can verify that $\tilde{A}$ is row-stochastic (see Lemma 1.31), and therefore, its spectral radius is 1 (see Exercise E1.4). In summary, we conclude that 1 is a simple eigenvalue of $\tilde{A}$, that 0 is a simple eigenvalue of $\tilde{L}$, that $\tilde{L}$ has rank $n-1$, and that $L(G)$ has rank $n-1$.

Regarding statement (iii), we first prove that $\mathrm{rank}(L(G)) = n - 1$ implies the existence of a globally reachable vertex. By contradiction, let G contain no globally reachable vertex. Then, by Lemma 1.27, there exist two nonempty disjoint subsets $U_1, U_2 \subset V(G)$ without any out-neighbor. After a permutation of the vertices, the adjacency matrix can be partitioned into the blocks

$$A(G) = \begin{bmatrix} A_{11} & 0 & 0 \\ 0 & A_{22} & 0 \\ A_{31} & A_{32} & A_{33} \end{bmatrix}.$$

Here, A_{12} and A_{13} vanish because U_1 does not have any out-neighbor, and A_{21} and A_{23} vanish because U_2 does not have any out-neighbor. Note that $D_{11} - A_{11}$ and $D_{22} - A_{22}$ are the Laplacian matrices of the graphs defined by restricting G to the vertices in U_1 and in U_2, respectively. Therefore, the eigenvalue 0 has geometric multiplicity at least 2 for the matrix $D_{\text{out}}(G) - A(G)$. This contradicts the assumption that $\text{rank}(L(G)) = n - 1$.

Next, still regarding statement (iii), we prove that the existence of a globally reachable vertex implies $\text{rank}(L(G)) = n-1$. Without loss of generality, we assume that G contains self-loops at each node (so that D_{out} is invertible). Let R be the set of globally reachable vertices; let $r \in \{1, \dots, n\}$ be its cardinality. If $r = n$, then the graph is strongly connected and statement (ii) implies that $\text{rank}(L(G)) = n - 1$. Therefore, assume that $r < n$. Renumber the vertices so that R is the set of the first r vertices. After this permutation, the adjacency matrix and the Laplacian matrix can be partitioned into the blocks

$$A(G) = \begin{bmatrix} A_{11} & 0 \\ A_{21} & A_{22} \end{bmatrix}, \quad \text{and} \quad L(G) = \begin{bmatrix} L_{11} & 0 \\ L_{21} & L_{22} \end{bmatrix}.$$

Here, $A_{12} \in \mathbb{R}^{r \times (n-r)}$ vanishes, because there can be no out-neighbor of R; otherwise that out-neighbor would be a globally reachable vertex in $V \setminus R$. Note that the rank of $L_{11} \in \mathbb{R}^{r \times r}$ is exactly $r-1$, since the digraph associated to A_{11} is strongly connected. To complete the proof it suffices to show that the rank of $L_{22} \in \mathbb{R}^{(n-r) \times (n-r)}$ is full. Note that the same block partition applies to the matrices $\tilde{A} = D_{\text{out}}^{-1} A$ and $\tilde{L} = D_{\text{out}}^{-1} L$ considered in the proof of statement (ii) above. With this block decomposition, we compute

$$\tilde{A}^{n-1} = \begin{bmatrix} \tilde{A}_{11}^{n-1} & 0 \\ \tilde{A}_{21}(n-1) & \tilde{A}_{22}^{n-1} \end{bmatrix},$$

for some matrix $\tilde{A}_{21}(n-1)$ that depends upon $\tilde{A}_{11}$, $\tilde{A}_{21}$ and $\tilde{A}_{22}$. Because a globally reachable node in G is globally reachable also in the digraph associated to $\tilde{A}$, Proposition 1.33(v) implies that $\tilde{A}_{21}(n-1)$ is positive. This fact, combined with the fact that $\tilde{A}$ and hence $\tilde{A}^{n-1}$ are row-stochastic, implies that $\tilde{A}_{22}^{n-1}$ has maximal row sum (that is, ∞-induced norm) strictly less than 1. Hence, the spectral radii of $\tilde{A}_{22}^{n-1}$ and of $\tilde{A}_{22}$ are strictly less than 1. Since $\tilde{A}_{22}$ has spectral radius strictly less than 1, the matrix $\tilde{L}_{22} = I_{n-r} - \tilde{A}_{22}$, and in turn the matrix L_{22}, have full rank.

Regarding statement (iv), the equivalence between (iv)a and (iv)b is proved as follows. Because $\sum_{j=1}^{n} l_{ij} = d_{\text{out}}(v_i) - d_{\text{in}}(v_i)$ for all $i \in \{1, \ldots, n\}$, it follows that $\mathbf{1}_n^T L(G) = \mathbf{0}_n^T$ if and only if $D_{\text{out}}(G) = D_{\text{in}}(G)$. Next, we prove that (iv)b implies (iv)c. Suppose that $L(G)^T \mathbf{1}_n = \mathbf{0}_n^T$ and consider the system $\dot{\gamma}(t) = -L(G)\gamma(t)$, $\gamma(0) = x_0$, together with the positive definite function $V : \mathbb{R}^n \to \mathbb{R}$ defined by $V(x) = x^T x$. We compute the Lie derivative of the function V along the vector field $x \mapsto -L(G)x$ as $\dot{V}(x) = -2x^T L(G)x$. Note that $\dot{V}(x) \leq 0$, for all $x \in \mathbb{R}^n$, is equivalent to $L(G) + L(G)^T \geq 0$. Because $\mathbf{1}_n^T L(G) = \mathbf{0}_n^T$ and $L(G)\mathbf{1}_n = \mathbf{0}_n$, it can immediately be established that $\exp(-L(G)t)$, $t \in \mathbb{R}$, is a doubly stochastic matrix. From Theorem 1.1, we know that if we let $\{P_\alpha\}$ be the set of $n \times n$ permutation matrices, then there exist time-dependent convex combination coefficients $\sum_\alpha \lambda_\alpha(t) = 1$, $\lambda_\alpha(t) \geq 0$, such that $\exp(-L(G)t) = \sum_\alpha \lambda_\alpha(t) P_\alpha$. By the convexity of V and its invariance under coordinate permutations, for any $x \in \mathbb{R}^n$, we have

$$V(\exp(-L(G)t)x) = V\left(\sum_\alpha \lambda_\alpha(t) P_\alpha x\right)$$

$$\leq \sum_\alpha \lambda_\alpha(t) V(P_\alpha x) = \sum_\alpha \lambda_\alpha(t) V(x) = V(x).$$

In other words, $V(\exp(-L(G)t)x) \leq V(x)$ for all $x \in \mathbb{R}^n$, which implies $\dot{V}(x) \leq 0$, for all $x \in \mathbb{R}^n$. Finally, we prove that (iv)c implies (iv)b. By assumption, $-x^T(L(G) + L(G)^T)x = -2x^T L(G)x \leq 0$ for all $x \in \mathbb{R}^n$. In particular, for any small $\varepsilon > 0$ and $x = \mathbf{1}_n - \varepsilon L(G)^T \mathbf{1}_n$,

$$-(\mathbf{1}_n^T - \varepsilon \mathbf{1}_n^T L(G))L(G)(\mathbf{1}_n - \varepsilon L(G)^T \mathbf{1}_n) = \varepsilon \|L(G)^T \mathbf{1}_n\|_2^2 + O(\varepsilon^2) \leq 0,$$

which is possible only if $L(G)^T \mathbf{1}_n = \mathbf{0}_n^T$. ∎

1.8.5 Proofs of Theorem 1.63 and Proposition 1.67

In this section, we prove Theorem 1.63. The exposition follows along the main lines of the original proof by Moreau (2005), with the variation of using the LaSalle Invariance Principle for set-valued dynamical systems, presented as Theorem 1.21. We begin with some preliminary results.

Lemma 1.84 (Union of digraphs and sums of adjacency matrices). *Let $G_1, \ldots, G_\delta$ be unweighted digraphs with common node set $\{1, \ldots, n\}$ and adjacency matrices $A_1, \ldots, A_\delta$. The unweighted digraph*

$$G_1 \cup \cdots \cup G_\delta = (\{1, \ldots, n\}, E(A_1) \cup \cdots \cup E(A_\delta))$$

is equal to the unweighted digraph associated to the nonnegative matrix $\sum_{k \in \{1, \ldots, \delta\}} A_k$, that is, the unweighted digraph $(\{1, \ldots, n\}, E(A_1 + \cdots + A_\delta))$.

Proof. If $(i, j) \in \cup_{k \in \{1, \ldots, \delta\}} E(G_k)$, then there exists $k_0 \in \{1, \ldots, \delta\}$ such that $(i,j) \in E(G_{k_0})$. Denoting the entries of the matrix A_k by $a_{ij}(k)$, this implies that $a_{ij}(k_0) > 0$, that $a_{ij}(1) + \cdots + a_{ij}(\delta) > 0$, and that (i,j) is an edge in $E(A_1 + \cdots + A_\delta)$. The converse statement is easily proved with an analogous reasoning. ∎

In what follows, for $\alpha \in {]0,1]}$, let $\mathcal{F}(\alpha)$ denote the set of $n \times n$ stochastic matrices that are non-degenerate with respect to α. Given $\alpha \in {]0,1]}$ and $\delta \in \mathbb{N}$, define the sets $\mathcal{F}_\delta(\alpha) \subset \mathbb{R}^{n \times n}$ by

$$\mathcal{F}_\delta(\alpha) = \big\{ F \in \mathcal{F}(\alpha^\delta) \mid \exists\, F_1, \ldots, F_\delta \in \mathcal{F}(\alpha) \text{ such that } F = F_\delta \cdots F_1$$
$$\text{and } G(F_1) \cup \cdots \cup G(F_\delta) \text{ contains a globally reachable node} \big\},$$

or, equivalently by Proposition 1.33,

$$\mathcal{F}_\delta(\alpha) = \big\{ F \in \mathcal{F}(\alpha^\delta) \mid \exists\, F_1, \ldots, F_\delta \in \mathcal{F}(\alpha) \text{ such that } F = F_\delta \ldots F_1$$
$$\text{and a column of } (F_1 + \cdots + F_\delta)^n \text{ has positive entries} \big\}.$$

Lemma 1.85 (Compact sets of stochastic matrices). *For $\alpha \in {]0,1]}$, the sets $\mathcal{F}(\alpha)$ and $\mathcal{F}_\delta(\alpha)$, $\delta \in \mathbb{N}$, are compact.*

Proof. All sets are clearly bounded. In Exercise E1.24, we invite the reader to prove that $\mathcal{F}(\alpha)$ is closed. Let us now prove that $\mathcal{F}_\delta(\alpha)$ is closed. Consider a matrix sequence $\{F(k) \mid k \in \mathbb{N}\} \subset \mathcal{F}_\delta(\alpha)$ convergent to some matrix F. Because $\mathcal{F}(\alpha^\delta)$ is closed, we establish that $F \in \mathcal{F}(\alpha^\delta)$. Because each matrix $F(k)$ belongs to $\mathcal{F}_\delta(\alpha)$, there exist matrices $F_1(k), \ldots, F_\delta(k) \in \mathcal{F}(\alpha)$ such that $F(k) = F_\delta(k) \cdots F_1(k)$. We claim that there exists a sequence $k_l \in \mathbb{N}$, for $l \in \mathbb{N}$, such that, for all $s \in \{1, \ldots, \delta\}$, the matrix sequences $F_s(k_l)$, $l \in \mathbb{N}$, are convergent. (To see this, note that $F_1(k)$ takes value in a compact set; hence it must have a convergent subsequence. Restrict $F_2(k)$ to the instants of time in the convergent subsequence for $F_1(k)$ and observe that it takes value in a compact set, etc.) Therefore, there exist matrices F_s, to which the matrix sequences $F_s(k_l)$, $l \in \mathbb{N}$, converge. Taking the limit as $l \to +\infty$ in the equality $F(k_l) = F_\delta(k_l) \cdots F_1(k_l)$, we establish that $F = F_\delta \cdots F_1$. Finally, it remains to be shown that a column of $B := (F_1 + \cdots + F_\delta)^n$ has positive entries. For $k \in \mathbb{N}$, define $B(k) = (F_1(k) + \cdots + F_\delta(k))^n$. Clearly, $B(k) \to B$ as $k \to +\infty$. By the definition of the sequence $F(k)$, each $B(k) = (b_{ij}(k))$ has the property that there exists $j_k \in \{1, \ldots, n\}$ such that $b_{ij_k}(k) > 0$ for all $i \in \{1, \ldots, n\}$. Since $\{1, \ldots, n\}$ is a finite set, there exists $j_0 \in \{1, \ldots, n\}$ that satisfies this property for an infinite subsequence of matrices $B(k_l)$, $l \in \mathbb{N}$. With some straightforward bookkeeping, we write

$$(B(k_l))_{ij_0} = \sum_{a_1,\ldots,a_n=1}^{\delta} \sum_{h_1=1}^{n} \cdots \sum_{h_{n-1}=1}^{n} (F_{a_1}(k_l))_{ih_1} \cdots (F_{a_n}(k_l))_{h_{n-1}j_0}.$$

Note that, because $F_s(k) \in \mathcal{F}(\alpha)$, for $s \in \{1, \ldots, \delta\}$, each nonzero entry $F_s(k)$ is lower bounded by $\alpha > 0$. Furthermore, each entry $(B(k_l))_{ij_0}$ is the sum of nonnegative terms, each of which is the product of n factors, each of which is lower bounded by α. Hence, because $(B(k_l))_{ij_0}$ is positive, it is also lower bounded by α^n. Since $\lim_{l \to +\infty} B(k_l) = B$, by the compactness of $[\alpha^n, 1] \cup \{0\}$, it must be that $B = (b_{ij})$ satisfies $b_{ji_0} \geq \alpha^n > 0$ for all $j \in \{1, \ldots, n\}$. In particular, this implies that $F \in \mathcal{F}_\delta(\alpha)$ and then $\mathcal{F}_\delta(\alpha)$ is closed.   ∎

Finally, we are able to prove the equivalences in Theorem 1.63.

Proof of Theorem 1.63. First, we prove that (i) implies (ii). Suppose that for all durations $\delta \in \mathbb{N}$, there exists some $\ell_0 \in \mathbb{N}$ such that the digraph with edges $\cup_{s \in [\ell_0, \ell_0 + \delta]} E(F(s))$ does not contain a globally reachable node. By Lemma 1.27, there must exist a set of nodes $U_1, U_2 \subset \{1, \ldots, n\}$ such that there are no out-going edges (i_1, j_1), with $i_1 \in U_1$, $i_1 \in \{1, \ldots, n\} \setminus U_1$ or (i_2, j_2), with $j_2 \in U_2$, $i_2 \in \{1, \ldots, n\} \setminus U_2$. Take any values $a, b \in \mathbb{R}$, $a \neq b$, and consider the initial conditions:

$$w_i(\ell_0) = \begin{cases} a, & i \in U_1, \\ b, & i \in U_2, \\ c \in \mathrm{co}(a, b), & i \in \{1, \ldots, n\} \setminus (U_1 \cup U_2). \end{cases}$$

Because of the properties of U_1 and U_2, for all $\delta \in \mathbb{N}$, we must have

$$w_j(\ell_0 + \delta + 1) = \begin{cases} a, & j \in U_1, \\ b, & j \in U_2, \\ c \in \mathrm{co}(a, b), & j \in \{1, \ldots, n\} \setminus (U_1 \cup U_2). \end{cases}$$

Because δ can be chosen arbitrarily large, one can easily construct a contradiction with the fact that $\mathrm{diag}(\mathbb{R}^n)$ is supposed to be uniformly globally attractive.

Next, we show that (ii) implies (i). Let $\alpha \in \,]0, 1]$ to be the scalar with respect to which the sequence is non-degenerate. Consider the set-valued discrete-time dynamical system $(\mathbb{R}^n, \mathbb{R}^n, T_{\alpha, \delta})$, with evolution map $T_{\alpha, \delta} : \mathbb{R}^n \rightrightarrows \mathbb{R}^n$ defined by

$$T_{\alpha, \delta}(w) = \{Fw \mid F \in \mathcal{F}_\delta(\alpha)\}.$$

Because of this definition, any trajectory $w : \mathbb{Z}_{\geq 0} \to \mathbb{R}^n$ of the averaging algorithm (1.6.2) satisfies

$$w((k+1)\delta) \in T_{\alpha, \delta}(w(k\delta)), \quad k \in \mathbb{Z}_{\geq 0}.$$

In what follows, we intend to use the LaSalle Invariance Principle for set-valued discrete dynamical systems, presented as Theorem 1.21, to prove that $\lim_{\ell \to +\infty} \mathrm{dist}(w(k\ell), \mathrm{diag}(\mathbb{R}^n)) = 0$. This will then imply, by Lemma 1.24, the uniform attractivity statement in the theorem. In the following, we check the conditions of the theorem.

Closedness of the set-valued dynamical system. Consider a pair of vector sequences $\{x_k \mid k \in \mathbb{N}\}$ and $\{y_k \mid k \in \mathbb{N}\}$ in $\mathbb{R}^n$ such that $\lim_{k\to+\infty}x_k = x$, $\lim_{k\to+\infty}y_k = y$, and $y_k \in T_{\alpha,\delta}(x_k)$, for all $k \in \mathbb{N}$. We need to show that $y \in T_{\alpha,\delta}(x)$. By definition of $T_{\alpha,\delta}$ and because $y_k \in T_{\alpha,\delta}(x_k)$, there exists a sequence $\{F(k) \mid k \in \mathbb{N}\} \subseteq \mathcal{F}_\delta(\alpha)$ such that $F(k)x_k = y_k$, for all $k \in \mathbb{N}$. Furthermore, since $\mathcal{F}_\delta(\alpha)$ is compact by Lemma 1.85, there exists a subsequence $\{F(k_l) \mid l \in \mathbb{N}\}$ that is convergent to some $F \in \mathcal{F}_\delta(\alpha)$. The desired conclusion follows from

$$y = \lim_{l\to+\infty} y_{k_\ell} = \lim_{l\to+\infty} F(k_\ell)x_{k_\ell} = Fx.$$

Non-increasing Lyapunov function. Define the function $V : \mathbb{R}^n \to \mathbb{R}_{\geq 0}$ by

$$V(x) = \max_{i\in\{1,\dots,n\}} x_i - \min_{i\in\{1,\dots,n\}} x_i.$$

Note that V is continuous. Pick any $x \in \mathbb{R}^n$ and any stochastic matrix $F \in \mathcal{F}_\delta(\alpha)$. Recall that $\|x\|_\infty = \max_{i\in\{1,\dots,n\}} |x_i|$, and that $\|F\|_\infty = 1$. Therefore, by the definition of the induced norm, $\|Fx\|_\infty \leq \|x\|_\infty$. Similarly, in components,

$$(Fx)_i = \sum_{j\in\{1,\dots,n\}} f_{ij}x_j \geq \left(\sum_{j\in\{1,\dots,n\}} f_{ij} \right) \min_{k\in\{1,\dots,n\}} x_k,$$

which implies $\min_{i\in\{1,\dots,n\}}(Fx)_i \geq \min_{k\in\{1,\dots,n\}} x_k$. Therefore, we have that $V(Fx) \leq V(x)$ for all $x \in \mathbb{R}^n$ and $F \in \mathcal{F}_\delta(\alpha)$. In other words, the function V is non-increasing along $T_{\alpha,\delta}$ in $\mathbb{R}^n$.

Boundedness. It can immediately be seen that, since $\|Fx\|_\infty \leq \|x\|_\infty$ for all stochastic matrices F and vectors x, the trajectory $k \mapsto w(k\delta)$ is bounded.

Invariant set. By Theorem 1.21, any trajectory of $T_{\alpha,\delta}$, and hence also the trajectory $w : \mathbb{Z}_{\geq 0} \to \mathbb{R}^n$ of the averaging algorithm (1.6.2), will converge to the largest weakly positively invariant set contained in a level set of the Lyapunov function V and in a set where the Lyapunov function does not decrease along T. In the following, we determine that this set must be contained in $\mathrm{diag}(\mathbb{R}^n)$.

For $k \in \mathbb{N}$ fixed, assume that $w(k\delta)$ satisfies $V(w(k\delta)) > 0$. Given the averaging algorithm (1.6.2) defined by the sequence $\{F(\ell) \mid \ell \in \mathbb{Z}_{\geq 0}\} \subset \mathcal{F}(\alpha)$, define $F_1(k) = F(k+1), \ldots, F_\delta(k) = F(k+\delta)$. Additionally, define $F(k) = F_\delta(k) \cdots F_1(k)$ and note that $F(k) \in \mathcal{F}_\delta(\alpha)$, by construction. With this notation, note that $w(k\delta + s) = F_s(k) \cdots F_1(k)w(k\delta)$ for $s \in \{1, \ldots, \delta\}$. Define $w_M = \max_{i \in \{1, \ldots, n\}} w_i(k\delta)$ and $w_m = \min_{i \in \{1, \ldots, n\}} w_i(k\delta)$; by hypothesis we know $w_M > w_m$. Define $U_M = \{i \in \{1, \ldots, n\} \mid w(k\delta) = w_M\}$ and $U_m = \{i \in \{1, \ldots, n\} \mid w_j(k\delta) = w_m\}$; by hypothesis we know $U_M \cap U_m = \emptyset$. Now, we are ready to use property (ii) in the theorem statement. Since $(\{1, \ldots, n\}, \cup_{s \in \{1, \ldots, \delta\}} E(F_s(k))$ contains a globally reachable node and since U_M and U_m are nonempty and disjoint, then Lemma 1.27 implies that there exists either

- (an out-neighbor of U_M) an edge $(i_M, j_M) \in E(F_s(k\delta))$ with $i_M \in U_M$, $j_M \in \{1, \ldots, n\} \setminus U_M$, and $s \in \{1, \ldots, \delta\}$; or

- (an out-neighbor of U_m) an edge $(i_m, j_m) \in E(F_s(k\delta))$ with $i_m \in U_m$, $j_m \in \{1, \ldots, n\} \setminus U_m$, and $s \in \{1, \ldots, \delta\}$.

Without loss of generality, suppose that an edge (i_M, j_M) exists and let $s_0 \in \{1, \ldots, \delta\}$ be the first time index for which this happens. We have the following two facts.

First, for every $s \in \{1, \ldots, s_0 - 1\}$, there does not exist any edge (i, h) with $i \in U_M$ and $h \notin U_M$, and, thus, for all $i \in U_M$;

$$w_i(k\delta + 1) = \sum_{j=1}^{n} (F_1(k))_{ij} w_j(k\delta) = \sum_{h \in U_M} (F_1(k))_{ih} w_h(k\delta)$$
$$= \Big(\sum_{h \in U_M} (F_1(k))_{ih}(k) \Big) w_M = w_M.$$

The same argument can be repeated for $F_2(k), \ldots, F_s(k)$, so that $w_i(k\delta + s) = w_M$ for all $i \in U_M$.

Second, if $i \notin U_M$ at time $k\delta$, then $w_i(k\delta + s) < w_M$ for all $s \in \{1, \ldots, s_0 - 1\}$. To see this, we compute

$$w_i(k\delta + 1) = \sum_{j=1}^{n} (F_1(k))_{ij} w_j(k\delta) = (F_1(k))_{ii} w_i(k\delta) + \sum_{j=1, j \neq i}^{n} (F_1(k))_{ij} w_j(k\delta)$$
$$\leq (F_1(k))_{ii} w_i(k\delta) + \Big(\sum_{j=1, j \neq i}^{n} (F_1(k))_{ij} \Big) w_M$$
$$\leq \alpha w_i(k\delta) + (1 - \alpha) w_M < w_M,$$

where we used the assumption of non-degeneracy with parameter $\alpha \in]0, 1]$.

The same argument can be repeated for the subsequent multiplications by the matrices $F_2(k), \ldots, F_s(k)$.

We finally reach time s_0 and compute

$$w_{i_M}(k\delta + s_0) = \sum_{j=1}^{n} (F_{s_0}(k))_{i_M j} w_j(k\delta + s_0 - 1)$$

$$= (F_{s_0}(k))_{i_M j_M} w_{j_M}(k\delta + s_0 - 1) + \sum_{j=1, j \neq j_M}^{n} (F_{s_0}(k))_{i_M j} w_j(k\delta + s_0 - 1)$$

$$< (F_{s_0}(k))_{i_M j_M} w_M + \sum_{j=1, j \neq j_M}^{n} (F_{s_0}(k))_{i_M j} w_j(k\delta + s_0 - 1) \leq w_M.$$

This implies that $w_{i_M}((k+1)\delta) < w_M$, so that i_M does not belong to U_M at time $(k+1)\delta$. That is, the cardinality of U_M decreases at least by one after $(k+1)\delta$. Since $\{1, \ldots, n\}$ is finite, after repeating this argument at most $n-1$ times, we have that U_M becomes empty at time $(k+n-1)\delta$. (Here we are assuming that the out-neighbor always exists for U_M; an analogous argument can be made for the general case.) This is enough to guarantee that $V(w((k+n)\delta)) < w_M - w_m = V(w(k\delta))$. This is what we need to conclude that $\lim_{k \to +\infty} \operatorname{dist}(w(k\delta), \operatorname{diag}(\mathbb{R})) = 0$. In summary, this concludes the proof of Theorem 1.63. ∎

We conclude this section by establishing convergence to an individual point, rather than a set of points.

Proof of Proposition 1.67. We adopt the same notation as above, that is, as in the proof of Theorem 1.63. Since $F(k) \in \mathcal{F}_\delta(\alpha)$, the set of sequence points $\{w(k\delta) \mid k \in \mathbb{N}\}$ belongs to the convex hull of all the components of the initial condition, that is, $[\min_i w_i(0), \max_i w_i(0)]^n$. Since $[\min_i w_i(0), \max_i w_i(0)]^n$ is compact, there exists a convergent subsequence $\{w(k_l\delta) \mid l \in \mathbb{N}\}$ to a point $c\mathbf{1}_n$. We also notice that for any $k_l \in \mathbb{N}$, we have $w_i((k_l + k)\delta) \in [\min_i w_i(k_l\delta), \max_i w_i(k_l\delta)]^n$, for all $i \in \{1, \ldots, n\}$ and $k \in \mathbb{N}$. Because $\lim_{l \to +\infty} w(k_l\delta) = c\mathbf{1}_n$ we know that $\lim_{l \to +\infty} [\min_i w_i(k_l\delta), \max_i w_i(k_l\delta)]^n = c\mathbf{1}_n$. Therefore, any sequence $\{w((k_l + k)\delta) \mid k \in \mathbb{N}\}$, for $l \in \mathbb{N}$, must converge to $c\mathbf{1}_n$. This implies that $\lim_{k \to +\infty} w(k\delta) = c\mathbf{1}_n$. ∎

1.8.6 Proofs of Theorems 1.79 and 1.80

Proof of Theorem 1.79. Let us prove fact (i). Because $\mathrm{Trid}_n(a, b, a)$ is a real symmetric matrix, $\mathrm{Trid}_n(a, b, a)$ is normal and its 2-induced norm—that is, its largest singular value—is equal to the magnitude of its eigenvalue with the largest magnitude. Based on this information and on the eigenvalue computation in Lemma 1.77, we compute

$$\| \mathrm{Trid}_n(a, b, a) \|_2 = \max_{i \in \{1, \ldots, n\}} \left| b + 2a \cos\left(\frac{i\pi}{n+1} \right) \right|$$

$$\leq |b| + 2|a| \max_{i \in \{1, \ldots, n\}} \left| \cos\left(\frac{i\pi}{n+1} \right) \right| \leq |b| + 2|a| \cos\left(\frac{\pi}{n+1} \right).$$

Because we assumed that $|b| + 2|a| = 1$ and because $\cos(\frac{\pi}{n+1}) < 1$ for any $n \geq 2$, the 2-induced norm of $\mathrm{Trid}_n(a, b, a)$ is strictly less than 1. Additionally, for $\ell > 0$, we bound from above the magnitude of the curve x, as

$$\|x(\ell)\|_2 = \| \mathrm{Trid}_n(a, b, a)^\ell x_0 \|_2 \leq \left(|b| + 2|a| \cos\left(\frac{\pi}{n+1} \right) \right)^\ell \|x_0\|_2.$$

In order to have $\|x(\ell)\|_2 < \varepsilon \|x_0\|_2$, it is sufficient to require that $\log \varepsilon > \ell \log \left(|b| + 2|a| \cos\left(\frac{\pi}{n+1} \right) \right)$, that is,

$$\ell > \frac{\log \varepsilon^{-1}}{- \log \left(|b| + 2|a| \cos\left(\frac{\pi}{n+1} \right) \right)}. \tag{1.8.2}$$

The upper bound now follows by noting that, as $t \to 0$, we have

$$-\frac{1}{\log(1 - 2|a|(1 - \cos t))} = \frac{1}{|a| t^2} + O(1).$$

Let us now show the lower bound. Assume, without loss of generality, that $ab > 0$ and consider the eigenvalue $b + 2a \cos(\frac{\pi}{n+1})$ of $\mathrm{Trid}_n(a, b, a)$. Note that $|b + 2a \cos(\frac{\pi}{n+1})| = |b| + 2|a| \cos(\frac{\pi}{n+1})$. (If $ab < 0$, then consider the eigenvalue $b + 2a \cos(\frac{n\pi}{n+1})$.) For $n > 2$, define the unit-length vector

$$\mathbf{v}_n = \sqrt{\frac{2}{n+1}} \begin{bmatrix} \sin \frac{\pi}{n+1} \\ \vdots \\ \sin \frac{n\pi}{n+1} \end{bmatrix} \in \mathbb{R}^n, \tag{1.8.3}$$

and note that, by Lemma 1.77(i), $\mathbf{v}_n$ is an eigenvector of $\mathrm{Trid}_n(a, b, a)$ with eigenvalue $b + 2a \cos(\frac{\pi}{n+1})$. The trajectory x with initial condition $\mathbf{v}_n$ satisfies $\|x(\ell)\|_2 = \left(|b| + 2|a| \cos\left(\frac{\pi}{n+1} \right) \right)^\ell \|\mathbf{v}_n\|_2$, and therefore, it will enter $B(\mathbf{1}_n, \varepsilon \|\mathbf{v}_n\|_2)$ only when ℓ satisfies equation (1.8.2). This completes the proof of fact (i).

Next, we prove fact (ii). Clearly, all eigenvalues of the matrix $\mathrm{Trid}_n(a, b, 0)$ are strictly inside the unit disk. For $\ell > 0$, we compute

$$\mathrm{Trid}_n(a, b, 0)^\ell$$

$$= b^\ell \left(I_n + \frac{a}{b} \mathrm{Trid}_n(1, 0, 0) \right)^\ell = b^\ell \sum_{j=0}^{n-1} \frac{\ell!}{j!(\ell - j)!} \left(\frac{a}{b} \right)^j \mathrm{Trid}_n(1, 0, 0)^j,$$

because of the nilpotency of $\mathrm{Trid}_n(1, 0, 0)$. Now, we can bound from above the magnitude of the curve x, as

$$\|x(\ell)\|_2 = \| \mathrm{Trid}_n(a, b, 0)^\ell x_0 \|_2$$

$$\leq |b|^\ell \sum_{j=0}^{n-1} \frac{\ell!}{j!(\ell - j)!} \left(\frac{a}{b} \right)^j \left\| \mathrm{Trid}_n(1, 0, 0)^j x_0 \right\|_2 \leq e^{a/b} \ell^{n-1} |b|^\ell \|x_0\|_2.$$

Here, we used $\| \mathrm{Trid}_n(1, 0, 0)^j x_0 \|_2 \leq \|x_0\|_2$ and $\max\{ \frac{\ell!}{(\ell-j)!} \mid j \in \{0, \dots, n-1\} \} \leq \ell^{n-1}$. Therefore, in order to have $\|x(\ell)\|_2 < \varepsilon \|x_0\|_2$, it suffices that $\log(e^{a/b}) + (n-1) \log \ell + \ell \log |b| \leq \log \varepsilon$, that is,

$$\ell - \frac{n-1}{-\log |b|} \log \ell > \frac{\frac{a}{b} - \log \varepsilon}{-\log |b|}.$$

A sufficient condition for $\ell - \alpha \log \ell > \beta$, for $\alpha, \beta > 0$, is that $\ell \geq 2\beta + 2\alpha \max\{1, \log \alpha\}$. For, if $\ell \geq 2\alpha$, then $\log \ell$ is bounded from above by the line $\ell/2\alpha + \log \alpha$. Furthermore, the line $\ell/2\alpha + \log \alpha$ is a lower bound for the line $(\ell - \beta)/\alpha$ if $\ell \geq 2\beta + 2\alpha \log \alpha$. In summary, it is true that $\|x(\ell)\|_2 \leq \varepsilon \|x(0)\|_2$ whenever

$$\ell \geq 2 \frac{\frac{a}{b} - \log \varepsilon}{-\log |b|} + 2 \frac{n-1}{-\log |b|} \max \left\{ 1, \log \frac{n-1}{-\log |b|} \right\}.$$

This completes the proof of the upper bound, that is, fact (ii).

The proof of fact (iii) is similar to that of fact (i). Because $\mathrm{Circ}_n(a, b, c)$ is circulant, it is also normal and each of its singular values corresponds to an eigenvector–eigenvalue pair. From Lemma 1.77(ii) and from the assumption $a + b + c = 1$, it is clear that the eigenvalue corresponding to $i = n$ is equal to 1; this is the largest singular value of $\mathrm{Circ}_n(a, b, c)$ and the corresponding eigenvector is $\mathbf{1}_n$. We now compute the second largest singular value:

$$\max_{i \in \{1, \dots, n-1\}} \left\| b + (a + c) \cos \left(\frac{i 2\pi}{n} \right) + \sqrt{-1}(c - a) \sin \left(\frac{i 2\pi}{n} \right) \right\|_{\mathbb{C}}$$

$$= \left\| 1 - (a + c) \left(1 - \cos \left(\frac{2\pi}{n} \right) \right) + \sqrt{-1}(c - a) \sin \left(\frac{2\pi}{n} \right) \right\|_{\mathbb{C}}.$$

Here, $\|\cdot\|_{\mathbb{C}}$ is the norm in $\mathbb{C}$. Because of the assumptions on a, b, c, the second largest singular value is strictly less than 1. In the orthogonal decomposition induced by the eigenvectors of $\mathrm{Circ}_n(a, b, c)$, we assume that the vector y_0 has a component y_{ave} along the eigenvector $\mathbf{1}_n$. For $\ell > 0$, we bound the distance of the curve $y(\ell)$ from $y_{\mathrm{ave}}\mathbf{1}_n$ as

$$\|y(\ell) - y_{\mathrm{ave}}\mathbf{1}_n\|_2$$
$$= \|\mathrm{Circ}_n(a, b, c)^\ell y_0 - y_{\mathrm{ave}}\mathbf{1}_n\|_2 = \|\mathrm{Circ}_n(a, b, c)^\ell (y_0 - y_{\mathrm{ave}}\mathbf{1}_n)\|_2$$
$$\leq \left\| 1 - (a + c)\left(1 - \cos\left(\frac{2\pi}{n}\right)\right) + \sqrt{-1}(c - a)\sin\left(\frac{2\pi}{n}\right) \right\|_{\mathbb{C}}^\ell \|y_0 - y_{\mathrm{ave}}\mathbf{1}_n\|_2.$$

This proves that $\lim_{\ell \to +\infty} y(\ell) = y_{\mathrm{ave}}\mathbf{1}_n$. Also, for $\alpha = a + c, \beta = c - a$ and as $t \to 0$, we have

$$-\frac{1}{\log\left(\left(1 - \alpha(1 - \cos t)\right)^2 + \beta^2 \sin^2 t\right)^{1/2}} = \frac{2}{(\alpha - \beta^2)t^2} + O(1).$$

Here, $\beta^2 < \alpha$ because $a, c \in \,]0, 1[$. From this, one deduces the upper bound in (iii).

Now, consider the eigenvalues $\lambda_n = b + (a+c)\cos\left(\frac{2\pi}{n}\right) + \sqrt{-1}(c-a)\sin\left(\frac{2\pi}{n}\right)$ and $\overline{\lambda}_n = b + (a+c)\cos\left(\frac{(n-1)2\pi}{n}\right) + \sqrt{-1}(c-a)\sin\left(\frac{(n-1)2\pi}{n}\right)$ of $\mathrm{Circ}_n(a, b, c)$, and its associated eigenvectors (cf. Lemma 1.77(ii))

$$\mathbf{v}_n = \begin{bmatrix} 1 \\ \omega \\ \vdots \\ \omega^{n-1} \end{bmatrix} \in \mathbb{C}^n, \quad \overline{\mathbf{v}}_n = \begin{bmatrix} 1 \\ \omega^{n-1} \\ \vdots \\ \omega \end{bmatrix} \in \mathbb{C}^n. \tag{1.8.4}$$

Note that the vector $\mathbf{v}_n + \overline{\mathbf{v}}_n$ belongs to $\mathbb{R}^n$. Moreover, its component y_{ave} along the eigenvector $\mathbf{1}_n$ is zero. The trajectory y with initial condition $\mathbf{v}_n + \overline{\mathbf{v}}_n$ satisfies $\|y(\ell)\|_2 = \|\lambda_n^\ell \mathbf{v}_n + \overline{\lambda}_n^\ell \overline{\mathbf{v}}_n\|_2 = |\lambda_n|^\ell \|\mathbf{v}_n + \overline{\mathbf{v}}_n\|_2$, and therefore it will enter $B(\mathbf{0}_n, \varepsilon \|\mathbf{v}_n + \overline{\mathbf{v}}_n\|_2)$ only when

$$\ell > \frac{\log \varepsilon^{-1}}{-\log\left\| 1 - (a+c)\left(1 - \cos\left(\frac{2\pi}{n}\right)\right) + \sqrt{-1}(c-a)\sin\left(\frac{2\pi}{n}\right) \right\|_{\mathbb{C}}}.$$

This completes the proof of fact (iii).   ∎

Proof of Theorem 1.80. We prove fact (i) and observe that the proof of fact (ii) is analogous. Consider the change of coordinates

$$x(\ell) = P_+ \begin{bmatrix} x'_{\mathrm{ave}}(\ell) \\ y(\ell) \end{bmatrix} = x'_{\mathrm{ave}}(\ell)\mathbf{1}_n + P_+ \begin{bmatrix} 0 \\ y(\ell) \end{bmatrix},$$

where $x'_{\mathrm{ave}}(\ell) \in \mathbb{R}$ and $y(\ell) \in \mathbb{R}^{n-1}$. A quick calculation shows that $x'_{\mathrm{ave}}(\ell) = \frac{1}{n}\mathbf{1}_n^T x(\ell)$, and the similarity transformation described in equation (1.6.7) implies

$$y(\ell + 1) = \mathrm{Trid}_{n-1}(a, b, a)\, y(\ell), \quad \text{and} \quad x'_{\mathrm{ave}}(\ell + 1) = (b + 2a)x'_{\mathrm{ave}}(\ell).$$

Therefore, $x_{\mathrm{ave}} = x'_{\mathrm{ave}}$. It is also clear that

$$x(\ell + 1) - x_{\mathrm{ave}}(\ell + 1)\mathbf{1}_n$$
$$= P_+ \begin{bmatrix} 0 \\ y(\ell + 1) \end{bmatrix} = \left(P_+ \begin{bmatrix} 0 & 0 \\ 0 & \mathrm{Trid}_{n-1}(a, b, a) \end{bmatrix} P_+^{-1} \right)(x(\ell) - x_{\mathrm{ave}}(\ell)\mathbf{1}_n).$$

Consider the matrix in parentheses determining the trajectory $\ell \mapsto (x(\ell) - x_{\mathrm{ave}}(\ell)\mathbf{1}_n)$. This matrix is symmetric, its singular values are 0 and the singular values of $\mathrm{Trid}_{n-1}(a, b, a)$, and its eigenvectors are $\mathbf{1}_n$ and the eigenvectors of $\mathrm{Trid}_{n-1}(a, b, a)$ (padded with an extra zero). These facts are sufficient to duplicate, step by step, the proof of fact (i) in Theorem 1.79. Therefore, the trajectory $\ell \mapsto (x(\ell) - x_{\mathrm{ave}}(\ell)\mathbf{1}_n)$ satisfies the stated properties. ∎

1.9 EXERCISES

E1.1   **(Orthogonal and permutation matrices).** Prove that

   (i)   the set of orthogonal matrices is a group;

   (ii) the set of permutation matrices is a group; and

   (iii) each permutation matrix is orthogonal.

E1.2   **(Doubly stochastic matrices).** Show that the set of doubly stochastic matrices is convex and that it contains the set of permutation matrices. Find in the literature as many distinct proofs of Theorem 1.1 as possible.
   Hint: A proof is contained in Horn and Johnson (1985). A second proof method is based on methods from combinatorics.

E1.3   **(Circulant matrices).** Given two $n \times n$ circulant matrices C_1 and C_2, show that the following hold:

   (i) C_1^T, $C_1 + C_2$ and $C_1 C_2$ are circulant; and

   (ii) $C_1 C_2 = C_2 C_1$.

E1.4   **(Spectral radius and ∞-induced norm of a row-stochastic matrix).** Show that the spectral radius and the ∞-induced norm of a row-stochastic matrix are 1.
   Hint: Let $A \in \mathbb{R}^{d \times d}$ be stochastic. First, show $\|A\|_\infty \le 1$ by direct algebraic manipulation. Second, use the bound in Lemma 1.5 to show that $\rho(A) \le 1$. Finally, conclude the proof by noting that 1 is an eigenvalue of A.
   Hint: An alternative proof that $\rho(A) = 1$ is as follows. First, use Geršgorin disks Theorem 1.2 to show that $\mathrm{spec}(A)$ is contained in the unit disk centered at the origin. Second, note that $\rho(A) \ge 1$, since 1 is an eigenvalue of A.

E1.5   **(Positive semidefinite matrix defined by a doubly stochastic and irreducible matrix).** Let $A \in \mathbb{R}^{n \times n}$ be doubly stochastic and irreducible. Show that the matrix

$$I_n - A^T A$$

is positive semidefinite and that its eigenvalue 0 is simple.

E1.6 **(M-matrices).** This exercise summarizes some properties of the so-called M-matrices (see Fiedler, 1986). A matrix $A \in \mathbb{R}^{n \times n}$ is an *M-matrix* (resp. an M_0-*matrix*) if

(i) all the off-diagonal elements of A are zero or negative; and

(ii) there exist a nonnegative matrix $C \in \mathbb{R}^{n \times n}$ and $k > \rho(C)$ (resp. $k \geq \rho(C)$) such that $A = kI_n - C$.

Show that:

(i) the matrix $B \in \mathbb{R}^{n \times n}$ is an M-matrix if

(a) all the off-diagonal elements of B are zero or negative; and

(b) there exists a vector $v \in \mathbb{R}^n$ with positive entries such that Bv has positive entries;

(ii) if A is an M_0-matrix, irreducible and singular, then there exists $x \in \mathbb{R}^n$ with positive entries such that $Ax = 0$ and $\operatorname{rank}(A) = n - 1$; and

(iii) if A is an M-matrix, then all eigenvalues of A have positive real parts.

E1.7 **(Decomposition of a stochastic matrix).** Consider the matrix

$$T = \begin{bmatrix} 1 & -1 & 0 & \dots & 0 \\ 0 & 1 & -1 & \dots & 0 \\ \vdots & & \ddots & \ddots & \vdots \\ 0 & \dots & 0 & 1 & -1 \\ \frac{1}{n} & \frac{1}{n} & \dots & \frac{1}{n} & \frac{1}{n} \end{bmatrix} \in \mathbb{R}^{n \times n}.$$

Show that:

(i) T is invertible.

(ii) For a stochastic matrix $F \in \mathbb{R}^{n \times n}$, there exist $F_{\mathrm{err}} \in \mathbb{R}^{(n-1) \times (n-1)}$ and $c_{\mathrm{err}} \in \mathbb{R}^{1 \times (n-1)}$ such that

$$TFT^{-1} = \begin{bmatrix} F_{\mathrm{err}} & 0_{(n-1) \times 1} \\ c_{\mathrm{err}} & 1 \end{bmatrix}.$$

Moreover, if F is symmetric, then $c_{\mathrm{err}} = 0_{1 \times (n-1)}$.

E1.8 This exercise establishes two extensions of the LaSalle Invariance Principle. Consider the same setup and assumptions as in Theorem 1.19, and remove the assumption that the set W is closed. Prove the following two conclusions.

(i) Each evolution with initial condition in W approaches a set of the form $V^{-1}(c) \cap (S \cup (\partial W \setminus W))$, where c is a real constant and S is the largest positively invariant set contained in $\{w \in W \mid V(f(w)) = V(w)\}$.

(ii) Each evolution $\gamma : \mathbb{Z}_{\geq 0} \to W$ with $\overline{\operatorname{image}(\gamma)} \subset W$ approaches a set of the form $V^{-1}(c) \cap S$, where c is a real constant and S is the largest positively invariant set contained in $\{w \in W \mid V(f(w)) = V(w)\}$.

Hint: Regarding part (i), follow the same steps as in the proof of Theorem 1.21 in Section 1.8.1 with the following difference: even though the set $\Omega(\gamma)$ is not a subset of W in general, the set $\Omega(\gamma) \cap W$ is a subset of W and is positively invariant.

E1.9 **(The closed map defined by a finite collection of continuous maps).** Let $f_1, \ldots, f_m : X \to X$ be continuous functions, where X is a d-dimensional space chosen among $\mathbb{R}^d$, $\mathbb{S}^d$, and the Cartesian products $\mathbb{R}^{d_1} \times \mathbb{S}^{d_2}$, for some $d_1 + d_2 = d$. Define the set-valued map $T : X \rightrightarrows X$ by

$$T(x) = \{f_1(x), \ldots, f_m(x)\}.$$

Show that T is closed on X.

E1.10 **(Overapproximation Lemma).** Prove Lemma 1.24.

E1.11 **(Acyclic digraphs).** Let G be an acyclic digraph. Show that:

    (i) G contains at least one sink, that is, a vertex without out-neighbors;

    (ii) G contains at least one source, that is, a vertex without in-neighbors; and

    (iii) in an appropriate ordering of the vertices of G, the adjacency matrix A is lower-triangular, that is, all its entries above the main diagonal vanish.
        Hint: Order the vertices of G according to their distance to a sink.

E1.12 **(A sufficient condition for a matrix to be primitive).** Show that if $A \in \mathbb{R}^{n \times n}$ is nonnegative, irreducible, and has a positive element on the diagonal, then A is primitive. Give an example that shows that this condition is sufficient but not necessary, that is, find a primitive matrix with no positive element on the diagonal.
Hint: See Exercise E1.23 below for a candidate matrix.

E1.13 **(Condensation digraph).** This exercise studies the decomposition of a digraph G in its strongly connected components. A subgraph H is a *strongly connected component* of G if H is strongly connected and any other subgraph of G strictly containing H is not strongly connected. The *condensation digraph* of G, denoted $C(G)$, is defined as follows: the nodes of $C(G)$ are the strongly connected components of G, and there exists a directed edge in $C(G)$ from node H_1 to node H_2 if and only if there exists a directed edge in G from a node of H_1 to a node of H_2. Show that:

    (i) every condensation digraph is acyclic;

    (ii) a digraph contains a globally reachable node if and only if its condensation digraph contains a globally reachable node; and

    (iii) a digraph contains a directed spanning tree if and only if its condensation digraph contains a directed spanning tree.

E1.14 **(Incidence matrix).** Given a weighted digraph G of order n, choose an arbitrary ordering of its edges. Define the *incidence matrix* $H(G) \in \mathbb{R}^{|E| \times n}$ of G by specifying that the row of $H(G)$ corresponding to edge (i, j) has an entry 1 in column i, an entry -1 in column j, and all other entries equal to zero. Show that

$$H(G)^T W H(G) = L(G) + L(\mathrm{rev}(G)),$$

where $W \in \mathbb{R}^{|E| \times |E|}$ is the diagonal matrix with a_{ij} in the entry corresponding to edge (i, j).

E1.15   **(From digraphs to stochastic matrices and back).** Let G be a weighted digraph of order n with adjacency matrix A, out-degree matrix D_{out}, and Laplacian matrix L. Define the following matrices:

$$F_1 = (\kappa I_n + D_{\text{out}})^{-1}(\kappa I_n + A), \quad \text{for } \kappa \in \mathbb{R}_{>0},$$

$$F_2 = I_n - \varepsilon L, \qquad\qquad\quad \text{for } \varepsilon \in [0, \min\{(D_{\text{out}})_{ii}^{-1} \mid i \in \{1, \ldots, n\}\}[.$$

Perform the following tasks:

(i) compute the entries of F_1 and F_2 as a function of the entries of $A(G)$;

(ii) show that the matrices F_1 and F_2 are row-stochastic;

(iii) identify the least restrictive conditions on G such that the matrices F_1 and F_2 are doubly stochastic; and

(iv) determine under what conditions a row-stochastic matrix can be written in the form F_1, or F_2 for some appropriate digraph (and for some appropriate scalars κ and ε).

E1.16   **(Metropolis–Hastings weights from the theory of Markov chains).** Given an undirected graph G of order n, define a weighted adjacency matrix A with entries

$$a_{ij} = \frac{1}{1 + \max\{|\mathcal{N}(i)|, |\mathcal{N}(j)|\}},$$

for $(i, j) \in E$. Perform the following tasks:

(i) show that the weighted degree of any vertex is strictly smaller than 1;

(ii) use (i) to justify that $\varepsilon = 1$ can be chosen in Exercise E1.15 for the construction of the matrix F_2; and

(iii) express the exponential convergence factor $r_{\exp}(F_2)$ as a function of the eigenvalues of the Laplacian of G.

E1.17   **(Some properties of products of stochastic matrices).** Show the following holds:

(i) If the matrices $A_1, \ldots, A_k$ are nonnegative, row-stochastic, or doubly stochastic, respectively, then their product $A_1 \cdots A_k$ is non-negative, row-stochastic, or doubly stochastic, respectively.

(ii) If the nonnegative matrices $A_1, \ldots, A_k$ have strictly positive diagonal elements, then their product $A_1 \cdots A_k$ has strictly positive diagonal elements.

(iii) Assume that $G_1, \ldots, G_k$ are digraphs associated with the nonnegative matrices $A_1, \ldots, A_k$ and that these matrices have strictly positive diagonal elements. If the digraph $G_1 \cup \ldots \cup G_k$ is strongly connected, then the matrix $A_1 \cdots A_k$ is irreducible.

E1.18   **(Disagreement function).** The *quadratic form* associated with a symmetric matrix $B \in \mathbb{R}^{n \times n}$ is the function $x \mapsto x^T B x$. Given a digraph G of order n, the *disagreement function* $\Phi_G : \mathbb{R}^n \to \mathbb{R}$ is defined by

$$\Phi_G(x) = \frac{1}{2} \sum_{i,j=1}^{n} a_{ij}(x_j - x_i)^2. \tag{E1.1}$$

Show that the following are true:

(i) the disagreement function is the quadratic form associated with the symmetric positive-semidefinite matrix

$$P(G) = \frac{1}{2}(D_{\mathrm{out}}(G) + D_{\mathrm{in}}(G) - A(G) - A(G)^T);$$

(ii) $P(G) = \frac{1}{2}\big(L(G) + L(\mathrm{rev}(G))\big).$

E1.19   **(Weight-balanced graphs and connectivity).** Let G be a weighted digraph and let A be a nonnegative $n \times n$ matrix. Show the following statements:

(i) if G is weight-balanced and contains a globally reachable node, then it is strongly connected;

(ii) if A is doubly stochastic and its associated weighted digraph contains a globally reachable node, then its associated weighted digraph is strongly connected; and

(iii) if A is doubly stochastic and a column of $\sum_{k=0}^{n-1} A^k$ is positive, then $\sum_{k=0}^{n-1} A^k$ is positive.

E1.20   **(The Laplacian matrix is positive semidefinite).** Without relying on the Geršgorin disks Theorem 1.2, show that if the weighted digraph G is undirected, then the matrix $L(G)$ is symmetric positive semidefinite. (Note that the proof of statement (i) in Theorem 1.37 relies on Geršgorin disks Theorem 1.2).

E1.21   **(Properties of the BFS algorithm).** Prove Lemma 1.28.

E1.22   **(LCR algorithm).** Consider the following LCR algorithm for leader election:

(i) Give a UID assignment to each processor for which $\Omega(n^2)$ messages are sent; and

(ii) give a UID assignment to each processor for which only $O(n)$ messages are sent.

(iii) Show that the average number of messages sent is $O(n \log n)$, where the average is taken over all possible ordering of the processors on the ring, each ordering assumed to be equally likely.

E1.23   **(Properties of a stochastic matrix and its associated digraph).** Consider the stochastic matrices

$$A_1 = \frac{1}{2}\begin{bmatrix} 0 & 1 & 1 \\ 1 & 0 & 1 \\ 1 & 1 & 0 \end{bmatrix} \quad \text{and} \quad A_2 = \frac{1}{2}\begin{bmatrix} 1 & 1 & 0 & 0 \\ 0 & 0 & 1 & 1 \\ 1 & 1 & 0 & 0 \\ 0 & 0 & 1 & 1 \end{bmatrix}.$$

Define and draw the associated digraphs G_1 and G_2. Without relying on the characterization in Propositions 1.33 and 1.35, perform the following tasks:

(i) show that the matrices A_1 and A_2 are irreducible and that the associated digraphs G_1 and G_2 are strongly connected;

(ii) show that the matrices A_1 and A_2 are primitive and that the associated digraphs G_1 and G_2 are strongly connected and aperiodic; and

(iii) show that the averaging algorithm associated with A_2 converges in a finite number of steps.

E1.24 **(Compactness of the set of non-degenerate matrices with respect to a parameter).** Show that, for any $\alpha \in \,]0,1]$, the set of non-degenerate matrices with respect to α is compact.

E1.25 **(Laplacian flow: Olfati-Saber and Murray, 2004).** Let G be a weighted directed graph with a globally reachable node. Define the *Laplacian flow* on $\mathbb{R}^n$ by

$$\dot{x} = -L(G)x,$$

or, equivalently in components,

$$\dot{x}_i = \sum_{j \in \mathcal{N}^{\text{out}}(i)} a_{ij}(x_j - x_i), \quad i \in \{1, \ldots, n\}.$$

Perform the following tasks:

(i)   Find the equilibrium points of the Laplacian flow.

(ii) Show that, if G is undirected, then the disagreement function (see Exercise E1.18) is monotonically non-increasing along the Laplacian flow.

(iii) Given $x_0 = ((x_0)_1, \ldots, (x_0)_n) \in \mathbb{R}^n$, show that the solution $t \mapsto x(t)$ of the Laplacian flow starting at x_0 verifies

$$\min\{(x_0)_1, \ldots, (x_0)_n\} \le x_i(t) \le \max\{(x_0)_1, \ldots, (x_0)_n\},$$

for all $t \in \mathbb{R}_{\ge 0}$. Use this fact to deduce that the solution $t \mapsto x(t)$ is bounded.

(iv) For G undirected, use (i)-(iii) to apply the LaSalle Invariance Principle in Theorem 1.20 and show that the solutions of the Laplacian flow converge to $\operatorname{diag}(\mathbb{R}^n)$.

(v)   Find an example G such that, with the notation in Exercise E1.18, the symmetric matrix $L(G)^T P(G) + P(G)L(G)$ is indefinite.
      Hint: *To show that the matrix is indefinite, it suffices to find $x_1, x_2 \in \mathbb{R}^n$ such that $x_1(L(G)^T P(G) + P(G)L(G))x_1 < 0$ and $x_2(L(G)^T P(G) + P(G)L(G))x_2 > 0$.*

(vi) Show that the Euler discretization of the Laplacian flow is the Laplacian-based averaging algorithm.

E1.26 **(Log–Sum–Exp consensus: Tahbaz-Salehi and Jadbabaie, 2006).** Pick $\alpha \in \mathbb{R} \setminus \{0\}$ and define the function $f_\alpha : \mathbb{R}^n \to \mathbb{R}$ by

$$f_\alpha(x) = \alpha \log\left(\frac{1}{n}\sum_{i=1}^{n} e^{x_i/\alpha}\right).$$

Show that:

(i) $\displaystyle\lim_{\alpha \to 0^-} f_\alpha(x) = \min\{x_1, \ldots, x_n\}$ and $\displaystyle\lim_{\alpha \to 0^+} f_\alpha(x) = \max\{x_1, \ldots, x_n\}$; and

(ii) $\displaystyle\lim_{\alpha \to +\infty} f_\alpha(x) = \lim_{\alpha \to -\infty} f_\alpha(x) = \frac{1}{n}(x_1 + \cdots + x_n).$

Next, let $A \in \mathbb{R}^{n \times n}$ be a non-degenerate, doubly stochastic matrix whose associated digraph contains a globally reachable node. Given such a matrix A, consider the discrete-time dynamical system

$$w_i(\ell + 1) = \alpha \log \Big(\sum_{j=1}^{n} a_{ij} \, e^{w_j(\ell)/\alpha} \Big).$$

(iii) Show that $w(\ell) \to f_\alpha(w(0))\mathbf{1}_n$ as $\ell \to +\infty$.

E1.27 **(The theory of Markov chains and random walks on graphs).** List as many connections as possible between the theory of averaging algorithms discussed in Section 1.6.2 and the theory of Markov chains. Some relevant references on Markov chains include Seneta (1981) and Lovász (1993).

Hint: There is a one-to-one correspondence between averaging algorithms and Markov chains. A homogeneous Markov chains corresponds precisely to a time-independent averaging algorithm. A reversible Markov chain corresponds precisely to a symmetric stochastic matrix.

E1.28 **(Distributed hypothesis testing: Rao and Durrant-Whyte, 1993; Olfati-Saber et al., 2006).** Let h_γ, for $\gamma \in \Gamma$ in a finite set Γ, be a set of alternative hypotheses about an uncertain event. Suppose that n nodes take measurements z_i, for $i \in \{1, \dots, n\}$, related to the event. Assume that each observation is conditionally independent of all other observations, given any hypothesis.

(i) Using Bayes' Theorem and the independence assumption, show that the *a posteriori probabilities* satisfy

$$p(h_\gamma | z_1, \dots, z_n) = \frac{p(h_\gamma)}{p(z_1, \dots, z_n)} \prod_{i=1}^{n} p(z_i | h_\gamma).$$

(ii) Suppose that the nodes form a undirected unweighted connected synchronous network with adjacency matrix A. Consider the discrete-time dynamical system

$$\pi_i(\ell + 1) = \Big(\pi_i(\ell) \prod_{j=1}^{n} \pi_j^{a_{ij}}(\ell) \Big)^{1/(1+d_{\mathrm{out}}(i))}.$$

Fix $\gamma \in \Gamma$, set $\pi_i(0) = p(z_i | h_\gamma)$, and show that $\pi(\ell) \to \sqrt[n]{\prod_{i=1}^{n} p(z_i | h_\gamma)} \mathbf{1}_n$ as $\ell \to +\infty$.

(iii) What information does each node need in order to compute the *maximum a posteriori estimate*, that is, to estimate the most likely hypothesis?
Hint: Can you compute $p(z_1, \dots, z_n)$, given knowledge of $p(h_\gamma)$ and of $\prod_{i=1}^{n} p(z_i | h_\gamma)$?

As a bibliographic note, the variable π_i is referred to as the *belief* in the seminal work by Pearl (1988).

E1.29 **(Bounds on vector norms).** Prove Lemma 1.82.

E1.30 **(The "n-bugs problem" and cyclic interactions).** The "n-bugs problem" related to the *pursuit curves* from mathematics, inquires about what the paths of n bugs, not aligned initially, are when they chase one another. Simple versions of the problem (e.g., for three bugs starting at the vertices of an equilateral triangle) were studied as early as the nineteenth century. It was in Watton and Kydon

(1969) that a general solution for the general n-bugs problem for non-collinear initial positions was given. The bugs trace out logarithmic spirals that eventually meet at the same point, and it is not necessary that they move with constant velocity. Surveys about cyclic pursuit problems are given in the papers in Watton and Kydon (1969) and Marshall et al. (2004). Cyclic pursuit, has also been studied recently in the multi-agent and control literature; see, for example Bruckstein et al. (1991), Marshall et al. (2004), and Smith et al. (2005). In particular, the paper Marshall et al. (2004) extends the n-bugs problem to the case of n kinematic unicycles evolving in continuous time.

Consider the simplified scenario of the n-bugs problem placed on a circle of radius r and suppose that the bugs' motion is constrained to be on that circle. Assume that agents are ordered counterclockwise with identities $i \in \{1, \ldots, n\}$, where, for convenience, we identify $n + 1$ with 1. Denote by $p_i(\ell) = (r, \theta_i(\ell))$ the sequence of positions of bug i, initially at $p_i(0) = (r, \theta_i(0))$. We illustrate two scenarios of interest in Figure E1.1 and we describe them in some detail below.

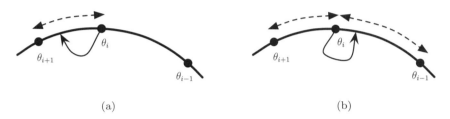

(a)   (b)

Figure E1.1 An illustration of the n-bugs problem. In (a), agent i looks at the position of agent $i+1$ and moves toward it by an amount proportional to their distance. In (b), agent i looks at the position of agents $i+1$ and $i-1$ and moves toward the one which is furthest by an amount proportional to the difference between the two distances. In both cases, the proportionality constant is k.

Cyclic pursuit. Suppose that each bug is chasing the closest counterclockwise neighbor (according to the order we have given them on the circle), see Figure E1.1(a). In other words, each bug feels an attraction toward the closest counterclockwise neighbor that can be described by the equation

$$\theta_i(\ell + 1) = (1 - k)\theta_i(\ell) + k\theta_{i+1}(\ell), \quad \ell \in \mathbb{Z}_{\geq 0},$$

where $k \in [0, 1]$. Determine for which values of k the bugs converge to a configuration for which $\text{dist}_c(\theta_{i+1}, \theta_i) = \text{dist}_c(\theta_i, \theta_{i-1})$ for all $i \in \{1, \ldots, n\}$. Observe that the bugs will approach this equally spaced configuration while moving around the circle indefinitely.

Cyclic balancing. Suppose that each bug makes a compromise between chasing its closest counterclockwise neighbor and the closest clockwise neighbor, see Figure E1.1(b). In other words, each bug feels an attraction towards the closest counterclockwise and clockwise neighbors that can be described by the equation

$$\theta_i(\ell + 1) = k\theta_{i+1}(\ell) + (1 - 2k)\theta_i(\ell) + k\theta_{i-1}(\ell), \quad \ell \in \mathbb{Z}_{\geq 0},$$

where $k \in [0, 1]$. Perform the following two tasks:

(i) Determine for which values of k the bugs converge to a configuration for which $\mathrm{dist}_{\mathrm{c}}(\theta_{i+1}, \theta_i) = \mathrm{dist}_{\mathrm{c}}(\theta_i, \theta_{i-1})$ for all $i \in \{1, \ldots, n\}$.

(ii) Show that the bugs will approach this equally spaced configuration while each of them converges to a stationary position on the circle.

Hint: *Rewrite the cyclic pursuit and cyclic balancing systems in terms of the inter-bug distances, that is, in terms of $d_i(\ell) = \mathrm{dist}_{\mathrm{c}}(\theta_{i+1}(\ell), \theta_i(\ell))$, $i \in \{1, \ldots, n\}$, $\ell \in \mathbb{Z}_{\geq 0}$. Find the matrices that describe the linear iterations in these new coordinates. Show that the agreement space, that is, the diagonal set in $\mathbb{R}^n$, is invariant under the dynamical systems. Finally, determine which values of k make each system converge to the agreement space. Lemma 1.77 might be of use in this regard. Regarding part (ii)b), recall that an exponentially decaying sequence is summable.*

Chapter Two

Geometric models and optimization

This chapter presents various geometric objects and geometric optimization problems that have strong connections with motion coordination. Basic geometric notions such as polytopes, centers, partitions, and distances are ubiquitous in cooperative strategies, coordination tasks, and the interaction of robotic networks with the physical environment. The notion of Voronoi partition finds application in diverse areas such as wireless communications, signal compression, facility location, and mesh optimization. Proximity graphs provide a natural way to mathematically model the network interconnection topology resulting from the agents' sensing and/or communication capabilities. Finally, multicenter functions play the role of aggregate objective functions in geometric optimization problems. We introduce these concepts here in preparation for the later chapters.

The chapter is organized as follows. We begin by presenting basic geometric constructions. This gives way to introduce the notion of proximity graphs along with numerous examples. The next section of the chapter presents geometric optimization problems and multicenter functions, paying special attention to the characterization of their smoothness properties and critical points. We end the chapter with three sections on, respectively, bibliographic notes, proofs of the results presented in the chapter, and exercises.

2.1 BASIC GEOMETRIC NOTIONS

In this section, we gather some classical geometric constructions that will be invoked regularly throughout the book.

2.1.1 Polygons and polytopes

For $p, q \in \mathbb{R}^d$, we let $]p, q[= \{\lambda p + (1 - \lambda)q \mid \lambda \in]0, 1[\}$ and $[p, q] = \{\lambda p + (1 - \lambda)q \mid \lambda \in [0, 1]\}$ denote the *open segment* and *closed segment*, with extreme points p and q, respectively. We let $H_{p,q} = \{x \in \mathbb{R}^d \mid \|x - p\|_2 \leq \|x - q\|_2\}$ denote the *closed halfspace* of $\mathbb{R}^d$ of points closer (in Euclidean distance) to p than to q. In the plane, we often refer to a halfspace as a *halfplane*.

As seen in Section 1.2, a set $S \subset \mathbb{R}^d$ is *convex* if, for any two points p, q in S, the closed segment $[p, q]$ is contained in S. The *convex hull* of a set is the smallest (with respect to the inclusion) convex set that contains it. We denote the convex hull of S by co(S). For $S = \{p_1, \ldots, p_n\}$ finite, the convex hull can be explicitly described as follows:

$$\mathrm{co}(S) = \big\{ \lambda_1 p_1 + \cdots + \lambda_n p_n \mid \lambda_i \geq 0 \text{ and } \sum_{i=1}^{n} \lambda_i = 1 \big\}.$$

Given p and q in $\mathbb{R}^d$ and a convex closed set $Q \subset \mathbb{R}^d$ with $p \in Q$ (see Figure 2.1), define the *from-to-inside* function by

$$\mathrm{fti}(p, q, Q) = \begin{cases} q, & \text{if } q \in Q, \\ [p, q] \cap \partial Q, & \text{if } q \notin Q. \end{cases}$$

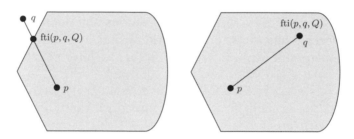

Figure 2.1 An illustration of the from-to-inside function fti.

The function fti selects the point in the closed segment $[p, q]$ which is at the same time closest to q and inside Q. Note that $\mathrm{fti}(p, q, Q)$ depends continuously on p and q.

A *polygon* is a set in $\mathbb{R}^2$ whose boundary is the union of a finite number of closed segments. A polygon is *simple* if its boundary, regarded as a curve, is not self-intersecting. We will only consider simple polygons. The closed segments composing the boundary of a polygon are called *edges*, and points resulting from the pairwise intersection between consecutive edges are called *vertices*. A convex polygon can be written as:

(i) the convex hull of its set of vertices; or

(ii) the intersection of halfplanes defined by its edges.

Two vertices whose open segment is contained in the interior of the polygon define a *diagonal*. To each vertex of a polygon we associate an *interior* and an *exterior angle*. A vertex is *strictly convex* (resp. *strictly nonconvex*) if its interior angle is strictly smaller (resp. greater) than π radians. A polygon is nonconvex if it has at least one strictly concave vertex. The *perimeter* of a polygon is the length of its boundary, that is, the sum of the lengths of its edges. A polytope is the generalization of the notion of polygon to $\mathbb{R}^d$, for $d \geq 3$. In this book, we will not consider nonconvex polytopes in dimension larger than 2. As for convex polygons, a *(convex) polytope* in $\mathbb{R}^d$ can be defined as either the convex hull of a finite set of points in $\mathbb{R}^d$ or the bounded intersection of a finite set of halfspaces. A $d - 1$ *face* (or a *facet*) of a polytope is the intersection between the polytope and the boundary of a closed halfspace that defines the polytope. A $d - 2$ face is a $d - 2$ face of a facet of the polytope. The faces of dimensions 0, 1, and $d - 1$ are called *vertices*, *edges*, and *faces*, respectively. For a convex polytope Q, we will refer to them as $\mathrm{Ve}(Q)$, $\mathrm{Ed}(Q)$, and $\mathrm{Fa}(Q)$, respectively.

2.1.2 Nonconvex geometry

In this section, we gather some basic notions on nonconvex geometry. We consider environments that include nonconvex polygons as a particular case.

We begin with some visibility notions. Given $S \subset \mathbb{R}^d$, two points $p, q \in S$ are *visible* to each other if the closed segment $[p, q]$ is contained in S. The *visibility set* $\mathrm{Vi}(p; S)$ is the set of all points in S visible from p. Given $r > 0$, the *range-limited visibility set* $\mathrm{Vi}_{\mathrm{disk}}(p; S) = \mathrm{Vi}(p; S) \cap \overline{B}(p, r)$ is the set of all points in S within a distance r and visible from p. The set S is *star-shaped* if there exists $p \in S$ such that $\mathrm{Vi}(p; S) = S$. The *kernel set* of S is comprised of all the points with this property, that is, $\mathrm{kernel}(S) = \{p \in S \mid \mathrm{Vi}(p; S) = S\}$. Trivially, any convex set is star-shaped. Given $\delta \in \mathbb{R}_{>0}$, the δ-contraction of S is the set $S_\delta = \{p \in S \mid \mathrm{dist}(p, \partial S) \geq \delta\}$. Note that if two points $p, q \in S$ are visible to each other in S_δ, then any point within distance δ of p and any point within distance δ of q are visible to each other. Figure 2.2 illustrates these visibility notions.

Next, we introduce various concavity notions. Given $S \subset \mathbb{R}^d$ connected and closed, $p \in \partial S$ is *strictly concave* if, for any $\varepsilon \in \mathbb{R}_{>0}$, there exist $q_1, q_2 \in B(p, \varepsilon) \cap \partial S$ such that $[q_1, q_2] \not\subset S$. This definition coincides with the notion of strictly concave vertex when the set S is a polygon. A *strict concavity* of S is either an isolated strictly concave point or a concave arc, that is, a connected set of strictly concave points. An *allowable environment* $S \subset \mathbb{R}^2$ is a set that satisfies the following properties: it is closed, simply connected, has a finite number of strict concavities, and its boundary can

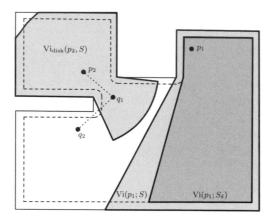

Figure 2.2 An illustration of various visibility notions. The visibility set $\mathrm{Vi}(p_1; S)$ from p_1 in S, the visibility set $\mathrm{Vi}(p_1; S_\delta)$ from p_1 in S_δ, and the range-limited visibility set $\mathrm{Vi}_{\mathrm{disk}}(p_2; S)$ from p_2 in S are depicted in light gray. The dashed curve in the interior of S corresponds to the boundary of the δ-contraction of S. The points p_2 and q_1 are visible to each other in S_δ. The points q_1 and q_2 are visible to each other in S, but they are not visible to each other in S_δ.

be described by a continuous and piecewise continuously differentiable curve which is not differentiable at most at a finite number of points. Figure 2.3 shows a sample allowable environment. Given an allowable environment S, let a point v belonging to a concave arc have the property that the boundary of S is continuously differentiable at v. The *internal tangent halfplane* $H_S(v)$ is the closed halfplane whose boundary is tangent to ∂S at v and whose interior does not contain any points of the strict concavity (see Figure 2.3).

The following result presents an interesting property of allowable environments. Its proof is left to the reader.

Lemma 2.1 (Contraction of allowable environments). *Given an allowable environment S, the δ-contraction S_δ is also allowable for sufficiently small $\delta \in \mathbb{R}_{>0}$ and does not have isolated strictly concave points. Furthermore, the boundary of S_δ is continuously differentiable at the concavities.*

Lemma 2.1 implies that the internal tangent halfplane is well-defined at any strict concavity of the δ-contraction S_δ.

A set $S \subset X$ is *relatively convex* in $X \subset \mathbb{R}^d$ if, for any two points p, q in S, the shortest curve in X that connects p and q is contained in S. Relatively convex sets in $\mathbb{R}^d$ are just convex sets. The *relative convex hull* of a set S in X is the smallest (with respect to the operation of inclusion)

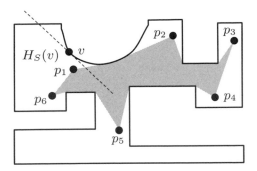

Figure 2.3 An allowable environment S. The curved portion of the boundary is a concave arc. The vertices whose interior angle is $3\pi/2$ radians are isolated strictly concave points. The relative convex hull of $\{p_1, \ldots, p_6\}$ in S is depicted in light gray. Finally, the dashed line represents the boundary of the internal tangent halfplane $H_S(v)$ tangent to ∂S at v.

relatively convex set in X that contains S (see Figure 2.3). We denote the relative convex hull of S in X by $\mathrm{rco}(S; X)$. The *(relative) perimeter* of S in X is the length of the shortest measurable closed curve contained in X that encloses S.

2.1.3 Geometric centers

Let $X = \mathbb{R}^d$, $X = \mathbb{S}^d$ or $X = \mathbb{R}^{d_1} \times \mathbb{S}^{d_2}$, $d = d_1 + d_2$. Recall our convention (cf., Section 1.1.2) that, unless otherwise noted, $\mathbb{R}^d$ is endowed with the Euclidean distance, $\mathbb{S}^d$ is endowed with the geodesic distance, and $\mathbb{R}^{d_1} \times \mathbb{S}^{d_2}$ is endowed with the Cartesian product distance $(\mathrm{dist}_2, \mathrm{dist}_g)$.

The *circumcenter* of a bounded set $S \subset X$, denoted by $\mathrm{CC}(S)$, is the center of the closed ball of minimum radius that contains S. The *circumradius* of S, denoted by $\mathrm{CR}(S)$, is the radius of this ball[1]. The circumcenter is always unique.

The computation of the circumcenter and the circumradius of a polytope $Q \subset \mathbb{R}^d$ is a strictly convex problem and, in particular, a quadratically constrained linear program in p (the center) and r (the radius). It consists of minimizing the radius r of the ball centered at p subject to the constraints that the distance between q and each of the polygon vertices is smaller than

---

[1]Note that the definition of circumcenter given here is in general different from the classical notion of circumcenter of a triangle, that is, the center of the circle passing through the three vertices of the triangle.

or equal to r. Formally, the problem can be expressed as

$$\text{minimize } r\,,$$

$$\text{subject to } \|q - p\|_2^2 \leq r^2, \text{ for all } q \in \text{Ve}(Q). \tag{2.1.1}$$

Next, we summarize some useful properties of the circumcenter in Euclidean space; see Exercise E2.1 for their proofs. In the following result, for $S \in \mathbb{F}(\mathbb{R}^d)$ with $d = 1$, we let $\text{Ve}(\text{co}(S))$ denote the set of extreme points of the interval $\text{co}(S)$.

Lemma 2.2 (Properties of the circumcenter in Euclidean space).
Let $S = \{p_1, \ldots, p_n\} \in \mathbb{F}(\mathbb{R}^d)$ with $n \geq 2$. The following properties hold:

(i) $\text{CC}(S) \in \text{co}(S) \setminus \text{Ve}(\text{co}(S))$; and

(ii) if $p \in \text{co}(S) \setminus \{\text{CC}(S)\}$ and $r \in \mathbb{R}_{>0}$ are such that $S \subset \overline{B}(p, r)$, then $]p, \text{CC}(S)[$ has a nonempty intersection with $\overline{B}(\frac{p+q}{2}, \frac{r}{2})$ for all $q \in \text{co}(S)$.

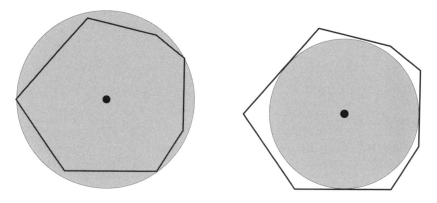

Figure 2.4 The circumcenter and circumradius (left), and incenter and inradius (right) of a convex polygon.

Given $X = \mathbb{R}^d$, $X = \mathbb{S}^d$ or $X = \mathbb{R}^{d_1} \times \mathbb{S}^{d_2}$, $d = d_1 + d_2$, the *incenter*, or *Chebyshev center* of a compact set $S \subset X$, denoted by $\text{IC}(S)$, is the set containing the centers of all closed balls of maximum radius contained in S. The *inradius* of S, denoted by $\text{IR}(S)$, is the common radius of any of these balls.

The computation of the incenter and the inradius of a polytope $Q \subset \mathbb{R}^d$ is a convex problem and, in particular, a linear program in p and r. It consists of maximizing the radius r of the ball centered at p subject to the constraints that the distance between p and each of the polytope facets is greater than or equal to r. Formally, the problem can be expressed as follows. For each

$f \in \mathrm{Fa}(Q)$, select a point $q_f \in Q$ belonging to f. Then, we set

$$\text{maximize } r,$$
$$\text{subject to } (q_f - p) \cdot \mathrm{n_{out}} \geq r, \text{ for all } f \in \mathrm{Fa}(Q), \qquad (2.1.2)$$

where $\mathrm{n_{out}}$ denotes the normal to the face f pointing toward the exterior of the polytope. The incenter of a polytope is not necessarily unique (consider, for instance, the case of a rectangle).

In Euclidean space, $X = \mathbb{R}^d$, we refer to a bounded measurable function $\phi : \mathbb{R}^d \to \mathbb{R}_{\geq 0}$ as a *density* on $\mathbb{R}^d$. The *(generalized) area* and the *centroid* (also called center of mass) of a bounded measurable set $S \subset \mathbb{R}^d$ with respect to ϕ, denoted by $\mathrm{A}_\phi(S)$ and $\mathrm{CM}_\phi(S)$ respectively, are given by

$$\mathrm{A}_\phi(S) = \int_S \phi(q)dq, \qquad \mathrm{CM}_\phi(S) = \frac{1}{\mathrm{A}_\phi(S)} \int_S q\phi(q)dq.$$

When the function ϕ that is being used is clear from the context, we simply refer to the area and the centroid of S. The centroid can alternatively be defined as follows. Define the *polar moment of inertia* of S about $p \in S$ by

$$\mathrm{J}_\phi(S, p) = \int_S \|q - p\|_2^2 \phi(q)dq.$$

Then, the centroid of S is precisely the point $p \in S$ that minimizes the polar moment of inertia of S about p. This can be easily seen from the Parallel Axis Theorem (Hibbeler, 2006), which states that

$$\mathrm{J}_\phi(S, p) = \mathrm{J}_\phi(S, \mathrm{CM}_\phi(S)) + \mathrm{A}_\phi(S)\|p - \mathrm{CM}_\phi(S)\|_2^2.$$

Remark 2.3 (Computation of geometric centers in the plane). The circumcenter, incenter, and centroid of a polygon can be computed in several ways. A simple procedure to compute the circumcenter consists of enumerating all pairs and triplets of vertices of the polygon, computing the centers and radiuses of the balls passing through them, and selecting the ball with the smallest radius that encloses the polygon. An alternative, more efficient, way of computing the circumcenter is to use the formulation (2.1.1). A convex quadratically constrained linear program is a particular case of a semidefinite-quadratic-linear program (SQLP). Several freely available numerical packages exist to solve SQLP problems; for example, SDPT3 (Tutuncu et al., 2003). The computation of the incenter set of a polygon can be performed via linear programming using the formulation (2.1.2). Finally, the centroid of a polygon can be computed with any numerical routine that accurately approximates the integral of a function over a planar domain. •

2.1.4 Voronoi and range-limited Voronoi partitions

A *partition* of a set S is a subdivision of S into connected subsets that are disjoint except for their boundary. Formally, a partition of S is a collection of closed connected sets $\{W_1, \ldots, W_m\} \subset \mathbb{P}(S)$ that verify

$$S = \cup_{i=1}^{m} W_i \quad \text{and} \quad \text{int}(W_j) \cap \text{int}(W_k) = \emptyset,$$

for $j, k \in \{1, \ldots, m\}$.

Definition 2.4 (Voronoi partition). Given a distance function dist : $X \times X \to \mathbb{R}_{\geq 0}$, a set $S \subset X$ and n distinct points $\mathcal{P} = \{p_1, \ldots, p_n\}$ in S, the *Voronoi partition* of S generated by $\mathcal{P}$ is the collection of sets $\mathcal{V}(\mathcal{P}) = \{V_1(\mathcal{P}), \ldots, V_n(\mathcal{P})\} \subset \mathbb{P}(S)$ defined by, for each $i \in \{1, \ldots, n\}$,

$$V_i(\mathcal{P}) = \{q \in S \mid \text{dist}(q, p_i) \leq \text{dist}(q, p_j), \text{ for all } p_j \in \mathcal{P} \setminus \{p_i\}\}. \qquad \bullet$$

In other words, $V_i(\mathcal{P})$ is the set of the points of S that are closer to p_i than to any of the other points in $\mathcal{P}$. We refer to $V_i(\mathcal{P})$ as the *Voronoi cell* of p_i. Unless explicitly noted otherwise, we compute the Voronoi partition according to the following conventions:

- for $X = \mathbb{R}^d$, with respect to the Euclidean distance;

- for $X = \mathbb{S}^d$, with respect to the geodesic distance; and

- for $X = \mathbb{R}^{d_1} \times \mathbb{S}^{d_2}$, $d_1 + d_2 = d$, with respect to the Cartesian product distance determined by dist_2 on $\mathbb{R}^{d_1}$ and dist_g on $\mathbb{S}^{d_2}$.

Figure 2.5 shows an example of the Voronoi partition of the circle generated by five points. In the Euclidean case, the Voronoi cell of p_i is equal to the intersection of half-spaces determined by p_i and the other locations in $\mathcal{P}$, and as such it is a convex polytope. The left plot in Figure 2.6 shows an example of the Voronoi partition of a convex polygon generated by 40 points.

Definition 2.5 (r-limited Voronoi partition). Given a distance function dist : $X \times X \to \mathbb{R}_{\geq 0}$, a set $S \subset X$, n distinct points $\mathcal{P} = \{p_1, \ldots, p_n\}$ in S, and a positive real number $r \in \mathbb{R}_{>0}$, the *r-limited Voronoi partition* inside S generated by $\mathcal{P}$ is the collection of sets $\mathcal{V}_r(\mathcal{P}) = \{V_{1,r}(\mathcal{P}), \ldots, V_{n,r}(\mathcal{P})\} \subset \mathbb{P}(S)$ defined by

$$V_{i,r}(\mathcal{P}) = V_i(\mathcal{P}) \cap \overline{B}(p_i, r), \quad i \in \{1, \ldots, n\}. \qquad \bullet$$

Note that the r-limited Voronoi partition inside S is precisely the Voronoi partition of the set $\cup_{i=1}^{n} \overline{B}(p_i, r) \cap S$. We will refer to $V_{i,r}(\mathcal{P})$ as the *r-limited*

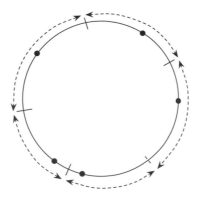

Figure 2.5 Voronoi partition of the circle generated by five points. The dashed segments correspond to the Voronoi cells of each individual point.

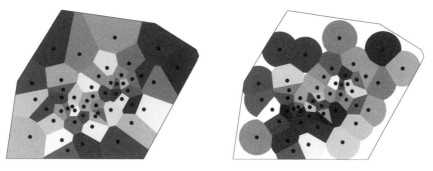

Figure 2.6 Voronoi partition of a convex polygon (left) and r-limited Voronoi partition inside a convex polygon (right) generated by 40 points.

Voronoi cell of p_i. The right-hand plot in Figure 2.6 shows an example of the r-limited Voronoi partition inside a convex polygon generated by 40 points.

Let $X = \mathbb{R}^d$, $X = \mathbb{S}^d$ or $X = \mathbb{R}^{d_1} \times \mathbb{S}^{d_2}$, $d = d_1 + d_2$. Given a density ϕ on X, a set of n distinct points $\mathcal{P} = \{p_1, \ldots, p_n\}$ in $S \subset X$ is:

(i) A *centroidal Voronoi configuration* if each point is the centroid of its own Voronoi cell, that is, $p_i = \mathrm{CM}_\phi(V_i(\mathcal{P}))$.

(ii) An *r-limited centroidal Voronoi configuration*, for $r \in \mathbb{R}_{>0}$, if each point is the centroid of its own r-limited Voronoi cell, that is, $p_i = \mathrm{CM}_\phi(V_{i,r}(\mathcal{P}))$. If $r \geq \mathrm{diam}(S)$, then an r-limited centroidal Voronoi configuration is a centroidal Voronoi configuration.

(iii) A *circumcenter Voronoi configuration* if each point is the circumcenter of its own Voronoi cell, that is, $p_i = \mathrm{CC}(V_i(\mathcal{P}))$.

(iv) An *incenter Voronoi configuration* if each point is an incenter of its own Voronoi cell, that is, $p_i \in \mathrm{IC}(V_i(\mathcal{P}))$.

Figure 2.7 illustrates of the various notions of center Voronoi configurations.

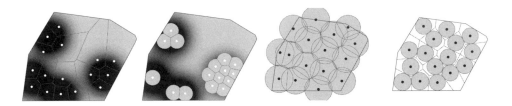

Figure 2.7 From left to right, centroidal, r-limited centroidal, circumcenter, and incenter
            Voronoi configurations composed by 16 points in a convex polygon. Darker
            colored areas correspond to higher values of the density ϕ.

2.2 PROXIMITY GRAPHS

Roughly speaking, a proximity graph is a graph whose vertex set is a set of
distinct points and whose edge set is a function of the relative locations of
the point set. Proximity graphs appear in computational geometry. In this
section, we study this important notion in detail following the presentation
by Cortés et al. (2005).

Definition 2.6 (Proximity graph). Assume that X is a d-dimensional
space chosen among $\mathbb{R}^d$, $\mathbb{S}^d$, and the Cartesian products $\mathbb{R}^{d_1} \times \mathbb{S}^{d_2}$, for some
$d_1 + d_2 = d$. For a set $S \subset X$, let $\mathbb{G}(S)$ be the set of all undirected graphs
whose vertex set is an element of $\mathbb{F}(S)$. A *proximity graph* $\mathcal{G} : \mathbb{F}(S) \to \mathbb{G}(S)$
associates to a set of distinct points $\mathcal{P} = \{p_1, \dots, p_n\} \subset S$ an undirected
graph with vertex set $\mathcal{P}$ and whose edge set is given by $\mathcal{E}_\mathcal{G}(\mathcal{P}) \subseteq \{(p, q) \in
\mathcal{P} \times \mathcal{P} \mid p \neq q\}$.   •

Note that in a proximity graph a point cannot be its own neighbor. From
this definition, we observe that the distinguishing feature of proximity graphs
is that their edge sets change with the location of their vertices. It is also
possible to define proximity graphs that associate to each point set a digraph,
but we will not consider them here.

Examples of proximity graphs on X, where we recall that $\mathrm{dist} = \mathrm{dist}_2$ if
$X = \mathbb{R}^d$, $\mathrm{dist} = \mathrm{dist}_g$ if $X = \mathbb{S}^d$, and $\mathrm{dist} = (\mathrm{dist}_2, \mathrm{dist}_g)$ if $X = \mathbb{R}^{d_1} \times \mathbb{S}^{d_2}$,
include the following:

(i) The *complete graph* $\mathcal{G}_{\mathrm{cmplt}}$ where any two points are neighbors.
    When convenient, we may view the complete graph as weighted
    by assigning the weight $\mathrm{dist}(p_i, p_j)$ to the edge $(p_i, p_j) \in \mathcal{E}_{\mathcal{G}_{\mathrm{cmplt}}}(\mathcal{P})$.

(ii) The *r-disk graph* $\mathcal{G}_{\text{disk}}(r)$, for $r \in \mathbb{R}_{>0}$, where two points are neighbors if they are located within a distance r, that is, $(p_i, p_j) \in \mathcal{E}_{\mathcal{G}_{\text{disk}}(r)}(\mathcal{P})$ if $\text{dist}(p_i, p_j) \leq r$.

(iii) The *Delaunay graph* $\mathcal{G}_{\text{D}}$, where two points are neighbors if their corresponding Voronoi cells intersect, that is, $(p_i, p_j) \in \mathcal{E}_{\mathcal{G}_{\text{D}}}(\mathcal{P})$ if $V_i(\mathcal{P}) \cap V_j(\mathcal{P}) \neq \emptyset$.

(iv) The *r-limited Delaunay graph* $\mathcal{G}_{\text{LD}}(r)$, for $r \in \mathbb{R}_{>0}$, where two points are neighbors if their corresponding $\frac{r}{2}$-limited Voronoi cells intersect, that is, $(p_i, p_j) \in \mathcal{E}_{\mathcal{G}_{\text{LD}}(r)}(\mathcal{P})$ if $V_{i,\frac{r}{2}}(\mathcal{P}) \cap V_{j,\frac{r}{2}}(\mathcal{P}) \neq \emptyset$.

(v) The *relative neighborhood graph* $\mathcal{G}_{\text{RN}}$, where two points are neighbors if their associated open lune (cf. Section 1.1.2) does not contain any point in $\mathcal{P}$, that is, $(p_i, p_j) \in \mathcal{E}_{\mathcal{G}_{\text{RN}}}(\mathcal{P})$ if, for all $p_k \in \mathcal{P}$, $k \notin \{i, j\}$

$$p_k \notin B(p_i, \text{dist}(p_i, p_j)) \cap B(p_j, \text{dist}(p_i, p_j)).$$

Figure 2.8 shows examples of these proximity graphs in the plane.

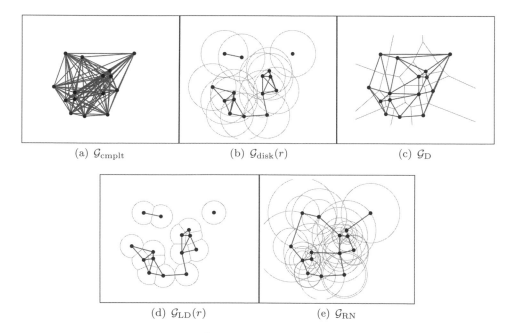

(a) $\mathcal{G}_{\text{cmplt}}$ (b) $\mathcal{G}_{\text{disk}}(r)$ (c) $\mathcal{G}_{\text{D}}$

(d) $\mathcal{G}_{\text{LD}}(r)$ (e) $\mathcal{G}_{\text{RN}}$

Figure 2.8 Proximity graphs in $\mathbb{R}^2$. From left to right, in the first row, complete, r-disk, and Delaunay, and in the second row, r-limited Delaunay and relative neighborhood for a set of 15 points. When appropriate, the geometric objects determining the edge relationship are plotted in lighter gray.

Additional examples of proximity graphs in the Euclidean space include the following:

(vi) The *Gabriel graph* $\mathcal{G}_{\mathrm{G}}$, where two points are neighbors if the ball centered at their midpoint and passing through both of them does not contain any point in $\mathcal{P}$, that is, $(p_i, p_j) \in \mathcal{E}_{\mathcal{G}_{\mathrm{G}}}(\mathcal{P})$ if $p_k \notin B\left(\frac{p_i + p_j}{2}, \frac{\mathrm{dist}(p_i, p_j)}{2}\right)$ for all $p_k \in \mathcal{P}$.

(vii) The r-∞-*disk graph* $\mathcal{G}_{\infty\text{-disk}}(r)$, for $r \in \mathbb{R}_{>0}$, where two points are neighbors if they are located within L^∞-distance r, that is, $(p_i, p_j) \in \mathcal{E}_{\mathcal{G}_{\infty\text{-disk}}(r)}(\mathcal{P})$ if $\mathrm{dist}_\infty(p_i, p_j) \le r$.

(viii) The *Euclidean minimum spanning tree* of a proximity graph $\mathcal{G}$, denoted by $\mathcal{G}_{\mathrm{EMST},\mathcal{G}}$, that assigns to each $\mathcal{P}$ a minimum-weight spanning tree (cf., Section 1.4.4.4) of $\mathcal{G}(\mathcal{P})$ with weighted adjacency matrix $a_{ij} = \|p_i - p_j\|_2$, for $(p_i, p_j) \in \mathcal{E}_{\mathcal{G}}(\mathcal{P})$. If $\mathcal{G}(\mathcal{P})$ is not connected, then $\mathcal{G}_{\mathrm{EMST},\mathcal{G}}(\mathcal{P})$ is the union of Euclidean minimum spanning trees of its connected components. When $\mathcal{G}$ is the complete graph, we simply denote the Euclidean minimum spanning tree by $\mathcal{G}_{\mathrm{EMST}}$.

(ix) the *visibility graph* $\mathcal{G}_{\mathrm{vis},Q}$ in an allowable environment Q in $\mathbb{R}^2$, where two points are neighbors if they are visible to each other, that is, $(p_i, p_j) \in \mathcal{E}_{\mathcal{G}_{\mathrm{vis},Q}}(\mathcal{P})$ if the closed segment $[p_i, p_j]$ from p_i to p_j is contained in Q.

(x) The *range-limited visibility graph* $\mathcal{G}_{\mathrm{vis\text{-}disk},Q}$ in an allowable environment Q in $\mathbb{R}^2$, where two points are neighbors if they are visible to each other and their distance is no more than r, that is, $(p_i, p_j) \in \mathcal{E}_{\mathcal{G}_{\mathrm{vis\text{-}disk},Q}}(\mathcal{P})$ if $(p_i, p_j) \in \mathcal{E}_{\mathcal{G}_{\mathrm{vis},Q}}(\mathcal{P})$ and $(p_i, p_j) \in \mathcal{E}_{\mathcal{G}_{\mathrm{disk}}(r)}(\mathcal{P})$.

Figure 2.9 shows examples of these proximity graphs in the plane; Figure 2.10 shows examples of these proximity graphs in a planar nonconvex environment; and Figure 2.11 shows example graphs in three-dimensions.

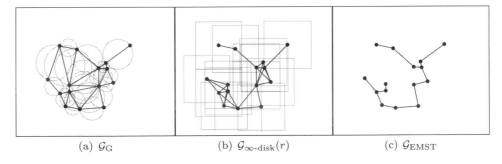

(a) $\mathcal{G}_{\mathrm{G}}$ (b) $\mathcal{G}_{\infty\text{-disk}}(r)$ (c) $\mathcal{G}_{\mathrm{EMST}}$

Figure 2.9 Proximity graphs in $\mathbb{R}^2$. From left to right, Gabriel graph, r-∞-disk graph, and Euclidean minimum spanning tree for 15 points. In two images, the geometric objects determining the edge relationship are plotted in light gray.

As for standard graphs, let us alternatively describe the edge set by means of the sets of neighbors of the individual graph vertices. To each proximity

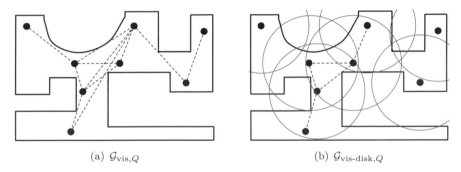

(a) $\mathcal{G}_{\mathrm{vis},Q}$ (b) $\mathcal{G}_{\mathrm{vis\text{-}disk},Q}$

Figure 2.10 The visibility and range-limited visibility graphs for 8 agents in an allowable
environment. The geometric objects determining the edge relationship are
plotted in light gray.

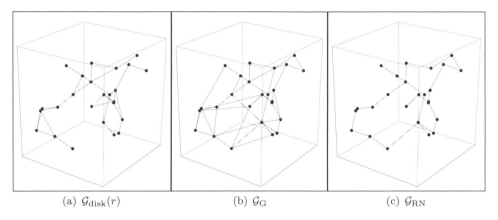

(a) $\mathcal{G}_{\mathrm{disk}}(r)$ (b) $\mathcal{G}_{\mathrm{G}}$ (c) $\mathcal{G}_{\mathrm{RN}}$

Figure 2.11 Proximity graphs in $\mathbb{R}^3$. From left to right, r-disk, relative neighborhood,
and Gabriel graphs for a set of 25 points.

graph $\mathcal{G}$, each $p \in X$ and each $\mathcal{P} = \{p_1, \ldots, p_n\} \in \mathbb{F}(X)$, we associate the
set of neighbors map $\mathcal{N}_{\mathcal{G}} : X \times \mathbb{F}(X) \to \mathbb{F}(X)$ defined by

$$\mathcal{N}_{\mathcal{G}}(p, \mathcal{P}) = \{q \in \mathcal{P} \mid (p, q) \in \mathcal{E}_{\mathcal{G}}(\mathcal{P} \cup \{p\})\}.$$

Typically, p is a point in $\mathcal{P}$, but the definition is well-posed for any $p \in$
X. Under the assumption that $\mathcal{P}$ does not contain repeated elements, the
definition will not lead to counterintuitive interpretations later. Given $p \in$
X, it is convenient to define the map $\mathcal{N}_{\mathcal{G},p} : \mathbb{F}(X) \to \mathbb{F}(X)$ by $\mathcal{N}_{\mathcal{G},p}(\mathcal{P}) =$
$\mathcal{N}_{\mathcal{G}}(p, \mathcal{P})$.

A proximity graph $\mathcal{G}_1$ is a *subgraph* of a proximity graph $\mathcal{G}_2$, denoted $\mathcal{G}_1 \subset$
$\mathcal{G}_2$, if $\mathcal{G}_1(\mathcal{P})$ is a subgraph of $\mathcal{G}_2(\mathcal{P})$ for all $\mathcal{P} \in \mathbb{F}(X)$. The following result,
whose proof is given in Section 2.5.1, summarizes the subgraph relationships
in the Euclidean case among the various proximity graphs introduced above.

Theorem 2.7 (Subgraph relationships among some standard proximity graphs on $\mathbb{R}^d$). *For $r \in \mathbb{R}_{>0}$, the following statements hold:*

(i) $\mathcal{G}_{\mathrm{EMST}} \subset \mathcal{G}_{\mathrm{RN}} \subset \mathcal{G}_{\mathrm{G}} \subset \mathcal{G}_{\mathrm{D}}$; and

(ii) $\mathcal{G}_{\mathrm{G}} \cap \mathcal{G}_{\mathrm{disk}}(r) \subset \mathcal{G}_{\mathrm{LD}}(r) \subset \mathcal{G}_{\mathrm{D}} \cap \mathcal{G}_{\mathrm{disk}}(r)$.

Note that the inclusion $\mathcal{G}_{\mathrm{LD}}(r) \subset \mathcal{G}_{\mathrm{D}} \cap \mathcal{G}_{\mathrm{disk}}(r)$ is in general strict; this counterintuitive fact is discussed in Exercise E2.3. Additionally, since $\mathcal{G}_{\mathrm{EMST}}$ is by definition connected, Theorem 2.7(i) implies that $\mathcal{G}_{\mathrm{RN}}$, $\mathcal{G}_{\mathrm{G}}$, and $\mathcal{G}_{\mathrm{D}}$ are connected. The connectivity properties of $\mathcal{G}_{\mathrm{disk}}(r)$ are characterized in the following result.

Theorem 2.8 (Connectivity properties of some standard proximity graphs on $\mathbb{R}^d$). *For $r \in \mathbb{R}_{>0}$, the following statements hold:*

(i) $\mathcal{G}_{\mathrm{EMST}} \subset \mathcal{G}_{\mathrm{disk}}(r)$ if and only if $\mathcal{G}_{\mathrm{disk}}(r)$ is connected; and

(ii) $\mathcal{G}_{\mathrm{EMST}} \cap \mathcal{G}_{\mathrm{disk}}(r)$, $\mathcal{G}_{\mathrm{RN}} \cap \mathcal{G}_{\mathrm{disk}}(r)$, $\mathcal{G}_{\mathrm{G}} \cap \mathcal{G}_{\mathrm{disk}}(r)$ and $\mathcal{G}_{\mathrm{LD}}(r)$ have the same connected components as $\mathcal{G}_{\mathrm{disk}}(r)$ (i.e., for all point sets $\mathcal{P} \in \mathbb{F}(\mathbb{R}^d)$, all graphs have the same number of connected components consisting of the same vertices).

The proof of this theorem is given in Section 2.5.1. Note that in Theorem 2.8, fact (ii) implies (i).

2.2.1 Spatially distributed proximity graphs

We now consider the following loosely stated question: When does a given proximity graph encode sufficient information to compute another proximity graph? For instance, if a node knows the position of its neighbors in the complete graph (i.e., of every other node in the graph), then it is clear that the node can compute its neighbors with respect to any proximity graph. Let us formalize this idea. A proximity graph $\mathcal{G}_1$ is *spatially distributed over* a proximity graph $\mathcal{G}_2$ if, for all $p \in \mathcal{P}$,

$$\mathcal{N}_{\mathcal{G}_1,p}(\mathcal{P}) = \mathcal{N}_{\mathcal{G}_1,p}\big(\mathcal{N}_{\mathcal{G}_2,p}(\mathcal{P})\big),$$

that is, any node informed about the location of its neighbors with respect to $\mathcal{G}_2$ can compute its set of neighbors with respect to $\mathcal{G}_1$.

Clearly, any proximity graph is spatially distributed over the complete graph. It is straightforward to deduce that if $\mathcal{G}_1$ is spatially distributed over $\mathcal{G}_2$, then $\mathcal{G}_1$ is a subgraph of $\mathcal{G}_2$. The converse is in general not true. For instance, $\mathcal{G}_{\mathrm{D}} \cap \mathcal{G}_{\mathrm{disk}}(r)$ is a subgraph of $\mathcal{G}_{\mathrm{disk}}(r)$, but $\mathcal{G}_{\mathrm{D}} \cap \mathcal{G}_{\mathrm{disk}}(r)$ is not spatially distributed over $\mathcal{G}_{\mathrm{disk}}(r)$; see Exercise E2.4.

The following result identifies proximity graphs which are spatially distributed over $\mathcal{G}_{\text{disk}}(r)$.

Proposition 2.9 (Spatially distributed graphs over the disk graph).
The proximity graphs $\mathcal{G}_{\text{RN}} \cap \mathcal{G}_{\text{disk}}(r)$, $\mathcal{G}_{\text{G}} \cap \mathcal{G}_{\text{disk}}(r)$, and $\mathcal{G}_{\text{LD}}(r)$ are spatially distributed over $\mathcal{G}_{\text{disk}}(r)$.

Remark 2.10 (Computation of the Delaunay graph over the r-disk graph). In general, for a fixed $r \in \mathbb{R}_{>0}$, $\mathcal{G}_{\text{D}}$ is not spatially distributed over $\mathcal{G}_{\text{disk}}(r)$. However, for a given $\mathcal{P} \in \mathbb{F}(X)$, it is always possible find r such that $\mathcal{G}_{\text{D}}(P)$ is spatially distributed over $\mathcal{G}_{\text{disk}}(r)(\mathcal{P})$. This is a consequence of the following observations. Given $\mathcal{P} \in \mathbb{F}(X)$, define the convex sets

$$W(p_i, r) = \overline{B}(p_i, r) \cap \left(\cap_{\|p_i - p_j\| \le r} H_{p_i, p_j} \right), \quad i \in \{1, \dots, n\},$$

where we recall that $H_{p,x}$ is the half-space of points q in $\mathbb{R}^d$ with the property that $\|q - p\|_2 \le \|q - x\|_2$. Note that the intersection $\overline{B}(p_i, r) \cap V_i$ is a subset of $W(p_i, r)$. Provided that r is twice as large as the maximum distance between p_i and the vertices of $W(p_i, r)$, then all Delaunay neighbors of p_i are within distance r from p_i. Equivalently, the half-space $H_{p_i, p}$ determined by p_i and a point p outside $\overline{B}(p_i, r)$ does not intersect $W(p_i, r)$. Therefore, the equality $V_i = W(p_i, r)$ holds. For node $i \in \{1, \dots, n\}$, the minimum adequate radius is then

$$r_{i,\min} = 2 \max\{\|p_i - q\|_2 \mid q \in W(p_i, r_{i,\min})\}.$$

The minimum adequate radius across the overall network is then $r_{\min} = \max_{i \in \{1,\dots,n\}} r_{i,\min}$. The algorithm presented in Cortés et al. (2004) builds on these observations to compute the Voronoi partition of a bounded set generated by a pointset in a distributed way.   •

2.2.2 The locally cliqueless graph of a proximity graph

Given a proximity graph, it is sometimes useful to construct another proximity graph that has fewer edges and the same number of connected components. This is certainly the case when optimizing multi-agent cost functions in which the proximity graph edges describe pairwise constraints between agents. Additionally, the construction of the new proximity graph should be spatially distributed over the original proximity graph. Here, we present the notion of locally cliqueless graph of a proximity graph.

Let $\mathcal{G}$ be a proximity graph in the Euclidean space. The *locally cliqueless graph $\mathcal{G}_{\text{lc},\mathcal{G}}$ of $\mathcal{G}$* is the proximity graph defined by: $(p_i, p_j) \in \mathcal{E}_{\mathcal{G}_{\text{lc},\mathcal{G}}}(\mathcal{P})$ if

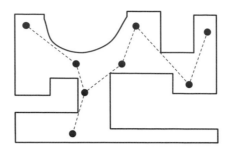

Figure 2.12 Locally cliqueless graph $\mathcal{G}_{\text{lc},\mathcal{G}_{\text{vis},Q}}$ of the visibility graph $\mathcal{G}_{\text{vis},Q}$ for the node configuration shown in Figure 2.9(d).

$(p_i, p_j) \in \mathcal{E}_{\mathcal{G}}(\mathcal{P})$ and

$$(p_i, p_j) \in \mathcal{E}_{\mathcal{G}_{\text{EMST}}}(\mathcal{P}'),$$

for any maximal clique $\mathcal{P}'$ of (p_i, p_j) in $\mathcal{G}$. Figure 2.12 shows an illustration of this notion. The properties of this construction are summarized in the following result; for the proof, see Ganguli et al. (2009).

Theorem 2.11 (Properties of the locally cliqueless graph). *Let $\mathcal{G}$ be a proximity graph in the Euclidean space. Then, the following statements hold:*

(i) $\mathcal{G}_{\text{EMST},\mathcal{G}} \subseteq \mathcal{G}_{\text{lc},\mathcal{G}} \subseteq \mathcal{G}$;

(ii) $\mathcal{G}_{\text{lc},\mathcal{G}}$ has the same connected components as $\mathcal{G}$; and

(iii) for $\mathcal{G} = \mathcal{G}_{\text{disk}}(r)$, $\mathcal{G}_{\text{vis},Q}$, and $\mathcal{G}_{\text{vis-disk},Q}$, where $r \in \mathbb{R}_{>0}$ and Q is an allowable environment, $\mathcal{G}_{\text{lc},\mathcal{G}}$ is spatially distributed over $\mathcal{G}$.

In general, the inclusions in Theorem 2.11(i) are strict.

2.2.3 Proximity graphs over tuples of points

The notion of proximity graph is defined for sets of distinct points $\mathcal{P} = \{p_1, \ldots, p_n\}$. However, we will be interested in considering tuples of elements of X of the form $P = (p_1, \ldots, p_n)$, where p_i corresponds to the position of an agent i of a robotic network. In principle, note that the tuple P might contain coincident points. In order to reconcile this mismatch between sets and tuples, we will do the following.

Let $i_{\mathbb{F}} : X^n \to \mathbb{F}(X)$ be the natural immersion of X^n into $\mathbb{F}(X)$, that is, $i_{\mathbb{F}}(P)$ is the point set that contains only the distinct points in $P = (p_1, \ldots, p_n)$. Note that $i_{\mathbb{F}}$ is invariant under permutations of its arguments and that the cardinality of $i_{\mathbb{F}}(p_1, \ldots, p_n)$ is in general less than or equal to n. In what follows, $\mathcal{P} = i_{\mathbb{F}}(P)$ will always denote the point set associated to $P \in X^n$. Using the natural inclusion, the notion of proximity graphs can be naturally extended as follows: given $\mathcal{G}$, we define (with a slight abuse of notation)

$$\mathcal{G} = \mathcal{G} \circ i_{\mathbb{F}} : X^n \to \mathbb{G}(X).$$

Additionally, we define the set of neighbors map $\mathcal{N}_{\mathcal{G}} : X \times X^n \to \mathbb{F}(X)$ by

$$\mathcal{N}_{\mathcal{G}}(p, (p_1, \ldots, p_n)) = \mathcal{N}_{\mathcal{G}}(p, i_{\mathbb{F}}(p_1, \ldots, p_n)).$$

According to this definition, coincident points in the tuple $(p_1, \ldots, p_n)$ will have the same set of neighbors. As before, it is convenient to define the shorthand notation $\mathcal{N}_{\mathcal{G},p} : X^n \to \mathbb{F}(X)$, $\mathcal{N}_{\mathcal{G},p}(P) = \mathcal{N}_{\mathcal{G}}(p, P)$ for $p \in X$.

2.2.4 Spatially distributed maps

Given a set Y and a proximity graph $\mathcal{G}$, a map $T : X^n \to Y^n$ is *spatially distributed over* $\mathcal{G}$ if there exists a map $\tilde{T} : X \times \mathbb{F}(X) \to Y$, with the property that, for all $(p_1, \ldots, p_n) \in X^n$ and for all $j \in \{1, \ldots, n\}$,

$$T_j(p_1, \ldots, p_n) = \tilde{T}(p_j, \mathcal{N}_{\mathcal{G},p_j}(p_1, \ldots, p_n)),$$

where T_j denotes the jth component of T. In other words, the jth component of a spatially distributed map at $(p_1, \ldots, p_n)$ can be computed with only knowledge of the vertex p_j and the neighboring vertices in the undirected graph $\mathcal{G}(P)$.

When studying coordination tasks and coordination algorithms, it will be relevant to characterize the spatially distributed features of functions, vector fields, and set-valued maps with respect to suitable proximity graphs.

Remark 2.12 (Relationship with the notion of spatially distributed graphs). Note that the proximity graph $\mathcal{G}_1$ is spatially distributed over the proximity graph $\mathcal{G}_2$ if and only if the map

$$P \in X^n \mapsto (\mathcal{N}_{\mathcal{G}_1,p_1}(P), \ldots, \mathcal{N}_{\mathcal{G}_1,p_n}(P)) \in \mathbb{F}(X)^n$$

is spatially distributed over $\mathcal{G}_2$.   •

2.3 GEOMETRIC OPTIMIZATION PROBLEMS AND MULTICENTER FUNCTIONS

In this section we consider various interesting geometric optimization problems. By geometric optimization, we mean an optimization problem induced by a collection of geometric objects (see Boltyanski et al., 1999). We shall pay particular attention to facility location problems, in which service sites are spatially allocated to fulfill a particular request.

2.3.1 Expected-value multicenter functions

Let $S \subset \mathbb{R}^d$ be a bounded environment of interest, and consider a density function $\phi : \mathbb{R}^d \to \mathbb{R}_{\geq 0}$. For the discussion of this section, only the value of ϕ restricted to S is of interest. One can regard ϕ as a function measuring the probability that some event takes place over the environment. The larger the value of $\phi(q)$, the more important the location q is. We refer to a non-increasing and piecewise continuously differentiable function $f : \mathbb{R}_{\geq 0} \to \mathbb{R}$, possibly with finite jump discontinuities, as a *performance*. Performance functions describe the utility of placing a node at a certain distance from a location in the environment. The smaller the distance, the larger the value of f, that is, the better the performance. For instance, in servicing problems, performance functions can encode the travel time or the energy expenditure required to service a specific destination. In sensing problems, performance functions can encode the signal-to-noise ratio between a source with an unknown location and a sensor attempting to locate it.

Given a bounded measurable set $S \subset \mathbb{R}^d$, a density function ϕ, and a performance function f, let us consider the expected value of the coverage over any point in S provided by a set of points $p_1, \ldots, p_n$. Formally, we define the *expected-value multicenter* function $\mathcal{H}_{\exp} : S^n \to \mathbb{R}$ by

$$\mathcal{H}_{\exp}(p_1, \ldots, p_n) = \int_S \max_{i \in \{1, \ldots, n\}} f(\|q - p_i\|_2) \phi(q) dq. \tag{2.3.1}$$

The definition of $\mathcal{H}_{\exp}$ can be interpreted as follows: for each location $q \in S$, consider the best coverage of q among those provided by each of the nodes $p_1, \ldots, p_n$, which corresponds to the value $\max_{i \in \{1, \ldots, n\}} f(\|q - p_i\|_2)$. Then, evaluate the performance by the importance $\phi(q)$ of the location q. Finally, sum the resulting quantity over all the locations of the environment S, to obtain $\mathcal{H}_{\exp}(p_1, \ldots, p_n)$ as a measure of the overall coverage provided by $p_1, \ldots, p_n$.

Given the meaning of $\mathcal{H}_{\exp}$, we seek to solve the following geometric optimization problem:

$$\text{maximize } \mathcal{H}_{\exp}(p_1, \ldots, p_n), \tag{2.3.2}$$

that is, we seek to determine a set of configurations $p_1, \ldots, p_n$ that maximize the value of the multicenter function $\mathcal{H}_{\exp}$. An equivalent formulation of this problem is referred to as a *continuous p-median problem* in the literature on facility location (see, e.g., Drezner, 1995). In our discussion, we will pay special attention to the case when $n = 1$, which we term the 1-*center* problem. For the purpose of solving (2.3.2), note that we can assume that the performance function satisfies $f(0) = 0$. This can be done without loss of generality, since for any $c \in \mathbb{R}$, one has

$$\int_S \max_{i \in \{1,\dots,n\}} (f(\|q - p_i\|_2) + c)\phi(q)dq = \mathcal{H}_{\exp}(p_1, \dots, p_n) + c\, A_\phi(S).$$

The expected-value multicenter function can be alternatively described in terms of the Voronoi partition of S generated by $\mathcal{P} = \{p_1, \dots, p_n\}$. Let us define the set

$$\mathcal{S}_{\text{coinc}} = \{(p_1, \dots, p_n) \in (\mathbb{R}^d)^n \mid p_i = p_j \text{ for some } i \neq j\},$$

consisting of tuples of n points, where some of them are repeated. Then, for $(p_1, \dots, p_n) \in S^n \setminus \mathcal{S}_{\text{coinc}}$, one has

$$\mathcal{H}_{\exp}(p_1, \dots, p_n) = \sum_{i=1}^{n} \int_{V_i(\mathcal{P})} f(\|q - p_i\|_2)\phi(q)dq. \qquad (2.3.3)$$

This expression of $\mathcal{H}_{\exp}$ is appealing because it clearly shows the result of the overall coverage of the environment as the aggregate contribution of all individual nodes. If $(p_1, \dots, p_n) \in \mathcal{S}_{\text{coinc}}$, then a similar decomposition of $\mathcal{H}_{\exp}$ can be written in terms of the distinct points $\mathcal{P} = i_{\mathbb{F}}(p_1, \dots, p_n)$.

Inspired by the expression (2.3.3), let us define a more general version of the expected-value multicenter function. Given $(p_1, \dots, p_n) \in S^n$ and a partition $\{W_1, \dots, W_n\} \subset \mathbb{P}(S)$ of S, let

$$\mathcal{H}_{\exp}(p_1, \dots, p_n, W_1, \dots, W_n) = \sum_{i=1}^{n} \int_{W_i} f(\|q - p_i\|_2)\phi(q)dq. \qquad (2.3.4)$$

Notice that $\mathcal{H}_{\exp}(p_1, \dots, p_n) = \mathcal{H}_{\exp}(p_1, \dots, p_n, V_1(\mathcal{P}), \dots, V_n(\mathcal{P}))$, for all $(p_1, \dots, p_n) \in S^n \setminus \mathcal{S}_{\text{coinc}}$. Moreover, one can establish the following optimality result (see Du et al., 1999).

Proposition 2.13 ($\mathcal{H}_{\exp}$-optimality of the Voronoi partition). *Let* $\mathcal{P} = \{p_1, \dots, p_n\} \in \mathbb{F}(S)$. *For any performance function* f *and for any partition* $\{W_1, \dots, W_n\} \subset \mathbb{P}(S)$ *of* S,

$$\mathcal{H}_{\exp}(p_1, \dots, p_n, V_1(\mathcal{P}), \dots, V_n(\mathcal{P})) \geq \mathcal{H}_{\exp}(p_1, \dots, p_n, W_1, \dots, W_n),$$

and the inequality is strict if any set in $\{W_1, \dots, W_n\}$ *differs from the corresponding set in* $\{V_1(\mathcal{P}), \dots, V_n(\mathcal{P})\}$ *by a set of positive measure. In other words, the Voronoi partition* $\mathcal{V}(\mathcal{P})$ *is optimal for* $\mathcal{H}_{\exp}$ *among all partitions of* S.

Proof. Assume that, for $i \neq j \in \{1, \ldots, n\}$, the set $\text{int}(W_i) \cap \text{int}(V_j(\mathcal{P}))$ has strictly positive measure. For all $q \in \text{int}(W_i) \cap \text{int}(V_j(\mathcal{P}))$, we know that $\|q-p_i\|_2 > \|q-p_j\|_2$. Because f is non-increasing, $f(\|q-p_i\|_2) < f(\|q-p_j\|_2)$ and, since $\text{int}(W_i) \cap \text{int}(V_j(\mathcal{P}))$ has strictly positive measure,

$$\int_{\text{int}(W_i) \cap \text{int}(V_j(\mathcal{P}))} f(\|q-p_i\|_2)\phi(q)dq < \int_{\text{int}(W_i) \cap \text{int}(V_j(\mathcal{P}))} f(\|q-p_j\|_2)\phi(q)dq.$$

Therefore, we deduce

$$\int_{W_i} f(\|q-p_i\|_2)\phi(q)dq < \sum_{j=1}^{n} \int_{W_i \cap V_j(\mathcal{P})} f(\|q-p_j\|_2)\phi(q)dq,$$

and the statements follow. ∎

Different performance functions lead to different expected-value multicenter functions. Let us examine some important cases.

Distortion problem: Consider the performance function $f(x) = -x^2$. Then, on $S^n \setminus S_{\text{coinc}}$, the expected-value multicenter function takes the form

$$\mathcal{H}_{\text{dist}}(p_1, \ldots, p_n) = -\sum_{i=1}^{n} \int_{V_i(P)} \|q - p_i\|_2^2 \phi(q)dq = -\sum_{i=1}^{n} \mathrm{J}_\phi(V_i(\mathcal{P}), p_i),$$

where recall that $\mathrm{J}_\phi(W, p)$ denotes the polar moment of inertia of the set W about the point p. In signal compression $-\mathcal{H}_{\text{dist}}$ is referred to as the *distortion function* and is relevant in many disciplines including vector quantization, signal compression, and numerical integration (see Gray and Neuhoff, 1998; Du et al., 1999). Here, distortion refers to the average deformation (weighted by the density ϕ) caused by reproducing $q \in S$ with the location p_i in $\mathcal{P} = \{p_1, \ldots, p_n\}$ such that $q \in V_i(\mathcal{P})$. It is interesting to note that

$$\mathcal{H}_{\text{dist}}(p_1, \ldots, p_n, W_1, \ldots, W_n) = -\sum_{i=1}^{n} \mathrm{J}_\phi(W_i, p_i)$$

$$= -\sum_{i=1}^{n} \mathrm{J}_\phi(W_i, \mathrm{CM}_\phi(W_i)) - \sum_{i=1}^{n} \mathrm{A}_\phi(W_i)\|p_i - \mathrm{CM}_\phi(W_i)\|_2^2, \quad (2.3.5)$$

where in the last equality we have used the Parallel Axis Theorem (Hibbeler, 2006). Note that the first term only depends on the partition of S, whereas the second term also depends on the location of the points. The following result is a consequence of this observation.

Proposition 2.14 ($\mathcal{H}_{\text{dist}}$-optimality of centroid locations). *Let* $\{W_1, \ldots, W_n\} \subset \mathbb{P}(S)$ *be a partition of* S. *Then, for any set points* $\mathcal{P} = \{p_1, \ldots, p_n\} \in \mathbb{F}(S)$,

$$\mathcal{H}_{\text{dist}}\big(\,\text{CM}_\phi(W_1), \ldots, \text{CM}_\phi(W_n), W_1, \ldots, W_n\big)$$
$$\geq \mathcal{H}_{\text{dist}}(p_1, \ldots, p_n, W_1, \ldots, W_n),$$

and the inequality is strict if there exists $i \in \{1, \ldots, n\}$ *for which* W_i *has non-vanishing area and* $p_i \neq \text{CM}_\phi(W_i)$. *In other words, the centroid locations* $\text{CM}_\phi(W_1), \ldots, \text{CM}_\phi(W_n)$ *are optimal for* $\mathcal{H}_{\text{dist}}$ *among all configurations in* S.

A consequence of this result is that for the 1-center problem, that is, when $n = 1$, the node location that optimizes $p \mapsto \mathcal{H}_{\text{dist}}(p) = -\text{J}_\phi(S, p)$ is the centroid of the set S, denoted by $\text{CM}_\phi(S)$.

Area problem: For $a \in \mathbb{R}_{>0}$, consider the performance function $f(x) = 1_{[0,a]}(x)$, that is, the indicator function of the closed interval $[0, a]$. Then, the expected-value multicenter function takes the form

$$\mathcal{H}_{\text{area},a}(p_1, \ldots, p_n) = \sum_{i=1}^{n} \int_{V_i(\mathcal{P})} 1_{[0,a]}(\|q - p_i\|_2)\phi(q)dq$$
$$= \sum_{i=1}^{n} \int_{V_i(\mathcal{P}) \cap \overline{B}(p_i, a)} \phi(q)dq$$
$$= \sum_{i=1}^{n} \text{A}_\phi(V_i(\mathcal{P}) \cap \overline{B}(p_i, a)) = \text{A}_\phi(\cup_{i=1}^{n} \overline{B}(p_i, a)),$$

that is, it corresponds to the area, measured according to ϕ, covered by the union of the n balls $\overline{B}(p_1, a), \ldots, \overline{B}(p_n, a)$. Exercise E2.5 discusses the 1-center area problem.

Mixed distortion-area problem: For $a \in \mathbb{R}_{>0}$ and $b \leq -a^2$, consider the performance function $f(x) = -x^2 \, 1_{[0,a]}(x) + b \cdot 1_{]a,+\infty[}(x)$. Then, on $S^n \setminus S_{\text{coinc}}$, the expected-value multicenter function takes the form

$$\mathcal{H}_{\text{dist-area},a,b}(p_1, \ldots, p_n) = -\sum_{i=1}^{n} \text{J}_\phi(V_{i,a}(\mathcal{P}), p_i) + b\,\text{A}_\phi(Q \setminus \cup_{i=1}^{n}\overline{B}(p_i, a)),$$

that is, it is a combination of the multicenter functions corresponding to the distortion problem and the area problem. Of special interest to us is the multicenter function that results from the choice $b = -a^2$. In this case, the performance function f is continuous, and we simply write $\mathcal{H}_{\text{dist-area},a}$. The extension of this function to sets of points and partitions of the space reads as follows:

$$\mathcal{H}_{\text{dist-area},a}(p_1, \ldots, p_n, W_1, \ldots, W_n)$$

$$= -\sum_{i=1}^{n} \left(\mathrm{J}_\phi(W_i \cap \overline{B}(p_i, a), p_i) + a^2 \, \mathrm{A}_\phi(W_i \cap (S \setminus \overline{B}(p_i, a))) \right).$$

We leave the proof of the following optimality result as a guided exercise for the reader (see Exercise E2.10).

Proposition 2.15 ($\mathcal{H}_{\text{dist-area},a}$-optimality of centroid locations).
Let $\{W_1, \ldots, W_n\} \subset \mathbb{P}(S)$ be a partition of S. Then, for any $\mathcal{P} = \{p_1, \ldots, p_n\} \in \mathbb{F}(S)$,

$$\mathcal{H}_{\text{dist-area},a}\left(q_1^*, \ldots, q_n^*, W_1, \ldots, W_n\right)$$

$$\geq \mathcal{H}_{\text{dist-area},a}(p_1, \ldots, p_n, W_1, \ldots, W_n),$$

where we have used the shorthands $q_i^* = \mathrm{CM}_\phi(W_i \cap \overline{B}(p_i, a))$, for $i \in \{1, \ldots, n\}$. Furthermore, the inequality is strict if there exists $i \in \{1, \ldots, n\}$ for which W_i has non-vanishing area and $p_i \neq q_i^*$.

A consequence of this result is that for the 1-center problem, that is, when $n = 1$—the node location that optimizes $p \mapsto \mathcal{H}_{\text{dist-area},a}(p) = \mathrm{J}_\phi(S \cap \overline{B}(p, a), p) + a^2 \, \mathrm{A}_\phi(S \setminus \overline{B}(p, a))$ is the centroid of the set $S \cap \overline{B}(p, a)$, denoted by $\mathrm{CM}_\phi(S \cap \overline{B}(p, a))$.

Next, we characterize the smoothness of the expected-value multicenter function. Before stating the precise result, let us introduce some useful notation. For a performance function f, let $\mathrm{Dscn}(f)$ denote the (finite) set of points where f is discontinuous. For each $a \in \mathrm{Dscn}(f)$, define the limiting values from the left and from the right, respectively, as

$$f_-(a) = \lim_{x \to a^-} f(x), \qquad f_+(a) = \lim_{x \to a^+} f(x).$$

We are now ready to characterize the smoothness of $\mathcal{H}_{\text{exp}}$, whose proof is given in Section 2.5.3. Before stating the result, recall that the line integral of a function $g : \mathbb{R}^2 \to \mathbb{R}$ over a curve C parameterized by a continuous and piecewise continuously differentiable map $\gamma : [0, 1] \to \mathbb{R}^2$ is defined by

$$\int_C g = \int_C g(\gamma) d\gamma := \int_0^1 g(\gamma(t)) \, \|\dot{\gamma}(t)\|_2 \, dt,$$

and is independent of the selected parameterization.

Theorem 2.16 (Smoothness properties of $\mathcal{H}_{\text{exp}}$). *Given a set $S \subset \mathbb{R}^d$ that is bounded and measurable, a density $\phi : \mathbb{R} \to \mathbb{R}_{\geq 0}$, and a performance function $f : \mathbb{R}_{\geq 0} \to \mathbb{R}$, the expected-value multicenter function $\mathcal{H}_{\text{exp}} : S^n \to \mathbb{R}$ is*

(i) *globally Lipschitz^2 on S^n; and*

(ii) *continuously differentiable on $S^n \setminus S_{\mathrm{coinc}}$, where for $i \in \{1, \ldots, n\}$*

$$\frac{\partial \mathcal{H}_{\exp}}{\partial p_i}(P) = \int_{V_i(\mathcal{P})} \frac{\partial}{\partial p_i} f(\|q - p_i\|_2)\phi(q)dq$$

$$+ \sum_{a \in \mathrm{Dscn}(f)} \left(f_-(a) - f_+(a)\right) \int_{V_i(\mathcal{P}) \cap \partial \overline{B}(p_i, a)} \mathrm{n}_{\mathrm{out}}(q)\phi(q)dq, \quad (2.3.6)$$

where $\mathrm{n}_{\mathrm{out}}$ is the outward normal vector to $\overline{B}(p_i, a)$. Therefore, the gradient of $\mathcal{H}_{\exp}$, interpreted as a map from S^n to $\mathbb{R}^n$, is spatially distributed (in the sense defined in Section 2.2.4) over the Delaunay graph $\mathcal{G}_D$.

Let us discuss how Theorem 2.16 particularizes to the distortion, area, and mixed distortion-area problems.

Distortion problem: In this case, the performance function does not have any discontinuities and, therefore, the second term in (2.3.6) vanishes. The gradient of $\mathcal{H}_{\mathrm{dist}}$ on $S^n \setminus S_{\mathrm{coinc}}$ then takes the form, for each $i \in \{1, \ldots, n\}$,

$$\frac{\partial \mathcal{H}_{\mathrm{dist}}}{\partial p_i}(P) = 2\, A_\phi(V_i(\mathcal{P}))(\mathrm{CM}_\phi(V_i(\mathcal{P})) - p_i),$$

that is, the ith component of the gradient points in the direction of the vector going from p_i to the centroid of its Voronoi cell. The critical points of $\mathcal{H}_{\mathrm{dist}}$ are therefore the set of centroidal Voronoi configurations in S (cf. Section 2.1.4). This is a natural generalization of the result for the 1-center case, where the optimal node location is the centroid $\mathrm{CM}_\phi(S)$.

Area problem: In this case, the performance function is differentiable everywhere except at a single discontinuity, and its derivative is identically zero. Therefore, the first term in (2.3.6) vanishes. The gradient of $\mathcal{H}_{\mathrm{area},a}$ on $S^n \setminus S_{\mathrm{coinc}}$ then takes the form, for each $i \in \{1, \ldots, n\}$,

$$\frac{\partial \mathcal{H}_{\mathrm{area},a}}{\partial p_i}(P) = \int_{V_i(\mathcal{P}) \cap \partial \overline{B}(p_i, a)} \mathrm{n}_{\mathrm{out}}(q)\phi(q)dq,$$

where $\mathrm{n}_{\mathrm{out}}$ is the outward normal vector to $\overline{B}(p_i, a)$. The gradient is an average of the normal at each point of $V_i(\mathcal{P}) \cap \partial \overline{B}(p_i, a)$, as illustrated in Figure 2.13. The critical points of $\mathcal{H}_{\mathrm{area},a}$ correspond to configurations with the property that each p_i is a local maximum for the area of $V_{i,a}(P) = V_i(P) \cap \overline{B}(p_i, a)$ at fixed $V_i(P)$. We refer to these

---

^{2}Given $S \subset \mathbb{R}^h$, a function $f : S \to \mathbb{R}^k$ is globally Lipschitz if there exists $K \in \mathbb{R}_{>0}$ such that $\|f(x - y)\|_2 \leq K\|x - y\|_2$ for all $x, y \in S$.

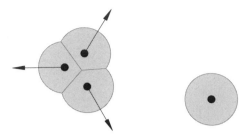

Figure 2.13 The gradient of the area function when the density function is constant. The component of the gradient corresponding to the rightmost node is zero; there is no incentive for this node to move in any particular direction. The component of the gradient for each of the three leftmost agents is non-zero; roughly speaking, by moving along the gradient directions, these agents decrease the overlapping among their respective disk and cover new regions of the space.

configurations as *a-limited area-centered* Voronoi configurations. This is a natural generalization of the result for the 1-center case, where the optimal node location maximizes $A_\phi(S \cap \overline{B}(p,a))$ (cf., Exercise E2.5).

Mixed distortion-area problem: In this case, the gradient of the multi-center function $\mathcal{H}_{\text{dist-area},a,b}$ is a combination of the gradients of $\mathcal{H}_{\text{dist}}$ and $\mathcal{H}_{\text{area},a}$. Specifically, one has for each $i \in \{1,\dots,n\}$,

$$\frac{\partial \mathcal{H}_{\text{dist-area},a,b}}{\partial p_i}(P) = 2\, A_\phi(V_{i,a}(\mathcal{P}))(CM_\phi(V_{i,a}(\mathcal{P})) - p_i)$$

$$- (a^2 + b) \int_{V_i(\mathcal{P}) \cap \partial \overline{B}(p_i,a)} n_{\text{out}}(q)\phi(q)dq,$$

where n_{out} is the outward normal vector to $\overline{B}(p_i,a)$. For the particular case when $b = -a^2$, the performance function is continuous, and the gradient of $\mathcal{H}_{\text{dist-area},a}$ takes the simpler form

$$\frac{\partial \mathcal{H}_{\text{dist-area},a}}{\partial p_i}(P) = 2\, A_\phi(V_{i,a}(\mathcal{P}))(CM_\phi(V_{i,a}(\mathcal{P})) - p_i),$$

which points in the direction of the vector from p_i to the centroid of its a-limited Voronoi cell. In this case, the critical points of $\mathcal{H}_{\text{dist-area},a}$ are therefore the set of a-limited centroidal Voronoi configurations in S (cf., Section 2.1.4). This is a natural generalization of the result for the 1-center case, where the optimal node location is the centroid $CM_\phi(S \cap \overline{B}(p,a))$.

We refer to $\mathcal{H}_{\text{dist}}$, $\mathcal{H}_{\text{area},a}$, and $\mathcal{H}_{\text{dist-area},a}$ as multicenter functions because, as the above discussion shows, their critical points correspond to various notions of center Voronoi configurations.

Note that the gradients of $\mathcal{H}_{\text{area},a}$ and $\mathcal{H}_{\text{dist-area},a,b}$ are spatially distributed over the $2a$-limited Delaunay graph $\mathcal{G}_{\text{LD}}(2a)$. This observation is important for practical considerations: robotic agents with range-limited interactions cannot in general compute the gradient of $\mathcal{H}_{\text{dist}}$ because, as we noted in Remark 2.10, for a given $r \in \mathbb{R}_{>0}$, $\mathcal{G}_{\text{D}}$ is not in general spatially distributed over $\mathcal{G}_{\text{disk}}(r)$. However, robotic agents with range-limited interactions can compute the gradients of $\mathcal{H}_{\text{area},a}$ and $\mathcal{H}_{\text{dist-area},a,b}$ as long as $r \geq 2a$ because, from Theorem 2.7(iii), $\mathcal{G}_{\text{LD}}(r)$ is spatially distributed over $\mathcal{G}_{\text{disk}}(r)$. The relevance of this fact is further justified by the following result.

Proposition 2.17 (Constant-factor approximation of $\mathcal{H}_{\text{dist}}$). *Let $S \subset \mathbb{R}^d$ be bounded and measurable. Consider the mixed distortion-area problem with $a \in]0, \text{diam}\, S]$ and $b = -\text{diam}(S)^2$. Then, for all $P \in S^n$,*

$$\mathcal{H}_{\text{dist-area},a,b}(P) \leq \mathcal{H}_{\text{dist}}(P) \leq \beta^2\, \mathcal{H}_{\text{dist-area},a,b}(P) < 0, \qquad (2.3.7)$$

where $\beta = \frac{a}{\text{diam}(S)} \in [0, 1]$.

In fact, similar constant-factor approximations of the expected-value multicenter function $\mathcal{H}_{\text{exp}}$ can also be established (see Cortés et al., 2005).

2.3.2 Worst-case and disk-covering multicenter functions

Given a compact set $S \subset \mathbb{R}^d$ and a performance function f, let us consider the point in S that is worst covered by a set of points $p_1, \ldots, p_n$. Formally, we define the *worst-case multicenter* function $\mathcal{H}_{\text{worst}} : S^n \to \mathbb{R}$ by

$$\mathcal{H}_{\text{worst}}(p_1, \ldots, p_n) = \min_{q \in S}\ \max_{i \in \{1,\ldots,n\}} f(\|q - p_i\|_2). \qquad (2.3.8)$$

The definition of $\mathcal{H}_{\text{worst}}$ can be read as follows: for each location $q \in S$, consider the best coverage of q among those provided by each of the nodes $p_1, \ldots, p_n$, which corresponds to the value $\max_{i \in \{1,\ldots,n\}} f(\|q - p_i\|_2)$. Then, compute the worst coverage $\mathcal{H}_{\text{worst}}(p_1, \ldots, p_n)$ by comparing the performance at all locations in S.

Given the interpretation of $\mathcal{H}_{\text{worst}}$, we seek to solve the following geometric optimization problem:

$$\text{maximize } \mathcal{H}_{\text{worst}}(p_1, \ldots, p_n), \qquad (2.3.9)$$

that is, we seek to determine configurations $p_1, \ldots, p_n$ that maximize the value of $\mathcal{H}_{\text{worst}}$. An equivalent formulation of this problem is referred to as a *continuous p-center problem* in the literature on facility location (see, e.g., Drezner, 1995).

In the present context, also relevant is the *disk-covering multicenter* function $\mathcal{H}_{\mathrm{dc}} : S^n \to \mathbb{R}$, defined by

$$\mathcal{H}_{\mathrm{dc}}(p_1, \ldots, p_n) = \max_{q \in S} \min_{i \in \{1, \ldots, n\}} \|q - p_i\|_2. \qquad (2.3.10)$$

The value of $\mathcal{H}_{\mathrm{dc}}$ can be interpreted as the largest possible distance from a point in S to one of the locations $p_1, \ldots, p_n$. Note that, by definition, the environment S is contained in the union of n closed balls centered at $p_1, \ldots, p_n$ with radius $\mathcal{H}_{\mathrm{dc}}(p_1, \ldots, p_n)$. The definition of $\mathcal{H}_{\mathrm{dc}}$ is illustrated in Figure 2.14(a).

The following result establishes the relationship between the worst-case and the disk-covering multicenter functions, and as byproduct, provides an elegant reformulation of the geometric optimization problem (2.3.9). Its proof is left to the reader.

Lemma 2.18 (Relationship between $\mathcal{H}_{\mathbf{worst}}$ and $\mathcal{H}_{\mathbf{dc}}$). *Given $S \subset \mathbb{R}^d$ compact and a performance function $f : \mathbb{R}_{\geq 0} \to \mathbb{R}$, one has $\mathcal{H}_{\mathrm{worst}} = f \circ \mathcal{H}_{\mathrm{dc}}$.*

Using Lemma 2.18 and the fact that f is non-increasing, we can reformulate the geometric optimization problem (2.3.9) as

$$\text{minimize } \mathcal{H}_{\mathrm{dc}}(p_1, \ldots, p_n), \qquad (2.3.11)$$

that is, find the minimum radius r such that the environment S is covered by n closed balls centered at $p_1, \ldots, p_n$ with equal radius r. Note the connection between this formulation and the classical disk-covering problem: how to cover a region with (possibly overlapping) disks of minimum radius. We shall comment more on this connection later.

Because of the equivalence between the geometric optimization problems (2.3.9) and (2.3.11), we focus our attention on $\mathcal{H}_{\mathrm{dc}}$. The disk-covering multicenter function can be alternatively described in terms of the Voronoi partition of S generated by $\mathcal{P} = \{p_1, \ldots, p_n\}$. For $(p_1, \ldots, p_n) \in S^n \setminus \mathcal{S}_{\mathrm{coinc}}$,

$$\begin{aligned}
\mathcal{H}_{\mathrm{dc}}(p_1, \ldots, p_n) &= \max_{i \in \{1, \ldots, n\}} \max_{q \in V_i(\mathcal{P})} \|q - p_i\|_2 \\
&= \max_{i \in \{1, \ldots, n\}} \max_{q \in \partial V_i(\mathcal{P})} \|q - p_i\|_2. \qquad (2.3.12)
\end{aligned}$$

This characterization of $\mathcal{H}_{\mathrm{dc}}$ is illustrated in Figure 2.14(b). The expression (2.3.12) is appealing because it clearly shows the value of $\mathcal{H}_{\mathrm{dc}}$ as the result of the aggregate contribution of all individual nodes. If $(p_1, \ldots, p_n) \in \mathcal{S}_{\mathrm{coinc}}$, then a similar decomposition of $\mathcal{H}_{\mathrm{dc}}$ can be written in terms of the distinct points $\mathcal{P} = i_{\mathbb{F}}(p_1, \ldots, p_n)$. A node $i \in \{1, \ldots, n\}$ is called *active* at $(p_1, \ldots, p_n)$ if $\max_{q \in \partial V_i(\mathcal{P})} \|q - p_i\|_2 = \mathcal{H}_{\mathrm{dc}}(p_1, \ldots, p_n)$. A node is *passive* at $(p_1, \ldots, p_n)$ if it is not active.

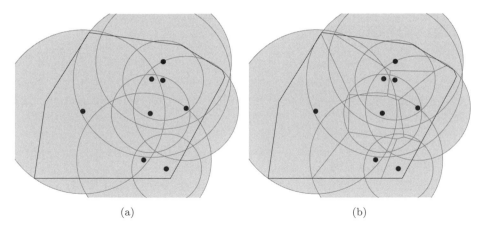

(a)   (b)

Figure 2.14 An illustration of the definition of $\mathcal{H}_{\mathrm{dc}}$: (a) and (b) show the same configuration, with and without the Voronoi configuration, respectively. For each
node, the disk is the minimum-radius disk centered at the node and enclosing
the Voronoi cell. The value of $\mathcal{H}_{\mathrm{dc}}$ is the radius of the disk centered at the
leftmost node.

Inspired by expression (2.3.12), let us define a more general version of the
disk-covering multicenter function. Given $(p_1, \ldots, p_n) \in S^n$ and a partition
$\{W_1, \ldots, W_n\} \subset \mathbb{P}(S)$ of S, let

$$\mathcal{H}_{\mathrm{dc}}(p_1, \ldots, p_n, W_1, \ldots, W_n) = \max_{i \in \{1, \ldots, n\}} \max_{q \in \partial W_i} \|q - p_i\|_2.$$

Note the relationship $\mathcal{H}_{\mathrm{dc}}(p_1, \ldots, p_n) = \mathcal{H}_{\mathrm{dc}}(p_1, \ldots, p_n, V_1(\mathcal{P}), \ldots, V_n(\mathcal{P}))$,
for all $(p_1, \ldots, p_n) \in S^n \setminus \mathcal{S}_{\mathrm{coinc}}$. Moreover, one can establish the following
optimality result, whose proof is given in Section 2.5.4.

**Proposition 2.19 ($\mathcal{H}_{\mathrm{dc}}$-optimality of the Voronoi partition and cir-
cumcenter locations).** *For any* $\mathcal{P} = \{p_1, \ldots, p_n\} \in \mathbb{F}(S)$ *and any partition* $\{W_1, \ldots, W_n\} \subset \mathbb{P}(S)$ *of* S,

$$\mathcal{H}_{\mathrm{dc}}(p_1, \ldots, p_n, V_1(\mathcal{P}), \ldots, V_n(\mathcal{P})) \le \mathcal{H}_{\mathrm{dc}}(p_1, \ldots, p_n, W_1, \ldots, W_n),$$

that is, the Voronoi partition $\mathcal{V}(\mathcal{P})$ *is optimal for* $\mathcal{H}_{\mathrm{dc}}$ *among all partitions
of* S, *and*

$$\mathcal{H}_{\mathrm{dc}}(\mathrm{CC}(W_1), \ldots, \mathrm{CC}(W_n), W_1, \ldots, W_n) \le \mathcal{H}_{\mathrm{dc}}(p_1, \ldots, p_n, W_1, \ldots, W_n),$$

that is, the circumcenter locations $\mathrm{CC}(W_1), \ldots, \mathrm{CC}(W_n)$ *are optimal for* $\mathcal{H}_{\mathrm{dc}}$
among all configurations in S.

As a corollary of this result, we have that the circumcenter of S is a global
optimum of $\mathcal{H}_{\mathrm{dc}}$ for the 1-center problem, that is, when $n = 1$. This comes
as no surprise since, in this case, the value $\mathcal{H}_{\mathrm{dc}}(p)$ corresponds to the radius
of the minimum-radius sphere centered at p that encloses S.

The following result characterizes the smoothness properties of the disk-covering multicenter function; for more details and for the proof, see Cortés and Bullo (2005).

Theorem 2.20 (Smoothness properties of $\mathcal{H}_{dc}$). *Given $S \subset \mathbb{R}^d$ compact, the disk-covering multicenter function $\mathcal{H}_{dc} : S^n \to \mathbb{R}$ is globally Lipschitz on S^n.*

The generalized gradient and the critical points of $\mathcal{H}_{dc}$ can be characterized, but require a careful study based on nonsmooth analysis (Clarke, 1983). In particular, two facts taken from Cortés and Bullo (2005) are of interest here. First, under certain technical conditions, one can show that the critical points of $\mathcal{H}_{dc}$ are circumcenter Voronoi configurations. This is why we refer to $\mathcal{H}_{dc}$ as a multicenter function. Second, the generalized gradient of $\mathcal{H}_{dc}$ is not spatially distributed over $\mathcal{G}_D$. This is essentially due to the inherent comparison among all agents that is embedded in the definition of $\mathcal{H}_{dc}$ (via the max function).

2.3.3 Sphere-packing multicenter functions

Given a compact connected set $S \subset \mathbb{R}^d$, imagine trying to fit inside S "maximally large" non-intersecting balls. Assuming that the balls are centered at a set of points $p_1, \ldots, p_n$, we aim to maximize their smallest radius. We define the *sphere-packing multicenter* function $\mathcal{H}_{sp} : S^n \to \mathbb{R}$ by

$$\mathcal{H}_{sp}(p_1, \ldots, p_n) = \min_{i \neq j \in \{1, \ldots, n\}} \left\{ \frac{1}{2}\|p_i - p_j\|_2, \mathrm{dist}(p_i, \partial S) \right\}. \qquad (2.3.13)$$

The definition of $\mathcal{H}_{sp}$ can be read as follows: consider the pairwise distances between any two points p_i, p_j (multiplied by a factor $1/2$ so that each point can fit a ball of equal radius and these balls do not intersect), and the individual distances from each point to the boundary of the environment. The value of $\mathcal{H}_{sp}$ is then the smallest of all distances, guaranteeing that the union of n open balls centered at $p_1, \ldots, p_n$ with radius $\mathcal{H}_{sp}(p_1, \ldots, p_n)$ is disjoint and contained in S. The definition of $\mathcal{H}_{sp}$ is illustrated in Figure 2.15(a).

Given the definition of $\mathcal{H}_{sp}$, we seek to solve the following geometric optimization problem:

$$\text{maximize } \mathcal{H}_{sp}(p_1, \ldots, p_n), \qquad (2.3.14)$$

that is, we seek to determine configurations $p_1, \ldots, p_n$ that maximize the value of $\mathcal{H}_{sp}$. Note the connection of this formulation with the classical sphere-packing problem: how to maximize the number of fixed-radius non-overlapping spheres inside a region.

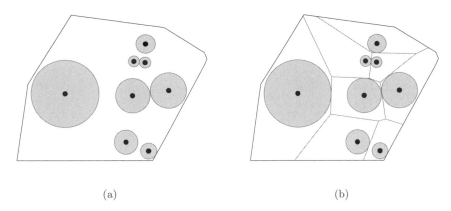

(a)   (b)

Figure 2.15 An illustration of the definition of $\mathcal{H}_{\mathrm{sp}}$: (a) and (b) show the same configuration, with and without the Voronoi configuration, respectively. For each node, the disk is the maximum-radius disk centered at the node and contained in the Voronoi cell. The value of $\mathcal{H}_{\mathrm{sp}}$ is the radius of the two equal-radius smallest disks.

The sphere-packing multicenter function can be alternatively described in terms of the Voronoi partition of S generated by $\mathcal{P} = \{p_1, \ldots, p_n\}$. For $(p_1, \ldots, p_n) \in S^n \setminus \mathcal{S}_{\mathrm{coinc}}$, one has

$$\mathcal{H}_{\mathrm{sp}}(p_1, \ldots, p_n) = \min_{i \in \{1, \ldots, n\}} \min_{q \in \partial V_i(\mathcal{P})} \|q - p_i\|_2. \tag{2.3.15}$$

This description is illustrated in Figure 2.15(b). As for the previous multicenter functions, expression (2.3.15) is appealing because it clearly shows the value of of $\mathcal{H}_{\mathrm{sp}}$ as the result of the aggregate contribution of all individual nodes. If $(p_1, \ldots, p_n) \in \mathcal{S}_{\mathrm{coinc}}$, then a similar decomposition of $\mathcal{H}_{\mathrm{sp}}$ exists in terms of the distinct points $\mathcal{P} = i_{\mathbb{F}}(p_1, \ldots, p_n)$. A node $i \in \{1, \ldots, n\}$ is called *active* at $(p_1, \ldots, p_n)$ if $\min_{q \in \partial V_i(\mathcal{P})} \|q - p_i\|_2 = \mathcal{H}_{\mathrm{sp}}(p_1, \ldots, p_n)$. A node is *passive* at $(p_1, \ldots, p_n)$ if it is not active.

Inspired by expression (2.3.15), let us define a more general version of the sphere-packing multicenter function. Given $(p_1, \ldots, p_n) \in S^n$ and a partition $\{W_1, \ldots, W_n\} \subset \mathbb{P}(S)$ of S, let

$$\mathcal{H}_{\mathrm{sp}}(p_1, \ldots, p_n, W_1, \ldots, W_n) = \min_{i \in \{1, \ldots, n\}} \min_{q \in \partial W_i} \|q - p_i\|_2.$$

Note the relationship $\mathcal{H}_{\mathrm{sp}}(p_1, \ldots, p_n) = \mathcal{H}_{\mathrm{sp}}(p_1, \ldots, p_n, V_1(\mathcal{P}), \ldots, V_n(\mathcal{P}))$, for all $(p_1, \ldots, p_n) \in S^n \setminus \mathcal{S}_{\mathrm{coinc}}$. Additionally, note that the quantity $\mathcal{H}_{\mathrm{sp}}(q_1, \ldots, q_n, W_1, \ldots, W_n)$ is the same for any $q_i \in \mathrm{IC}(W_i)$, $i \in \{1, \ldots, n\}$. With a slight abuse of notation, we refer to this common value using the symbol $\mathcal{H}_{\mathrm{sp}}(\mathrm{IC}(W_1), \ldots, \mathrm{IC}(W_n), W_1, \ldots, W_n)$. Moreover, one can establish the following optimality result (for the proof, see Section 2.5.5).

Proposition 2.21 ($\mathcal{H}_{sp}$-optimality of the Voronoi partition and incenter locations). *For any $\mathcal{P} = \{p_1, \ldots, p_n\} \in \mathbb{F}(S)$ and any partition $\{W_1, \ldots, W_n\} \subset \mathbb{P}(S)$ of S,*

$$\mathcal{H}_{sp}(p_1, \ldots, p_n, V_1(\mathcal{P}), \ldots, V_n(\mathcal{P})) \geq \mathcal{H}_{sp}(p_1, \ldots, p_n, W_1, \ldots, W_n),$$

that is, the Voronoi partition $\mathcal{V}(\mathcal{P})$ is optimal for $\mathcal{H}_{sp}$ among all partitions of S, and

$$\mathcal{H}_{sp}(\mathrm{IC}(W_1), \ldots, \mathrm{IC}(W_n), W_1, \ldots, W_n) \geq \mathcal{H}_{sp}(p_1, \ldots, p_n, W_1, \ldots, W_n),$$

that is, the incenters $\mathrm{IC}(W_1), \ldots, \mathrm{IC}(W_n)$ are optimal for $\mathcal{H}_{sp}$ among all configurations in S.

As a corollary of this result, we have that the incenter set of S is composed of global optima of $\mathcal{H}_{sp}$ for the 1-center problem, that is, when $n = 1$. This comes as no surprise since, in this case, the value $\mathcal{H}_{sp}(p)$ corresponds to the radius of the maximum-radius sphere centered at p enclosed in S. The following result characterizes the smoothness properties of the sphere-packing multicenter function (see Cortés and Bullo, 2005).

Theorem 2.22 (Smoothness properties of $\mathcal{H}_{sp}$). *Given $S \subset \mathbb{R}^d$ compact, the sphere-packing multicenter function $\mathcal{H}_{sp} : S^n \to \mathbb{R}$ is globally Lipschitz on S^n.*

We conclude this section with some remarks that are analogous to those for the function $\mathcal{H}_{dc}$. The generalized gradient and the critical points of $\mathcal{H}_{sp}$ can be characterized, but require a careful study based on nonsmooth analysis (Clarke, 1983). In particular, two facts taken from Cortés and Bullo (2005) are of interest here. First, under certain technical conditions, one can show that the critical points of $\mathcal{H}_{sp}$ are incenter Voronoi configurations. This is why we refer to $\mathcal{H}_{sp}$ as a multicenter function. Second, the generalized gradient of $\mathcal{H}_{sp}$ is not spatially distributed over $\mathcal{G}_D$. This is essentially due to the inherent comparison among all agents that is embedded in the definition of $\mathcal{H}_{sp}$ (via the min function).

2.4 NOTES

A thorough introduction to computational geometric concepts can be found in Preparata and Shamos (1993), de Berg et al. (2000), and O'Rourke (2000). The handbooks Goodman and O'Rourke (2004) and Sack and Urrutia (2000) present a comprehensive overview of computational geometric problems and their applications. Among the numerous topics that we do not discuss in this chapter, we mention distance geometry and rigidity theory (Whiteley, 1997), which are notable for their applications to network localization and formation control.

The notion of Voronoi partition, and generalizations of it, have been applied in numerous areas, including spatial interpolation, pattern analysis, spatial processes modeling, and optimization, to name a few. The survey Aurenhammer (1991) and the book by Okabe et al. (2000) discuss the history, properties, and applications of Voronoi partitions. The nearest-neighbor and natural-neighbor interpolations based on Voronoi partitions (see, for example Sibson, 1981; Boissonnat and Cazals, 2002) are of particular interest to the treatment of this chapter because of their spatially distributed computation character. Spatially distributed maps for motion coordination are discussed in Martínez et al. (2007c) and adopted in later chapters.

Proximity graphs (Jaromczyk and Toussaint, 1992) are a powerful tool to capture the structure and shape of geometric objects, and therefore have applications in multiple areas, including topology control of wireless networks (Santi, 2005), computer graphics (Langetepe and Zachmann, 2006), and geographic analysis (Radke, 1988). The connectivity properties of certain proximity graphs (including those stated in Theorem 2.8) are taken from Cortés et al. (2005, 2006). In cooperative control, a closely related notion is that of state-dependent graph (Mesbahi, 2005). Random geometric graphs (Penrose, 2003) and percolation theory (Bollobás and Riordan, 2006; Meester and Roy, 2008) study the properties of proximity graphs associated to the random deployment of points according to some specified density function.

Locational optimization problems (Drezner, 1995; Drezner and Hamacher, 2001) are spatial resource-allocation problems (e.g., where to place mailboxes in a city, or where to place cache serves on the internet) that pervade a broad spectrum of scientific disciplines. Computational geometry plays an important role in locational optimization (Robert and Toussaint, 1990; Okabe et al., 2000). The field of geometric optimization (Mitchell, 1997; Agarwal and Sharir, 1998; Boltyanski et al., 1999) blends the geometric and locational optimization aspects to study a wide variety of optimization problems induced by geometric objects. The smoothness properties of the cost function $\mathcal{H}_{\mathrm{exp}}$ are taken from Cortés et al. (2005).

2.5 PROOFS

This section gathers the proofs of the main results presented in the chapter.

2.5.1 Proofs of Theorem 2.7 and Theorem 2.8

Proof of Theorem 2.7. The inclusions in fact (i) are taken from Jaromczyk and Toussaint (1992), and de Berg et al. (2000). The proof of the first inclusion in fact (ii) is as follows. Let $(p_i, p_j) \in \mathcal{E}_{\mathcal{G}_{\mathrm{G}} \cap \mathcal{G}_{\mathrm{disk}(r)}}(\mathcal{P})$. From the definition of the Gabriel graph, we deduce that $\|\frac{p_i+p_j}{2} - p_i\|_2 = \|\frac{p_i+p_j}{2} - p_j\|_2 \leq \|\frac{p_i+p_j}{2} - p_k\|_2$, for all $k \in \{1,\dots,n\} \setminus \{i,j\}$, and therefore, $\frac{p_i+p_j}{2} \in V_i(\mathcal{P}) \cap V_j(\mathcal{P})$. Since $(p_i, p_j) \in \mathcal{E}_{\mathcal{G}_{\mathrm{disk}(r)}}(\mathcal{P})$, we deduce that $\frac{p_i+p_j}{2} \in \overline{B}(p_i, \frac{r}{2}) \cap \overline{B}(p_j, \frac{r}{2})$, and hence $(p_i, p_j) \in \mathcal{E}_{\mathcal{G}_{\mathrm{LD}(r)}}(\mathcal{P})$. The second inclusion in (ii) is straightforward: if $(p_i, p_j) \in \mathcal{E}_{\mathcal{G}_{\mathrm{LD}(r)}}(\mathcal{P})$, then $V_i(\mathcal{P}) \cap V_j(\mathcal{P}) \neq \emptyset$, that is, $(p_i, p_j) \in \mathcal{E}_{\mathcal{G}_{\mathrm{D}}}(\mathcal{P})$. Since clearly $(p_i, p_j) \in \mathcal{E}_{\mathcal{G}_{\mathrm{disk}(r)}}(\mathcal{P})$, we conclude (ii). ∎

Proof of Theorem 2.8. The proof of fact (i) is as follows. Let $\mathcal{P} \in \mathbb{F}(\mathbb{R}^d)$. If $\mathcal{G}_{\mathrm{EMST}}(\mathcal{P}) \subseteq \mathcal{G}_{\mathrm{disk}(r)}(\mathcal{P})$, then clearly $\mathcal{G}_{\mathrm{disk}(r)}(\mathcal{P})$ is connected. To prove the other implication, we reason by contradiction. Assume $\mathcal{G}_{\mathrm{disk}(r)}(\mathcal{P})$ is connected and let $\mathcal{G}_{\mathrm{EMST}}(\mathcal{P}) \not\subseteq \mathcal{G}_{\mathrm{disk}(r)}(\mathcal{P})$, that is, there exist p_i and p_j with $(p_i, p_j) \in \mathcal{E}_{\mathcal{G}_{\mathrm{EMST}}}(\mathcal{P})$ and $\|p_i - p_j\|_2 > r$. If we remove this edge from $\mathcal{E}_{\mathcal{G}_{\mathrm{EMST}}}(\mathcal{P})$, then the tree becomes disconnected into two connected components T_1 and T_2, with $p_i \in T_1$ and $p_j \in T_2$. Now, since by hypothesis $\mathcal{G}_{\mathrm{disk}(r)}(\mathcal{P})$ is connected, there must exist $k, l \in \{1,\dots,n\}$ such that $p_k \in T_1$, $p_l \in T_2$ and $\|p_k - p_l\|_2 \leq r$. If we add the edge (p_k, p_l) to the set of edges of $T_1 \cup T_2$, then the resulting graph G is acyclic, connected, and contains all the vertices $\mathcal{P}$, that is, G is a spanning tree. Moreover, since $\|p_k - p_l\|_2 \leq r < \|p_i - p_j\|_2$ and T_1 and T_2 are induced subgraphs of $\mathcal{G}_{\mathrm{EMST}}(\mathcal{P})$, we conclude that G has smaller length than $\mathcal{G}_{\mathrm{EMST}}(\mathcal{P})$, which is a contradiction with the definition of the Euclidean minimum spanning tree.

Next, we prove fact (ii). For $r \in \mathbb{R}_+$, it suffices for us to show that $\mathcal{G}_{\mathrm{EMST}} \cap \mathcal{G}_{\mathrm{disk}}(r)$ has the same connected components as $\mathcal{G}_{\mathrm{disk}}(r)$, since this implies that the same result holds for $\mathcal{G}_{\mathrm{RN}} \cap \mathcal{G}_{\mathrm{disk}}(r)$, $\mathcal{G}_{\mathrm{G}} \cap \mathcal{G}_{\mathrm{disk}}(r)$, and $\mathcal{G}_{\mathrm{LD}}(r)$. Since $\mathcal{G}_{\mathrm{EMST}} \cap \mathcal{G}_{\mathrm{disk}}(r)$ is a subgraph of $\mathcal{G}_{\mathrm{disk}}(r)$, it is clear that vertices belonging to the same connected component of $\mathcal{G}_{\mathrm{EMST}} \cap \mathcal{G}_{\mathrm{disk}}(r)$ must also belong to the same connected component of $\mathcal{G}_{\mathrm{disk}}(r)$. To prove the converse, let $\mathcal{P} \in \mathbb{F}(\mathbb{R}^d)$, and assume that p_i and p_j in $\mathcal{P}$ verify $\|p_i - p_j\|_2 \leq r$. Let C be the connected component of $\mathcal{G}_{\mathrm{disk}}(r)(\mathcal{P})$ to which they belong. With a slight abuse of notation, we also denote by C the vertices of the connected component. Since C is connected, then $\mathcal{G}_{\mathrm{EMST}}(C) \subset C$ by fact (i). Moreover, since all the nodes in $\mathcal{P} \setminus C$ are at a distance strictly larger than r from any node of C, we deduce from the definition of the Euclidean minimum spanning tree that $\mathcal{G}_{\mathrm{EMST}}(C)$ is equal to the subgraph of $\mathcal{G}_{\mathrm{EMST}}(\mathcal{P})$ induced by C. Therefore, $\mathcal{G}_{\mathrm{EMST}}(C) \subset \mathcal{G}_{\mathrm{EMST}} \cap \mathcal{G}_{\mathrm{disk}}(r)(\mathcal{P})$, and p_i and

p_j belong to the same component of $\mathcal{G}_{\text{EMST}} \cap \mathcal{G}_{\text{disk}}(r)(\mathcal{P})$. This implies the result. ∎

2.5.2 Proof of Proposition 2.9

Proof. Regarding the statement on $\mathcal{G}_{\text{RN}} \cap \mathcal{G}_{\text{disk}}(r)$, note that

$$B(p_i, \|p_i - p_j\|_2) \cap B(p_j, \|p_i - p_j\|_2) \subset B(p_i, \|p_i - p_j\|_2).$$

Therefore, if $\|p_i - p_j\|_2 \leq r$, then any node contained in the intersection $B(p_i, \|p_i - p_j\|_2) \cap B(p_j, \|p_i - p_j\|_2)$ must necessarily be within a distance r of p_i. From here, we deduce that $\mathcal{G}_{\text{RN}} \cap \mathcal{G}_{\text{disk}}(r)$ is spatially distributed over $\mathcal{G}_{\text{disk}}(r)$. Regarding the statement on $\mathcal{G}_{\text{G}} \cap \mathcal{G}_{\text{disk}}(r)$, note that

$$B\left(\frac{p_i + p_j}{2}, \frac{\|p_i - p_j\|_2}{2}\right) \subset B(p_i, \|p_i - p_j\|_2).$$

Therefore, if $\|p_i - p_j\|_2 \leq r$, then any node contained in $B\left(\frac{p_i + p_j}{2}, \frac{\|p_i - p_j\|_2}{2}\right)$ must necessarily be within a distance r of p_i. From here, we deduce that $\mathcal{G}_{\text{G}} \cap \mathcal{G}_{\text{disk}}(r)$ is spatially distributed over $\mathcal{G}_{\text{disk}}(r)$. Finally, note that if $\|p_i - p_j\|_2 > r$, then the half-plane $\{q \in \mathbb{R}^2 \mid \|q - p_i\|_2 \leq \|q - p_j\|_2\}$ contains the ball $\overline{B}(p_i, \frac{r}{2})$. Accordingly,

$$
\begin{aligned}
V_{i, \frac{r}{2}}(\mathcal{P}) &= V_i(\mathcal{P}) \cap \overline{B}(p_i, \tfrac{r}{2}) \\
&= \{q \in \mathbb{R}^2 \mid \|q - p_i\|_2 \leq \|q - p_j\|_2, \text{ for all } p_j \in \mathcal{P}\} \cap \overline{B}(p_i, \tfrac{r}{2}) \\
&= \{q \in \mathbb{R}^2 \mid \|q - p_i\|_2 \leq \|q - p_j\|_2, \text{ for all } p_j \in \mathcal{N}_{\mathcal{G}_{\text{disk}}(r), p_i}(\mathcal{P})\} \cap \overline{B}(p_i, \tfrac{r}{2}),
\end{aligned}
$$

from which we deduce that $\mathcal{G}_{\text{LD}}(r)$ is spatially distributed over $\mathcal{G}_{\text{disk}}(r)$. ∎

2.5.3 Proof of Theorem 2.16

We begin with some preliminary notions. In the following, a set $\Omega \subset \mathbb{R}^2$ is *piecewise continuously differentiable* if its boundary, $\partial\Omega$, is a not self-intersecting closed curve that admits a continuous and piecewise continuously differentiable parameterization $\gamma : [0, 1] \to \mathbb{R}^2$. Likewise, a collection of sets $\{\Omega(x) \subset \mathbb{R}^2 \mid x \in (a, b)\}$ is a *piecewise continuously differentiable family* if $\Omega(x)$ is piecewise continuously differentiable for all $x \in (a, b)$, and there exists a continuous function $\gamma : [0, 1] \times (a, b) \to \mathbb{R}^2$, $(t, x) \mapsto \gamma(t, x)$, continuously differentiable with respect to its second argument, such that for each $x \in (a, b)$, the map $t \mapsto \gamma_x(t) = \gamma(t, x)$ is a continuous and piecewise continuously differentiable parameterization of $\partial\Omega(x)$. We refer to γ as a *parameterization for the family* $\{\Omega(x) \subset \mathbb{R}^2 \mid x \in (a, b)\}$.

The following result is an extension of the Law of Conservation of Mass in fluid mechanics (Chorin and Marsden, 1994) and of the classic divergence theorem in differential geometry (Chavel, 1984).

Proposition 2.23 (Generalized conservation of mass). *Let $\{\Omega(x) \subset \mathbb{R}^2 \mid x \in (a,b)\}$ be a family of star-shaped sets with piecewise continuously differentiable boundary. Let the function $\phi : \mathbb{R}^2 \times (a,b) \to \mathbb{R}$ be continuous on $\mathbb{R}^2 \times (a,b)$ that is continuously differentiable with respect to its second argument for all $x \in (a,b)$ and almost all $q \in \Omega(x)$, and such that for each $x \in (a,b)$, the maps $q \mapsto \phi(q,x)$ and $q \mapsto \frac{\partial \phi}{\partial x}(q,x)$ are measurable, and integrable on $\Omega(x)$. Then, the function*

$$(a,b) \ni x \mapsto \int_{\Omega(x)} \phi(q,x)dq \qquad (2.5.1)$$

is continuously differentiable and

$$\frac{d}{dx} \int_{\Omega(x)} \phi(q,x)dq = \int_{\Omega(x)} \frac{\partial \phi}{\partial x}(q,x)dq + \int_{\partial\Omega(x)} \phi(\gamma,x)\left(n(\gamma) \cdot \frac{\partial \gamma}{\partial x}\right)d\gamma \,,$$

where $n : \partial\Omega(x) \to \mathbb{R}^2$, $q \mapsto n(q)$, denotes the unit outward normal to $\partial\Omega(x)$ at $q \in \partial\Omega(x)$, and $\gamma : [0,1] \times (a,b) \to \mathbb{R}^2$ is a parameterization for the family $\{\Omega(x) \subset \mathbb{R}^2 \mid x \in (a,b)\}$.

We interpret the proposition as follows: in the fluid mechanics interpretation, as the parameter x changes, the total mass variation inside the region can be decomposed into two terms. The first term is the amount of mass created inside the region, whereas the second term is the amount of mass that crosses the moving boundary of the region.

Proof of Proposition 2.23. Let $x_0 \in (a,b)$. Using the fact that the map γ is continuous and that $\Omega(x_0)$ is star-shaped, one can show that there exist an interval around x_0 of the form $(x_0 - \varepsilon, x_0 + \varepsilon)$, a continuously differentiable function $u_{x_0} : [0,1] \times \mathbb{R}_{\geq 0} \to \mathbb{R}^2$ and a function $r_{x_0} : [0,1] \times (x_0 - \varepsilon, x_0 + \varepsilon) \to \mathbb{R}_{\geq 0}$ continuously differentiable in its second argument and piecewise continuously differentiable in its first argument, such that for all $x \in (x_0 - \varepsilon, x_0 + \varepsilon)$, one has

$$\Omega(x) = \cup_{t \in [0,1]}\{u_{x_0}(t,s) \mid 0 \leq s \leq r_{x_0}(t,x)\},$$
$$\gamma(t,x) = u_{x_0}(t, r_{x_0}(t,x)), \quad \text{for all } t \in [0,1].$$

For simplicity, we denote by r and u the functions r_{x_0} and u_{x_0}, respectively. By definition, the function in (2.5.1) is continuously differentiable at x_0 if the following limit exists:

$$\lim_{h \to 0} \frac{1}{h}\left(\int_{\Omega(x_0+h)} \phi(q,x_0+h)dq - \int_{\Omega(x_0)} \phi(q,x_0)dq \right),$$

and depends continuously on x_0. Now, we can rewrite the previous limit as

$$\lim_{h \to 0} \frac{1}{h} \int_0^1 \Big(\int_0^{r(t,x_0+h)} \phi(u(t,s), x_0 + h) \Big\| \frac{\partial u}{\partial t} \times \frac{\partial u}{\partial s} \Big\|_2 ds$$

$$- \int_0^{r(t,x_0)} \phi(u(t,s), x_0) \Big\| \frac{\partial u}{\partial t} \times \frac{\partial u}{\partial s} \Big\|_2 ds \Big) dt$$

$$= \lim_{h \to 0} \frac{1}{h} \int_0^1 \Big(\int_{r(t,x_0)}^{r(t,x_0+h)} \phi(u(t,s), x_0 + h) \Big\| \frac{\partial u}{\partial t} \times \frac{\partial u}{\partial s} \Big\|_2 ds$$

$$+ \int_0^{r(t,x_0)} (\phi(u(t,s), x_0 + h) - \phi(u(t,s), x_0)) \Big\| \frac{\partial u}{\partial t} \times \frac{\partial u}{\partial s} \Big\|_2 ds \Big) dt, \quad (2.5.2)$$

where $\times$ denotes the vector product and for brevity we omit the fact that the partial derivatives $\frac{\partial u}{\partial t}$ and $\frac{\partial u}{\partial s}$ are evaluated at (t,s) in the integrals. Regarding the second integral in the last equality of (2.5.2), since

$$\lim_{h \to 0} \frac{1}{h} \Big((\phi(u(t,s), x_0 + h) - \phi(u(t,s), x_0)) \Big\| \frac{\partial u}{\partial t} \times \frac{\partial u}{\partial s} \Big\|_2 \Big)$$

$$= \frac{\partial \phi}{\partial x_0}(u(t,s), x_0) \Big\| \frac{\partial u}{\partial t} \times \frac{\partial u}{\partial s} \Big\|_2,$$

almost everywhere, and this function is measurable and its integral over the bounded set $\Omega(x_0)$ is finite by hypothesis, the Lebesgue Dominated Convergence Theorem (Bartle, 1995) implies that

$$\lim_{h \to 0} \frac{1}{h} \int_0^1 \int_0^{r(t,x_0)} (\phi(u(t,s), x_0 + h) - \phi(u(t,s), x_0)) \Big\| \frac{\partial u}{\partial t} \times \frac{\partial u}{\partial s} \Big\|_2 ds dt$$

$$= \int_0^1 \int_0^{r(t,x_0)} \frac{\partial \phi}{\partial x}(u(t,s), x_0) \Big\| \frac{\partial u}{\partial t} \times \frac{\partial u}{\partial s} \Big\|_2 ds dt$$

$$= \int_{\Omega(x_0)} \frac{\partial \phi}{\partial x}(q, x_0) dq. \quad (2.5.3)$$

On the other hand, regarding the first integral in the last equality of (2.5.2), using the continuity of ϕ, one can deduce that

$$\lim_{h \to 0} \frac{1}{h} \int_0^1 \int_{r(t,x_0)}^{r(t,x_0+h)} \phi(u(t,s), x_0 + h) \Big\| \frac{\partial u}{\partial t}(t,s) \times \frac{\partial u}{\partial s}(t,s) \Big\|_2 ds \, dt$$

$$= \lim_{h \to 0} \frac{1}{h} \int_0^1 \int_{x_0}^{x_0+h} \phi(u(t,r(t,z)), x_0 + h)$$

$$\cdot \Big\| \frac{\partial u}{\partial t}(t,r(t,z)) \times \frac{\partial u}{\partial s}(t,r(t,z)) \Big\|_2 \frac{\partial r}{\partial x}(t,z) \, dz \, dt$$

$$= \int_0^1 \phi(u(t,r(t,x_0)), x_0) \Big\| \frac{\partial u}{\partial t}(t,r(t,x_0)) \times \frac{\partial u}{\partial s}(t,r(t,x_0)) \Big\|_2 \frac{\partial r}{\partial x_0}(t,x_0) \, dt.$$

Since $\gamma(t,x) = u(t, r(t,x))$ for all $t \in [0,1]$ and $x \in (x_0 - \varepsilon, x_0 + \varepsilon)$, one has

$$\frac{\partial \gamma}{\partial t}(t, x_0) = \frac{\partial u}{\partial t}(t, r(t, x_0)) + \frac{\partial u}{\partial s}(t, r(t, x_0))\frac{\partial r}{\partial t}(t, x_0),$$

$$\frac{\partial \gamma}{\partial x}(t, x_0) = \frac{\partial u}{\partial s}(t, r(t, x_0))\frac{\partial r}{\partial x}(t, x_0).$$

Let χ denote the angle formed by $\frac{\partial \gamma}{\partial t}(t, x_0)$ and $\frac{\partial u}{\partial s}(t, r(t, x_0))$. Then (omitting the expression $(t, r(t,x))$ for brevity),

$$\left\| \frac{\partial u}{\partial t} \times \frac{\partial u}{\partial s} \right\|_2 = \left\| \left(\frac{\partial u}{\partial t} + \frac{\partial u}{\partial s}\frac{\partial r}{\partial t} \right) \times \frac{\partial u}{\partial s} \right\|_2$$

$$= \left\| \frac{d\gamma}{dt} \right\|_2 \left\| \frac{\partial u}{\partial s} \right\|_2 \sin\chi = \left\| \frac{\partial\gamma}{\partial t} \right\|_2 n^T(\gamma)\frac{\partial u}{\partial s},$$

where in the last inequality we have used the fact that, since γ_{x_0} is a parameterization of $\partial\Omega(x_0)$, then $\sin\chi = \cos\psi$, where ψ is the angle formed by n, the outward normal to $\partial\Omega(x_0)$, and $\frac{\partial u}{\partial s}$. Therefore, we finally arrive at

$$\int_0^1 \phi(\gamma(t), x_0)\left\| \frac{\partial u}{\partial t}(t, r(t, x_0)) \times \frac{\partial u}{\partial s}(t, r(t, x_0)) \right\|_2 \frac{\partial r}{\partial x}(t, x_0)dt$$

$$= \int_0^1 \phi(\gamma(t), x_0)\left\| \frac{\partial\gamma}{\partial t}(t, x_0) \right\|_2 n^T(\gamma(t, x_0))\frac{\partial\gamma}{\partial x}(t, x_0)dt$$

$$= \int_{\partial\Omega(x_0)} \phi(\gamma, x_0)n^T(\gamma)\frac{\partial\gamma}{\partial x}d\gamma. \qquad (2.5.4)$$

Given the hypothesis of Proposition 2.23, both terms in (2.5.3) and (2.5.4) have a continuous dependence on $x_0 \in (a, b)$. This concludes the proof. ∎

We are finally ready to state the proof of the main result of Section 2.3.

Proof of Theorem 2.16. We prove the theorem statement when the performance function is continuously differentiable and we refer to Cortés et al. (2005) for the complete proof for the case when the performance function is piecewise continuously differentiable. Specifically, we show that if f is continuously differentiable, then for $P \in S^n \setminus S_{\text{coinc}}$,

$$\frac{\partial \mathcal{H}_{\exp}}{\partial p_i}(P) = \int_{V_i(\mathcal{P})} \frac{\partial}{\partial p_i} f(\|q - p_i\|_2)\phi(q)dq.$$

From Proposition 2.23, we have

$$\frac{\partial}{\partial p_i}\Big(\sum_{j=1}^{n}\int_{V_j(\mathcal{P})}f(\|q-p_j\|_2)\phi(q)dq\Big) = \int_{V_i(\mathcal{P})}\frac{\partial}{\partial p_i}f(\|q-p_i\|_2)\phi(q)dq$$

$$+\sum_{j=1}^{n}\int_{\partial V_j(\mathcal{P})}\varphi(p_j,q)\Big(n(\gamma_j)\cdot\frac{\partial\gamma_j}{\partial p_i}\Big)d\gamma_j,$$

where γ_j is a parametrization of $V_j(\mathcal{P})$ and where we abbreviate $\varphi(p_j,q) = f(\|q-p_j\|_2)\phi(q)$. Next, we show that the second term vanishes. Note that the motion of p_i affects the Voronoi cell $V_i(\mathcal{P})$ and the cells of all its neighbors in $\mathcal{N}_{\mathcal{G}_{\mathrm{D}},p_i}(\mathcal{P})$. Therefore, the second term equals

$$\int_{\partial V_i(\mathcal{P})}\varphi(p_i,q)\Big(n(\gamma_i)\cdot\frac{\partial\gamma_i}{\partial p_i}\Big)d\gamma_i$$

$$+\sum_{p_j\in\mathcal{N}_{\mathcal{G}_{\mathrm{D}},p_i}(\mathcal{P})}\int_{\partial V_j(\mathcal{P})}\varphi(p_j,q)\Big(n(\gamma_j)\cdot\frac{\partial\gamma_j}{\partial p_i}\Big)d\gamma_j.$$

Without loss of generality, assume that $V_i(\mathcal{P})$ does not share any face with ∂S. Since the boundary of $V_i(\mathcal{P})$ satisfies $\partial V_i(\mathcal{P}) = \bigcup_j \Delta_{ij}$, where $\Delta_{ij} = \Delta_{ji}$ is the edge between $V_i(\mathcal{P})$ and $V_j(\mathcal{P})$, for all neighbors p_j, we compute

$$\int_{\partial V_i(\mathcal{P})}\varphi(p_i,q)\Big(n(\gamma_i)\cdot\frac{\partial\gamma_i}{\partial p_i}\Big)d\gamma_i = \sum_{p_j\in\mathcal{N}_{\mathcal{G}_{\mathrm{D}},p_i}(\mathcal{P})}\int_{\Delta_{ij}}\varphi(p_i,q)\Big(n_{ij}(\gamma_j)\cdot\frac{\partial\gamma_j}{\partial p_i}\Big)d\gamma_j,$$

$$\int_{\partial V_j(\mathcal{P})}\varphi(p_j,q)\Big(n(\gamma_j)\cdot\frac{\partial\gamma_j}{\partial p_i}\Big)d\gamma_j = \int_{\Delta_{ji}}\varphi(p_j,q)\Big(n_{ji}(\gamma_j)\cdot\frac{\partial\gamma_j}{\partial p_i}\Big)d\gamma_j,$$

where n_{ij} denotes the unit normal along Δ_{ij} outward of $V_i(P)$. Noting that $n_{ji} = -n_{ij}$ and collecting the results obtained so far, we write

$$\sum_{j=1}^{n}\int_{\partial V_j(\mathcal{P})}\varphi(p_j,q)\Big(n(\gamma_j)\cdot\frac{\partial\gamma_j}{\partial p_i}\Big)d\gamma_j$$

$$= \sum_{p_j\in\mathcal{N}_{\mathcal{G}_{\mathrm{D}},p_i}(\mathcal{P})}\int_{\Delta_{ij}}\big(\varphi(p_i,q)-\varphi(p_j,q)\big)\Big(n_{ij}(\gamma_j)\cdot\frac{\partial\gamma_j}{\partial p_i}\Big)d\gamma_j.$$

This quantity vanishes because $f(\|q-p_i\|_2) = f(\|q-p_j\|_2)$, and therefore $\varphi(p_i,q) = \varphi(p_j,q)$ for any q belonging to the edge Δ_{ij}. ∎

2.5.4 Proof of Proposition 2.19

Proof. Recall that $\mathcal{H}_{\mathrm{dc}}(p_1,\ldots,p_n) = \mathcal{H}_{\mathrm{dc}}(p_1,\ldots,p_n,V_1(\mathcal{P}),\ldots,V_n(\mathcal{P}))$. To show the first inequality, let $j\in\{1,\ldots,n\}$ and $q_*\in V_j(\mathcal{P})$ be such that

$\mathcal{H}_{\mathrm{dc}}(p_1, \ldots, p_n) = \|q_* - p_j\|_2$. By definition, given a partition $\{W_1, \ldots, W_n\}$ of S, there exists k such that $q_* \in W_k$. Therefore,

$$\begin{aligned} \mathcal{H}_{\mathrm{dc}}(p_1, \ldots, p_n) = \|q_* - p_j\|_2 &\le \|q_* - p_k\|_2 \\ &\le \max_{q \in W_k} \|q - p_j\|_2 \le \mathcal{H}_{\mathrm{dc}}(p_1, \ldots, p_n, W_1, \ldots, W_n). \end{aligned}$$

To show the second inequality, note that the definition of circumcenter implies that, for each $i \in \{1, \ldots, n\}$,

$$\max_{q \in \partial W_i} \|q - \mathrm{CC}(W_i)\|_2 \le \max_{q \in \partial W_i} \|q - p_i\|_2.$$

Taking the maximum over all nodes, we deduce that

$$\mathcal{H}_{\mathrm{dc}}(\mathrm{CC}(W_1), \ldots, \mathrm{CC}(W_n), W_1, \ldots, W_n) \le \mathcal{H}_{\mathrm{dc}}(p_1, \ldots, p_n, W_1, \ldots, W_n),$$

as claimed. ∎

2.5.5 Proof of Proposition 2.21

Proof. Recall that $\mathcal{H}_{\mathrm{sp}}(p_1, \ldots, p_n) = \mathcal{H}_{\mathrm{sp}}(p_1, \ldots, p_n, V_1(\mathcal{P}), \ldots, V_n(\mathcal{P}))$. To show the first inequality, let $j \in \{1, \ldots, n\}$ and $q_* \notin \mathrm{int}(V_j(\mathcal{P}))$ be such that $\mathcal{H}_{\mathrm{sp}}(p_1, \ldots, p_n) = \|q_* - p_j\|_2$. Since $q_* \notin \mathrm{int}(V_j(\mathcal{P}))$, there exists $i \in \{1, \ldots, n\}$ such that $\|q_* - p_j\|_2 \ge \|q_* - p_i\|_2$. On the other hand, there must exist $k \in \{1, \ldots, n\}$ such that $q_* \in W_k$. Now, if $k = j$, then $q_* \notin \mathrm{int}(W_i)$. Therefore,

$$\begin{aligned} \mathcal{H}_{\mathrm{sp}}(p_1, \ldots, p_n) = \|q_* - p_j\|_2 &\ge \|q_* - p_i\|_2 \\ &\ge \min_{q \notin \mathrm{int}(W_i)} \|q - p_i\|_2 \ge \mathcal{H}_{\mathrm{sp}}(p_1, \ldots, p_n, W_1, \ldots, W_n). \end{aligned}$$

Now, if $k = i$, then $q_* \notin \mathrm{int}(W_j)$. Therefore,

$$\mathcal{H}_{\mathrm{sp}}(P) = \|q_* - p_j\|_2 \ge \min_{q \notin \mathrm{int}(W_j)} \|q - p_i\|_2 \ge \mathcal{H}_{\mathrm{sp}}(p_1, \ldots, p_n, W_1, \ldots, W_n).$$

Finally, if $k \ne i, j$, then $q_* \notin \mathrm{int}(W_i) \cup \mathrm{int}(W_j)$, and a similar argument guarantees $\mathcal{H}_{\mathrm{sp}}(p_1, \ldots, p_n) \ge \mathcal{H}_{\mathrm{sp}}(p_1, \ldots, p_n, W_1, \ldots, W_n)$.

To show the second inequality, let $i \in \{1, \ldots, n\}$ and select $q_i \in \mathrm{IC}(W_i)$. The definition of the incenter set implies that,

$$\min_{q \in \partial W_i} \|q - q_i\|_2 \ge \min_{q \in \partial W_i} \|q - p_i\|_2.$$

The expression on the left does not depend on the specific point selected in the incenter set. Taking the minimum over all nodes, we deduce that

$$\mathcal{H}_{\mathrm{sp}}(\mathrm{IC}(W_1), \ldots, \mathrm{IC}(W_n), W_1, \ldots, W_n) \ge \mathcal{H}_{\mathrm{sp}}(p_1, \ldots, p_n, W_1, \ldots, W_n),$$

as claimed. ∎

2.6 EXERCISES

E2.1   **(Proof of Lemma 2.2).** For $S = \{p_1, \ldots, p_n\} \in \mathbb{F}(\mathbb{R}^d)$ with $n \geq 2$, prove the following statements:

   (i) $\mathrm{CC}(S) \in \mathrm{co}(S) \setminus \mathrm{Ve}(\mathrm{co}(S))$;

   (ii) if $p \in \mathrm{co}(S) \setminus \{\mathrm{CC}(S)\}$ and $r \in \mathbb{R}_{>0}$ are such that $S \subset \overline{B}(p, r)$, then the segment $]p, \mathrm{CC}(S)[$ has a nonempty intersection with $\overline{B}(\frac{p+q}{2}, \frac{r}{2})$ for all $q \in \mathrm{co}(S)$.

   Hint: To show (i), invoke the definition of circumcenter. To show (ii), distinguish between the case when $\|p - q\|_2 < r$ and $\|p - q\|_2 = r$. A proof is contained in Cortés et al. (2006).

E2.2   **(The centroid of a convex set is an interior point).** Let S be a bounded measurable convex set in $\mathbb{R}^d$ and let $\phi : S \to \mathbb{R}_{>0}$ be a bounded measurable density function that is positive over S. Show that

$$\mathrm{CM}_\phi(S) \in \mathrm{int}(S).$$

E2.3   **(The inclusion $\mathcal{G}_{\mathrm{LD}}(r) \subset \mathcal{G}_{\mathrm{D}} \cap \mathcal{G}_{\mathrm{disk}}(r)$ is in general strict).** Consider the nodes $p_1 = (0, 0)$, $p_2 = (1, 0)$, and $p_3 = (2, \frac{1}{10})$. Pick $r = 3$ and perform the following tasks:

   (i) draw the three points, their Voronoi polygons and the disks centered at the points with radius r; and

   (ii) show that p_1 and p_3 are neighbors in the graph $\mathcal{G}_{\mathrm{D}} \cap \mathcal{G}_{\mathrm{disk}}(r)$, but not in the graph $\mathcal{G}_{\mathrm{LD}}(r)$.

E2.4   **(The proximity graph $\mathcal{G}_{\mathrm{D}} \cap \mathcal{G}_{\mathrm{disk}}(r)$ is not spatially distributed over $\mathcal{G}_{\mathrm{disk}}(r)$).** Consider the nodes $p_1 = (0, 0)$, $p_2 = (1, 0)$, $p_3 = (2, \frac{1}{10})$, and $p_4 = (0, \frac{31}{10})$. Compute the Voronoi partitions of the plane generated by $\{p_1, p_2, p_3\}$ and $\{p_1, p_2, p_3, p_4\}$. For $r = 3$, show that p_1 and p_3 are neighbors in the graph $\mathcal{G}_{\mathrm{D}} \cap \mathcal{G}_{\mathrm{disk}}(r)(\{p_1, p_2, p_3\})$ but not in the graph $\mathcal{G}_{\mathrm{D}} \cap \mathcal{G}_{\mathrm{disk}}(r)(\{p_1, p_2, p_3, p_4\})$. Why does this exercise illustrate that $\mathcal{G}_{\mathrm{D}} \cap \mathcal{G}_{\mathrm{disk}}(r)$ is not spatially distributed over $\mathcal{G}_{\mathrm{disk}}(r)$?

E2.5   **(1-center area problem).** Let $W \subset \mathbb{R}^2$ be a convex polygon, let ϕ be a density function on $\mathbb{R}^2$, and let $a \in \mathbb{R}_{>0}$. Assume that the a-contraction of W is nonempty. Consider the area function $\mathcal{H}_1 : W \to \mathbb{R}$, defined by

$$\mathcal{H}_1(p) = \int_{W \cap \overline{B}(p,a)} \phi(q)dq = \mathrm{A}_\phi(W \cap \overline{B}(p, a)).$$

Justify informally why, at points in the boundary of a convex polygon W, the gradient of $\mathcal{H}_1$ is non-vanishing, and points toward the interior of the polygon. (Note that it is not known whether the function $\mathcal{H}_1$ is concave and how to characterize critical points of $\mathcal{H}_1$ in geometric terms.)

E2.6   **(Concavity of performance function and 1-center function).** Given a performance function f, define the 1-center function $\mathcal{H}_{\mathrm{exp},1} : S \to \mathbb{R}$ by

$$\mathcal{H}_{\mathrm{exp},1}(p) = \int_S f(\|q - p\|_2)\phi(q)dq.$$

Prove the following facts:

(i) if f is concave, then $\mathcal{H}_{\mathrm{exp},1}$ is concave; and

(ii) if f is concave and decreasing and S has positive measure, then $\mathcal{H}_{\mathrm{exp},1}$ is strictly concave.

E2.7 **(Fermat–Weber center).** Let $S \subset \mathbb{R}^2$ be a convex polygon and let ϕ be a density function on S. Define the *Fermat–Weber function* $\mathcal{H}_{\mathrm{FW}} : \mathbb{R}^2 \to \mathbb{R}$ by

$$\mathcal{H}_{\mathrm{FW}}(p) = \int_S \|p - q\|_2 \phi(q) dq.$$

(i) Prove that $\mathcal{H}_{\mathrm{FW}}$ is strictly convex.

(ii) Show that $\mathcal{H}_{\mathrm{FW}}$ has a unique global minimum point inside S.

(iii) Compute the derivative of $\mathcal{H}_{\mathrm{FW}}$ and propose an algorithm to compute the global minimum point.

(iv) Is the function strictly convex even if the polygon S is not convex?

The unique minimum of $\mathcal{H}_{\mathrm{FW}}$ is called the *Fermat–Weber point* or, alternatively, the *median point* of the region S. Further details on this problem are available in Fekete et al. (2005) and references therein.

E2.8 **(Proof of Proposition 2.14).** In this exercise, you are asked to prove a statement that is slightly more general than Proposition 2.14. Let $\{W_1, \ldots, W_n\} \subset \mathbb{P}(S)$ be a partition of $S \subset \mathbb{R}^d$ and let ϕ be a density function on $\mathbb{R}^d$. Select $\{p_1, \ldots, p_n\}, \{\overline{p}_1, \ldots, \overline{p}_n\} \in \mathbb{F}(S)$ with the property that, for all $i \in \{1, \ldots, n\}$,

$$\|\overline{p}_i - \mathrm{CM}_\phi(W_i)\|_2 \leq \|p_i - \mathrm{CM}_\phi(W_i)\|_2.$$

Show that

$$\mathcal{H}_{\mathrm{dist}}(\overline{p}_1, \ldots, \overline{p}_n, W_1, \ldots, W_n) \geq \mathcal{H}_{\mathrm{dist}}(p_1, \ldots, p_n, W_1, \ldots, W_n),$$

and that the inequality is strict if there exists $i \in \{1, \ldots, n\}$ such that $\|\overline{p}_i - \mathrm{CM}_\phi(W_i)\|_2 < \|p_i - \mathrm{CM}_\phi(W_i)\|_2$ and such that W_i has positive area. *Hint: Use the expression of $\mathcal{H}_{\mathrm{dist}}$ in (2.3.5).*

E2.9 **(Mixed distortion-area multicenter function).** Show that the expected multicenter function $\mathcal{H}_{\mathrm{exp}}$ takes the form of $\mathcal{H}_{\mathrm{dist\text{-}area},a,b}$ stated in Section 2.3.1 when the performance function is

$$f(x) = -x^2 \, 1_{[0,a]}(x) + b \cdot 1_{]a,+\infty[}(x),$$

with $a \in \mathbb{R}_{>0}$ and $b \leq -a^2$.

Hint: As an intermediate step, show that for $P = (p_1, \ldots, p_n) \in S^n$, one has $V_i(P) \cap (S \setminus \overline{B}(p_i, a)) = V_i(P) \cap (S \setminus \cup_{k=1}^n \overline{B}(p_k, a))$ for all $i \in \{1, \ldots, n\}$.

E2.10 **(Proof of Proposition 2.15).** This exercise is a guided proof of Proposition 2.15. Let $W \subset \mathbb{R}^d$ be a connected set, let ϕ be a density function on $\mathbb{R}^d$, and let $a \in \mathbb{R}_{>0}$. For $p \in W$ and $\overline{B}$ a closed ball centered at a point in W with radius a, define $(p, \overline{B}) \mapsto \mathcal{H}_W(p, \overline{B})$ by

$$\mathcal{H}_W(p, \overline{B}) = -\int_{W \cap \overline{B}} \|q - p\|_2^2 \phi(q) dq - \int_{W \cap (S \setminus \overline{B})} a^2 \phi(q) dq.$$

Do the following:

(i) Show that the multicenter function $\mathcal{H}_{\text{dist-area},a}$ admits the expression

$$\mathcal{H}_{\text{dist-area},a}(p_1,\ldots,p_n,W_1,\ldots,W_n) = \sum_{i=1}^{n} \mathcal{H}_{W_i}(p_i,\overline{B}(p_i,a)).$$

(ii) Given a closed ball $\overline{B}$ centered at a point in W with radius a, show that for any $p \in W$,

$$\mathcal{H}_W(\text{CM}_\phi(W \cap \overline{B}),\overline{B}) \geq \mathcal{H}_W(p,\overline{B}),$$

with strict inequality unless $p = \text{CM}_\phi(W \cap \overline{B})$.
Hint: *Use the Parallel Axis Theorem (Hibbeler, 2006).*

(iii) Given $p \in W$, show that for any closed ball $\overline{B}$ centered at a point in W with radius a,

$$\mathcal{H}_W(p,\overline{B}(p,a)) \geq \mathcal{H}_W(p,\overline{B}).$$

Hint: *Consider the decomposition of W given by the union of the disjoint sets $\overline{B}(p,a) \cap \overline{B}$, $\overline{B}(p,a) \cap (W \setminus \overline{B})$, $(W \setminus \overline{B}(p,a)) \cap \overline{B}$ and $(W \setminus \overline{B}(p,a)) \cap (W \setminus \overline{B})$, and compare the integrals over each set.*

(iv) Deduce, using (ii) and (iii), that

$$\mathcal{H}_W(\text{CM}_\phi(W \cap \overline{B}(p,a)),\overline{B}(\text{CM}_\phi(W \cap \overline{B}(p,a)),a)) \geq \mathcal{H}_W(p,\overline{B}(p,a)),$$

with strict inequality unless $p = \text{CM}_\phi(W \cap \overline{B})$.

(v) Combine (i) and (iv) to prove Proposition 2.15.

E2.11   **(Locally cliqueless proximity graph).** Give an example of an allowable environment Q and a configuration of points such that the following inclusions (taken from Theorem 2.11(i)) are strict for $\mathcal{G} = \mathcal{G}_{\text{vis},Q}$:

$$\mathcal{G}_{\text{EMST},\mathcal{G}} \subseteq \mathcal{G}_{\text{lc},\mathcal{G}} \subseteq \mathcal{G},$$

E2.12   **(Properties of the locally cliqueless graph).** Prove Theorem 2.11.
Hint: *This exercise has notable theoretical content. To prove Theorem 2.11(i), use an argument by contradiction to show that the first inclusion holds, and use the definition of locally cliqueless graph to show that the second inclusion holds.*

E2.13   **(When are the total derivative and the partial derivative of a function equal?).** Assume that $f : \mathbb{R} \times \mathbb{R} \to \mathbb{R}$ is continuously differentiable in its both of its arguments and let $\partial_1 f$ be its partial derivative with respect to its first argument. Assume that the function $y^* : \mathbb{R} \to \mathbb{R}$ satisfies, for each $x \in \mathbb{R}$,

$$f(x,y^*(x)) = \max\{f(x,z) \mid z \in \mathbb{R}\},$$

and is continuously differentiable. Perform the following tasks:

(i) Show that

$$\frac{d}{dx} f(x,y^*(x)) = \partial_1 f(x,y^*(x)). \tag{E2.1}$$

(ii) Explain how this result gives an insight into the expression of the gradient of $\mathcal{H}_{\text{exp}}$ in Theorem 2.16(ii) for a continuously differentiable performance function. Also, explain why this formula is not directly applicable to the function $\mathcal{H}_{\text{exp}}$.

Note that equation (E2.1) is referred to as the *envelope theorem* in the economics literature.

E2.14 **(Distortion gradient ascent flow).** Given a (convex) polytope $S \subset \mathbb{R}^d$ and a density function ϕ, consider n nodes $p_1, \ldots, p_n$ evolving under the continuous-time gradient ascent flow of the multicenter function $\mathcal{H}_{\text{dist}}$,

$$\dot{p}_i = 2 \, \mathrm{A}_\phi(V_i(\mathcal{P}))(\mathrm{CM}_\phi(V_i(\mathcal{P})) - p_i), \quad i \in \{1, \ldots, n\}.$$

(i) What are the equilibrium points?

(ii) Show that $\mathcal{H}_{\text{dist}}$ is monotonically non-decreasing along the flow.

(iii) Show that the set S^N is invariant, i.e., that the trajectories of all nodes remain in S.

(iv) Use (i)–(iii) to apply the LaSalle Invariance Principle and show that the solutions of the flow converge to the set of centroidal Voronoi configurations in S.

(v) Implement numerically the flow in the software of your choice. Select the unit square $S = [0, 1] \times [0, 1]$ and the density function

$$\phi = \exp\left(-\left(x - \frac{1}{8}\right)^2 - \left(y - \frac{1}{8}\right)^2\right) + \exp\left(-\left(x - \frac{7}{8}\right)^2 - \left(y - \frac{7}{8}\right)^2\right).$$

Run simulations from different initial conditions and with different numbers of nodes. Show by illustration that multiple local maxima exist.

Hint: To perform step (iv), one should also prove that any two nodes never converge to the same location (in finite or infinite time); this property needs to be established because the function $\mathcal{H}_{\text{dist}}$ is not differentiable on such configurations. For this and the next exercise, do not worry about proving this property and instead refer to Cortés et al. (2005, Proposition 3.1).

E2.15 **(Area gradient ascent flow).** Given a (convex) polytope $S \subset \mathbb{R}^d$, a density function ϕ, and a radius $a \in \mathbb{R}_{>0}$, consider n nodes $p_1, \ldots, p_n$ evolving under the continuous-time gradient ascent flow of the multicenter function $\mathcal{H}_{\text{area},a}$,

$$\dot{p}_i = \int_{V_i(\mathcal{P}) \cap \partial \overline{B}(p_i, a)} \mathrm{n}_{\text{out}}(q)\phi(q)dq, \quad i \in \{1, \ldots, n\},$$

where n_{out} is the outward normal vector to the ball $\overline{B}(p_i, a)$.

(i) What are the equilibrium points?

(ii) Show that $\mathcal{H}_{\text{area},a}$ is monotonically non-decreasing along the flow.

(iii) Show that the set S^N is invariant, i.e., that the trajectories of all nodes remain in S.

(iv) Use (i-)-(iii) to apply the LaSalle Invariance Principle and show that the solutions of the flow converge to the set of a-limited area-centered Voronoi configurations in S.

(v) Implement numerically the flow in the software of your choice. Select the unit square $S = [0, 1] \times [0, 1]$, the density function

$$\phi(x, y) = \exp\left(-\left(x - \frac{1}{8}\right)^2 - \left(y - \frac{1}{8}\right)^2\right) + \exp\left(-\left(x - \frac{7}{8}\right)^2 - \left(y - \frac{7}{8}\right)^2\right),$$

and the parameter $a = \frac{1}{8}$. Run simulations from different initial conditions and with different numbers of nodes. Show by illustration that multiple local maxima exist.

Chapter Three

Robotic network models and complexity notions

This chapter introduces the main subject of study of this book, namely a model for groups of robots that sense their own position, exchange messages according to a geometric communication topology, process information, and control their motion. We refer to such systems as robotic networks. The content of this chapter has evolved from Martínez et al. (2007a).

The chapter is organized as follows. The first section contains the formal model. We begin by presenting the physical components of a network, that is, the mobile robots and the communication service connecting them. We then present the notion of control and communication law, and how a law is executed by a robotic network. These notions subsume the notions of synchronous network and distributed algorithm described in Section 1.5. As an example of these notions, we introduce a simple law, called the agree and pursue law, which combines ideas from leader election algorithms and from cyclic pursuit (i.e., a game in which robots chase each other in a circular environment). In the second section, we propose a model of groups of robots that interact through sensing, rather than communication. The third section discusses time, space, and communication complexity notions for robotic networks as extensions of the corresponding notions for distributed algorithms. The complexity notions rely on the basic concept of coordination task and task achievement. The fourth and last section establishes the time, space, and communication required by the agree and pursue law to steer a group of robots to a uniformly spaced rotating configuration. We end the chapter with three sections on, respectively, bibliographical notes, proofs of the results presented in the chapter, and exercises.

3.1 A MODEL FOR SYNCHRONOUS ROBOTIC NETWORKS

Here, we introduce a model for a synchronous robotic network. This model is an extension of the synchronous network model presented in Section 1.5.1. We start by detailing the physical components of the network, which include the robots themselves as well as the communication service among them.

3.1.1 Physical components

We start by providing a basic definition of a robot and a model for how each robot moves in space.

A *mobile robot* is a continuous-time continuous-space dynamical system as defined in Section 1.3, that is, a tuple (X, U, X_0, f), where

(i) X is d-dimensional space chosen among $\mathbb{R}^d$, $\mathbb{S}^d$, and the Cartesian products $\mathbb{R}^{d_1} \times \mathbb{S}^{d_2}$, for some $d_1 + d_2 = d$, called the *state space*;

(ii) U is a compact subset of $\mathbb{R}^m$ containing $\mathbf{0}_m$, called the *input space*;

(iii) X_0 is a subset of X, called the *set of allowable initial states*; and

(iv) $f : X \times U \to \mathbb{R}^d$ is a continuously differentiable control vector field on X, that is, f determines the robot motion $x : \mathbb{R}_{\geq 0} \to X$ via the differential equation, or control system,

$$\dot{x}(t) = f(x(t), u(t)), \tag{3.1.1}$$

subject to the control $u : \mathbb{R}_{\geq 0} \to U$.

We will use the terms "robot" and "agent" interchangeably. We refer to $x \in X$ and $u \in U$ as a *physical state* and an *input* of the mobile robot, respectively. Most often, the physical state will have the interpretation of a location, or a location and velocity. We will often consider control-affine vector fields. In such a case, we represent f as the ordered family of continuously differentiable vector fields $(f_0, f_1, \ldots, f_m)$ on X. In general, the control signal u will not depend only on time but also on x and possible other variables in the system. Note that there is no additional difficulty in modeling mobile robots using dynamical systems defined on manifolds (Bullo and Lewis, 2004), but we avoid it here in the interest of simplicity.

Example 3.1 (Planar vehicle models). The following models of control systems are commonly used in robotics, beginning with the early works of Dubins (1957), and Reeds and Shepp (1990). Figures 3.1(a) and (b) show a two-wheeled vehicle and a four-wheeled vehicle, respectively. The two-wheeled planar vehicle is described by the dynamical system

$$\dot{x} = v \cos\theta, \quad \dot{y} = v \sin\theta, \quad \dot{\theta} = \omega, \tag{3.1.2}$$

with state variables $x \in \mathbb{R}$, $y \in \mathbb{R}$, and $\theta \in \mathbb{S}^1$, describing the planar position and orientation of the vehicle, and with controls v and ω, describing the forward linear velocity and the angular velocity of the vehicle. Depending on which set the controls are restricted to, we define the following models:

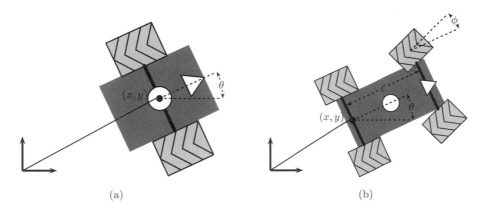

(a)   (b)

Figure 3.1 A two-wheeled vehicle (a) and four-wheeled vehicle (b). In each case, the orientation of the vehicle is indicated by the small triangle.

The unicycle. The controls v and ω take value in $[-1, 1]$ and $[-1, 1]$, respectively.

The differential drive robot. Set $v = (\omega_{\mathrm{right}} + \omega_{\mathrm{left}})/2$ and $\omega = (\omega_{\mathrm{right}} - \omega_{\mathrm{left}})/2$ and assume that both ω_{right} and ω_{left} take value in $[-1, 1]$.

The Reeds–Shepp car. The control v takes values in $\{-1, 0, 1\}$ and the control ω takes values in $[-1, 1]$.

The Dubins vehicle. The control v is set equal to 1 and the control ω takes value in $[-1, 1]$.

Finally, the four-wheeled planar vehicle, composed of a front and a rear axle separated by a distance ℓ, is described by the same dynamical system (3.1.2) with the following distinctions: $(x, y) \in \mathbb{R}^2$ is the position of the midpoint of the rear axle, $\theta \in \mathbb{S}^1$ is the orientation of the rear axle, the control v is the forward linear velocity of the rear axle, and the angular velocity satisfies $\omega = \dfrac{v}{\ell} \tan \phi$, where the control ϕ is the steering angle of the vehicle.   •

Next, we generalize the notion of synchronous network introduced in Definition 1.38 and introduce a corresponding notion of robotic network.

Definition 3.2 (Robotic network). The physical components of a *robotic network* $\mathcal{S}$ consist of a tuple $(I, \mathcal{R}, E_{\mathrm{cmm}})$, where

(i) $I = \{1, \dots, n\}$, I is called the *set of unique identifiers (UIDs)*;

(ii) $\mathcal{R} = \{R^{[i]}\}_{i \in I} = \{(X^{[i]}, U^{[i]}, X_0^{[i]}, f^{[i]})\}_{i \in I}$ is a set of mobile robots;

(iii) E_{cmm} is a map from $\prod_{i \in I} X^{[i]}$ to the subsets of $I \times I$—this map is called the *communication edge map*.

Additionally, if all mobile robots are identical, that is, if $R^{[i]} = (X, U, X_0, f)$ for all $i \in \{1, \ldots, n\}$, then the robotic network is *uniform*.   •

Remarks 3.3 (Notational conventions and meaning of the communication edge map).

(i) Following the convention established in Section 1.5, we let the superscript $[i]$ denote the variables and spaces which correspond to the robot with unique identifier i; for instance, $x^{[i]} \in X^{[i]}$ and $x_0^{[i]} \in X_0^{[i]}$ denote the physical state and the initial physical state of robot $R^{[i]}$, respectively. We refer to $x = (x^{[1]}, \ldots, x^{[n]}) \in \prod_{i \in I} X^{[i]}$ as a *state* of the network.

(ii) The map $x \mapsto (I, E_{\mathrm{cmm}}(x))$ models the topology of the communication service among the robots: at a physical state $x = (x^{[1]}, \ldots, x^{[n]})$, two robots at locations $x^{[i]}$ and $x^{[j]}$ can communicate if and only if the pair (i, j) is an edge in $E_{\mathrm{cmm}}(x) = E_{\mathrm{cmm}}(x^{[1]}, \ldots, x^{[n]})$. Accordingly, we refer to $(I, E_{\mathrm{cmm}}(x))$ as the *communication graph* at x. When and which robots communicate is discussed in Section 3.1.2. As communication graphs, we will often adopt one of the proximity graphs discussed in Section 2.2, and in particular the (undirected) disk graph.   •

To make things concrete, let us present some examples of robotic networks that will be commonly used later.

Example 3.4 (First-order robots with range-limited communication). Consider a group of robots moving in $\mathbb{R}^d$, $d \geq 1$. As in Chapter 2, we let p denote a point in $\mathbb{R}^d$ and we let $\{p^{[1]}, \ldots, p^{[n]}\}$ denote the robot locations. Assume that the robots move according to

$$\dot{p}^{[i]}(t) = u^{[i]}(t), \tag{3.1.3}$$

with $u^{[i]} \in [-u_{\max}, u_{\max}]^d$; for an illustration, see Figure 3.2. According to our mobile robot notation, these are identical robots of the form

$$(\mathbb{R}^d, [-u_{\max}, u_{\max}]^d, \mathbb{R}^d, (\mathbf{0}_d, e_1, \ldots, e_d)).$$

We assume that each robot can sense its own position and can communicate with any other robot within distance r, that is, we adopt the r-disk graph $\mathcal{G}_{\mathrm{disk}}(r)$ defined in Section 2.2 as communication graph. These data define the uniform robotic network $\mathcal{S}_{\mathrm{disk}}$.

It will also be interesting to consider first-order robots with communication graphs other than the disk graph; important examples include the Delaunay graph $\mathcal{G}_{\mathrm{D}}$, the limited Delaunay graph $\mathcal{G}_{\mathrm{LD}}(r)$, and the ∞-disk

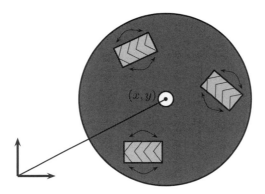

Figure 3.2 An omnidirectional vehicle. In addition to controlling the rotation speed of the wheels, the vehicle can also actuate the direction in which they point. This allows the vehicle to move in any direction according to the first-order dynamics (3.1.3).

graph $\mathcal{G}_{\infty\text{-disk}}(r)$, discussed in Section 2.2. These three graphs, adopted as communication models, give rise to three robotic networks denoted $\mathcal{S}_\text{D}$, and $\mathcal{S}_\text{LD}$, $\mathcal{S}_{\infty\text{-disk}}$, respectively.   •

Example 3.5 (Planar vehicle robots with Delaunay communication). We consider a group of vehicle robots moving in an allowable environment $Q \subset \mathbb{R}^2$ according to the planar vehicle dynamics introduced in Example 3.1. We let $\{(p^{[1]}, \theta^{[1]}), \dots, (p^{[n]}, \theta^{[n]})\}$ denote the robot physical states, where $p^{[i]} = (x^{[i]}, y^{[i]}) \in Q$ corresponds to the position and $\theta^{[i]} \in \mathbb{S}^1$ corresponds to the orientation of the robot $i \in I$. As the communication graph, we adopt the Delaunay graph $\mathcal{G}_\text{D}$ on Q introduced in Section 2.2. These data define the uniform robotic network $\mathcal{S}_\text{vehicles}$.   •

Example 3.6 (Robots with line-of-sight communication). We consider a group of robots moving in an allowable environment $Q \subset \mathbb{R}^2$. As in Example 3.4, we let $\{p^{[1]}, \dots, p^{[n]}\}$ denote the robot locations and we assume that the robots move according to the motion model (3.1.3). Each robot can sense its own position and the boundary of ∂Q, and can communicate with any other robot within distance r and within line of sight, that is, we adopt the range-limited visibility graph $\mathcal{G}_{\text{vis-disk},Q}$ in Q defined in Section 2.2 as the communication graph. These data define the uniform robotic network $\mathcal{S}_\text{vis-disk}$.   •

Example 3.7 (First-order robots in $\mathbb{S}^1$). Consider a group of n robots $\{\theta^{[1]}, \dots, \theta^{[n]}\}$ in $\mathbb{S}^1$, moving along on the unit circle with an angular velocity equal to the control input. Each identical robot is described by the tuple $(\mathbb{S}^1, [-u_\text{max}, u_\text{max}], \mathbb{S}^1, (0, e))$, where e is the vector field on $\mathbb{S}^1$ describing unit-speed counterclockwise rotation. As in the previous examples, we assume that each robot can sense its own position and can communicate with any other robot within distance r along the circle, that is, we adopt the r-disk graph $\mathcal{G}_{\text{disk}}(r)$ on $\mathbb{S}^1$ defined in Section 2.2 as the communication graph. These data define the uniform robotic network $\mathcal{S}_{\text{circle}}$.   •

We conclude this section with a remark.

Remark 3.8 (Congestion models in robotic networks). The behavior of a robotic network might be affected by communication and physical congestion problems.

Communication congestion: Omnidirectional wireless transmissions interfere. Clear reception of a signal requires that no other signals are present at the same point in time and space. In an *ad hoc* network, node i receives a message transmitted by node j only if all other neighbors of i are silent. In other words, the transmission medium is shared among the agents. As the density of agents increases, so does wireless communication congestion. The following asymptotic and optimization results are known.

First, for *ad hoc* networks with n uniformly randomly placed nodes, it is known (Gupta and Kumar, 2000) that the maximum-throughput communication range $r(n)$ of each node decreases as the density of nodes increases; in d dimensions, the appropriate scaling law is $r(n) \in \Theta\big((\log(n)/n)^{1/d}\big)$. This is referred to as the *connectivity regime* in percolation theory and statistical mechanics. Using the k-nearest-neighbor graph over uniformly placed nodes, the analysis in Xue and Kumar (2004) suggests that the minimal number of neighbors in a connected network grows with $\log(n)$.

Second, a growing body of literature (Santi, 2005; Lloyd et al., 2005) is available on *topology control*, that is, on how to compute transmission power values in an *ad hoc* network so as to minimize energy consumption and interference (due to multiple sources), while achieving various graph topological properties, such as connectivity or low network diameter.

Physical congestion. Robots can collide: it is clearly important to avoid "simultaneous access to the same physical area" by multiple robots. It is reasonable to assume that, as the number of robots increases, so should the area available for their motion. A convenient alternative approach is the one taken by Sharma et al. (2007), where robots' safety zones decrease with decreasing robot speed. This suggests that, in a fixed environment, individual nodes of a large ensemble have to move at a speed decreasing with n, and in particular, at a speed proportional to $n^{-1/d}$. Roughly speaking, if the overall volume V in which the groups of agents move is constant, and there are n robots, then the speed v at which they can move goes approximately as $v^d \approx \frac{V}{n}$.

In summary, one way to incorporate congestion effects into the robotic network model is to assume that the parameters of the network physical components depend upon the number of robots n. In the limit as $n \to +\infty$, we will sometimes assume that r and $u_{\max}$, the communication range and the velocity upper bound in Examples 3.4 and 3.7, are of order $n^{-1/d}$. •

3.1.2 Control and communication laws

Here, we present a discrete-time communication, continuous-time motion model for the evolution of a robotic network subject to a communication and control law. In our model, each robot evolves in the physical domain in continuous time, senses its position in continuous time, and, in discrete time, exchanges information with other robots and executes a state machine, which we shall refer to as a processor. The following definition is a generalization of the concept of distributed algorithm introduced in Definition 1.39 and of the classical notion of dynamical feedback controller.

Definition 3.9 (Control and communication law). A *control and communication law* $\mathcal{CC}$ for a robotic network $\mathcal{S}$ consists of the sets:

(i) $\mathbb{A}$, a set containing the `null` element, called the *communication alphabet*—elements of $\mathbb{A}$ are called *messages*;

(ii) $W^{[i]}$, $i \in I$, called the *processor state sets*; and

(iii) $W_0^{[i]} \subseteq W^{[i]}$, $i \in I$, sets of *allowable initial values*;

and of the following maps:

(i) $\mathrm{msg}^{[i]} : X^{[i]} \times W^{[i]} \times I \to \mathbb{A}$, $i \in I$, called *message-generation functions*;

(ii) $\mathrm{stf}^{[i]} : X^{[i]} \times W^{[i]} \times \mathbb{A}^n \to W^{[i]}$, $i \in I$, called *(processor) state-transition functions*; and

(iii) $\mathrm{ctl}^{[i]} : X^{[i]} \times X^{[i]} \times W^{[i]} \times \mathbb{A}^n \to U^{[i]}$, $i \in I$, called *(motion) control functions*.

If $\mathcal{S}$ is uniform and if $W^{[i]} = W$, $\mathrm{msg}^{[i]} = \mathrm{msg}$, $\mathrm{stf}^{[i]} = \mathrm{stf}$, and $\mathrm{ctl}^{[i]} = \mathrm{ctl}$, for all $i \in I$, then $\mathcal{CC}$ is said to be *uniform* and is described by a tuple $(\mathbb{A}, W, \{W_0^{[i]}\}_{i \in I}, \mathrm{msg}, \mathrm{stf}, \mathrm{ctl})$. •

We will sometimes refer to a control and communication law as a *distributed motion coordination algorithm*. Roughly speaking, the rationale behind Definition 3.9 is as follows (see Figure 3.3). The state of robot i in-

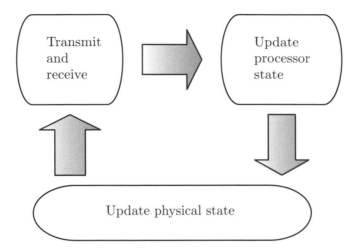

Figure 3.3 The execution of a control and communication law by a robotic network.

cludes both the physical state $x^{[i]} \in X^{[i]}$ and the *processor state* $w^{[i]} \in W^{[i]}$ of the state machine that robot i implements. These states are initialized with values in their corresponding allowable initial sets $X_0^{[i]}$ and $W_0^{[i]}$. We assume that the robot can sense it own physical position $x^{[i]}$. At each time instant $\ell \in \mathbb{Z}_{\geq 0}$, robot i sends to each of its out-neighbors j in the communication digraph $(I, E_{\mathrm{cmm}}(x))$ a message (possibly the `null` message) computed by applying the message-generation function $\mathrm{msg}^{[i]}$ to the current values of its physical state $x^{[i]}$ and processor state $w^{[i]}$, and to the identity j. Subsequently, but still at the time instant $\ell \in \mathbb{Z}_{\geq 0}$, robot i updates the value of its processor state $w^{[i]}$ by applying the state-transition function $\mathrm{stf}^{[i]}$ to the current value of its physical state $x^{[i]}$, processor state $w^{[i]}$ and to the messages it receives from its in-neighbors. Between communication instants, that is, for $t \in [\ell, \ell + 1)$ for some $\ell \in \mathbb{Z}_{\geq 0}$, the motion of the ith robot is determined by applying the control function to the current value of $x^{[i]}$, the value of $x^{[i]}$ at time ℓ, the current value of $w^{[i]}$, and the messages received at time ℓ. This evolution model is very similar to the one that we introduced for synchronous networks in Definition 1.40: in each communication round, the first step is transmission and the second one is computation and, except for the dependence on the physical state x, the communication and state transition processes are identical.

These ideas are formalized in the following definition.

Definition 3.10 (Evolution of a robotic network). Let $\mathcal{CC}$ be a control and communication law for the robotic network $\mathcal{S}$. The *evolution* of $(\mathcal{S}, \mathcal{CC})$ from initial conditions $x_0^{[i]} \in X_0^{[i]}$ and $w_0^{[i]} \in W_0^{[i]}$, $i \in I$, is the collection of curves $x^{[i]} : \mathbb{R}_{\geq 0} \to X^{[i]}$ and $w^{[i]} : \mathbb{Z}_{\geq 0} \to W^{[i]}$, $i \in I$, defined by

$$\dot{x}^{[i]}(t) = f\left(x^{[i]}(t),\, \mathrm{ctl}^{[i]}\left(x^{[i]}(t), x^{[i]}(\lfloor t \rfloor), w^{[i]}(\lfloor t \rfloor), y^{[i]}(\lfloor t \rfloor)\right)\right),$$

where $\lfloor t \rfloor = \max\{\ell \in \mathbb{Z}_{\geq 0} \mid \ell < t\}$, and

$$w^{[i]}(\ell) = \mathrm{stf}^{[i]}(x^{[i]}(\ell), w^{[i]}(\ell - 1), y^{[i]}(\ell)),$$

with $x^{[i]}(0) = x_0^{[i]}$ and $w^{[i]}(-1) = w_0^{[i]}$, $i \in I$. In the previous equations, $y^{[i]} : \mathbb{Z}_{\geq 0} \to \mathbb{A}^n$ (describing the messages received by processor i) has components $y_j^{[i]}(\ell)$, for $j \in I$, defined by

$$y_j^{[i]}(\ell) = \begin{cases} \mathrm{msg}^{[j]}(x^{[j]}(\ell), w^{[j]}(\ell - 1), i), & \text{if } (j, i) \in E_{\mathrm{cmm}}\left(x^{[1]}(\ell), \dots, x^{[n]}(\ell)\right), \\ \texttt{null}, & \text{otherwise.} \end{cases}$$

$\bullet$

For convenience, we define $w(t) = w(\lfloor t \rfloor)$ for all $t \in \mathbb{R}_{\geq 0}$, and let $\mathbb{R}_{\geq 0} \ni t \mapsto (x(t), w(t))$ denote the curves $x^{[i]}$ and $w^{[i]}$, for $i \in \{1, \dots, n\}$.

Remarks 3.11 (Simplifications of control and communication laws).

(i) A control and communication law $\mathcal{CC}$ is *static* if the processor state set $W^{[i]}$ is a singleton for all $i \in I$. This means that there is no meaningful evolution of the processor state. In this case, $\mathcal{CC}$ can be described by a tuple $(\mathbb{A}, \{\mathrm{msg}^{[i]}\}_{i \in I}, \{\mathrm{ctl}^{[i]}\}_{i \in I})$, with $\mathrm{msg}^{[i]} : X^{[i]} \times I \to \mathbb{A}$, and $\mathrm{ctl}^{[i]} : X^{[i]} \times X^{[i]} \times \mathbb{A}^n \to U^{[i]}$, for $i \in I$.

(ii) A control and communication law $\mathcal{CC}$ is *data-sampled* if the control functions are independent of the current position of the robot and depend only upon the robot's position at the last sample time. Specifically, the control functions have the following property: given a processor state $w^{[i]} \in W^{[i]}$, an array of messages $y^{[i]} \in \mathbb{A}^n$, a current state $x^{[i]}$, and a state at last sample time $x_{\mathrm{smpld}}^{[i]}$, the control input $\mathrm{ctl}^{[i]}(x^{[i]}, x_{\mathrm{smpld}}^{[i]}, w^{[i]}, y^{[i]})$ is independent of $x^{[i]}$, for all $i \in I$. In this case, the control functions can be described by maps of the form $\mathrm{ctl}^{[i]} : X^{[i]} \times W^{[i]} \times \mathbb{A}^n \to U^{[i]}$, for $i \in I$.

(iii) In many control and communication laws, the robots exchange full information about their states, including both their processor and their physical states. For such laws, we identify the communication alphabet with $\mathbb{A} = (X \times W) \cup \{\texttt{null}\}$ and we refer to the corresponding message-generation function $\mathrm{msg}_{\mathrm{std}}(x, w, j) = (x, w)$ as the *standard message-generation function*.

Note that we allow the processor state set and the communication alphabet to contain an infinite number of symbols. In other words, we assume that a robot can store and transmit a (finite number of) integer and real numbers, among other things. This is equivalent to assuming that we neglect any inaccuracies due to quantization, as we did in Section 1.6.   •

Remark 3.12 (Extensions of control and communication laws). Here, we briefly discuss alternative models and extensions of the proposed models.

Asynchronous sensor-based interactions. In the early network model proposed by Suzuki and Yamashita (1999), robots are referred to as "anonymous" and "oblivious" in precisely the same way in which we defined control and communication laws to be uniform and static, respectively. As compared with our notion of robotic network, the model in Suzuki and Yamashita (1999) is more general in that the robots' activation schedules do not necessarily coincide (i.e., this model is asynchronous), and at the same time it is less general in that (1) robots cannot communicate any information other than their respective positions, and (2) each robot observes every other robot's position (i.e., the complete communication graph is adopted). In the Section 3.2 below, we present a model in which robots rely on sensing rather than communication for their interaction.

Discrete-time motion models. For some algorithms in later chapters, it will be convenient to consider discrete-time motion models; for example, we present discrete-time motion models for first-order agents in Section 4.1. In some other cases, it will be convenient to consider dynamical interactions between agents taking place in continuous time.

Stochastic link models. Although we do not present any results on this topic in this notes, it is possible to develop robotic networks models over random graphs and random geometric graphs, as studied by Bollobás (2001) and Penrose (2003). Furthermore, it is of interest to consider communication links with time-varying rates.   •

3.1.3 The agree and pursue control and communication law

We conclude this section with an example of a dynamic control and communication law. The problem is described as follows: a collection of robots with range-limited communication are placed on the unit circle; the robots move and communicate with the objectives of (1) agreeing on a direction of motion (clockwise or counterclockwise) and (2) achieving an equidistant configuration where all robots are equally angularly spaced. To achieve these two objectives, we combine ideas from leader election algorithms for synchronous networks (see Section 1.5.4) and from cyclic pursuit problems (see Exercise E1.30): the robots move a distance proportional to an appropriate inter-robot separation, and they repeatedly compare their identifiers to discover the direction of motion of the robot with the largest identifier. In other words, the robots run a leader election task in their processor states and a uniform robotic deployment task in their physical state—these are among the most basic tasks in distributed algorithms and cooperative control. We present the algorithm here and characterize its correctness and performance later in the chapter.

From Example 3.7, we consider the uniform network $\mathcal{S}_{\text{circle}}$ of locally connected first-order robots on $\mathbb{S}^1$. For $r, u_{\max}, k_{\text{prop}} \in]0, \frac{1}{2}[$ with $k_{\text{prop}} r \leq u_{\max}$, we define the AGREE & PURSUE law, denoted by $\mathcal{CC}_{\text{AGREE \& PURSUE}}$, as the *uniform data-sampled* law loosely described as follows:

> *[Informal description]* The processor state consists of dir (the robot's direction of motion) taking values in $\{c, cc\}$ (meaning clockwise and counterclockwise) and max-id (the largest UID received by the robot, initially set to the robot's UID) taking values in I. In each communication round, each robot transmits its position and its processor state. Among the messages received from agents moving toward its position, each agent picks the message with the largest value of max-id. If this value is larger than its own value, the agent resets its processor state with the selected message. Between communication rounds, each robot moves in the clockwise or counterclockwise direction depending on whether its processor state dir is c or cc. Each robot moves k_{prop} times the distance to the immediately next neighbor in the chosen direction, or, if no neighbors are detected, k_{prop} times the communication range r.

Note that the processor state with the largest UID will propagate throughout the network as in the FLOODMAX ALGORITHM for leader election. Also, note that the assumption $k_{\text{prop}} r \leq u_{\max}$ guarantees that the desired control is always within the allowable range $[-u_{\max}, u_{\max}]$. Next, we define the law formally:

Robotic Network: $\mathcal{S}_{\text{circle}}$, first-order agents in $\mathbb{S}^1$
with absolute sensing of own position, and
with communication range r

Distributed Algorithm: AGREE & PURSUE

Alphabet: $\mathbb{A} = \mathbb{S}^1 \times \{c, cc\} \times I \cup \{\texttt{null}\}$

`Processor State:` $w = (\texttt{dir}, \texttt{max-id})$, where

  `dir` $\in \{\texttt{c}, \texttt{cc}\}$,   initially: $\texttt{dir}^{[i]}$ unspecified

  `max-id` $\in I$,   initially: $\texttt{max-id}^{[i]} = i$ for all i

% Standard message-generation function

`function` $\mathrm{msg}(\theta, w, i)$

  1: **return** (θ, w)

`function` $\mathrm{stf}(\theta, w, y)$

  1: **for** each non-**null** message $(\theta_{\mathrm{rcvd}}, (\texttt{dir}_{\mathrm{rcvd}}, \texttt{max-id}_{\mathrm{rcvd}}))$ in y **do**

  2:     **if** $(\texttt{max-id}_{\mathrm{rcvd}} > \texttt{max-id})$ AND $(\mathrm{dist}_{\texttt{cc}}(\theta, \theta_{\mathrm{rcvd}}) \leq r$ AND $\texttt{dir}_{\mathrm{rcvd}} =$
       c) OR $(\mathrm{dist}_{\texttt{c}}(\theta, \theta_{\mathrm{rcvd}}) \leq r$ AND $\texttt{dir}_{\mathrm{rcvd}} = \texttt{cc})$ **then**

  3:       $\texttt{new-dir} := \texttt{dir}_{\mathrm{rcvd}}$

  4:       $\texttt{new-id} := \texttt{max-id}_{\mathrm{rcvd}}$

  5: **return** $(\texttt{new-dir}, \texttt{new-id})$

`function` $\mathrm{ctl}(\theta_{\mathrm{smpld}}, w, y)$

  1: $d_{\mathrm{tmp}} := r$

  2: **for** each non-**null** message $(\theta_{\mathrm{rcvd}}, (\texttt{dir}_{\mathrm{rcvd}}, \texttt{max-id}_{\mathrm{rcvd}}))$ in y **do**

  3:     **if** $(\texttt{dir} = \texttt{cc})$ AND $(\mathrm{dist}_{\texttt{cc}}(\theta_{\mathrm{smpld}}, \theta_{\mathrm{rcvd}}) < d_{\mathrm{tmp}})$ **then**

  4:       $d_{\mathrm{tmp}} := \mathrm{dist}_{\texttt{cc}}(\theta_{\mathrm{smpld}}, \theta_{\mathrm{rcvd}})$

  5:       $u_{\mathrm{tmp}} := k_{\mathrm{prop}} d_{\mathrm{tmp}}$

  6:     **if** $(\texttt{dir} = \texttt{c})$ AND $(\mathrm{dist}_{\texttt{c}}(\theta_{\mathrm{smpld}}, \theta_{\mathrm{rcvd}}) < d_{\mathrm{tmp}})$ **then**

  7:       $d_{\mathrm{tmp}} := \mathrm{dist}_{\texttt{c}}(\theta_{\mathrm{smpld}}, \theta_{\mathrm{rcvd}})$

  8:       $u_{\mathrm{tmp}} := -k_{\mathrm{prop}} d_{\mathrm{tmp}}$

  9: **return** u_{tmp}

    An implementation of this control and communication law is shown in Figure 3.4. As parameters, we select $n = 45$, $r = 2\pi/40$, $u_{\max} = 1/4$ and $k_{\mathrm{prop}} = 7/16$. Along the evolution, all robots agree upon a common direction of motion and, after a suitable time, they reach a uniform distribution.

Figure 3.4 The AGREE & PURSUE law. Disks and circles correspond to robots moving counterclockwise and clockwise, respectively. The initial positions and the initial directions of motion are randomly generated. The five pictures depict the state of the network at times 0, 9, 20, 100, and 800.

3.2 ROBOTIC NETWORKS WITH RELATIVE SENSING

The model presented above assumes the ability of each robot to know its own absolute position. Here, we treat the alternative setting in which the robots do not communicate amongst themselves, but instead detect and measure each other's relative position through appropriate sensors. Additionally, we assume that the robots will perform measurements of the environment without having any *a priori* knowledge of it. We assume that robots do not have the ability to perform measurements expressed in a common reference frame. An early reference in which relative information is adopted is Lin et al. (2005).

3.2.1 Kinematics notions

Because the robots do not have a common reference frame, all the measurements generated by their on-board sensors are expressed in a local reference frame. To formalize this fact, it is useful to review some basic kinematics conventions. We let $\Sigma^{\text{fixed}} = (p^{\text{fixed}}, \{\boldsymbol{x}^{\text{fixed}}, \boldsymbol{y}^{\text{fixed}}, \boldsymbol{z}^{\text{fixed}}\})$ be a fixed reference frame in $\mathbb{R}^3$. A point q, a vector v, and a set of points S expressed with respect to the frame Σ^{fixed} are denoted by q_{fixed}, v_{fixed} and S_{fixed}, respectively. Next, let $\Sigma^{\text{b}} = (p^{\text{b}}, \{\boldsymbol{x}^{\text{b}}, \boldsymbol{y}^{\text{b}}, \boldsymbol{z}^{\text{b}}\})$ be a reference frame fixed to

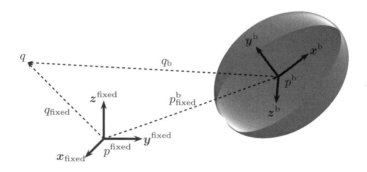

Figure 3.5 Inertially fixed and body-fixed frames in $\mathbb{R}^3$.

a moving body. The origin of Σ^{b} is the point p^{b}, denoted by $p^{\text{b}}_{\text{fixed}}$ when expressed with respect to Σ^{fixed}. The orientation of Σ^{b} is characterized by the d-dimensional rotation matrix $R^{\text{b}}_{\text{fixed}}$, whose columns are the frame vectors $\{\boldsymbol{x}^{\text{b}}, \boldsymbol{y}^{\text{b}}, \boldsymbol{z}^{\text{b}}\}$ of Σ^{b} expressed with respect to Σ^{fixed}. We recall here the definition of the group of rotation matrices in d-dimensions:

$$\mathrm{SO}(d) = \{R \in \mathbb{R}^{d\times d} \mid RR^T = I_d,\ \det(R) = +1\}.$$

With these notations, changes of reference frames are described by

$$q_{\text{fixed}} = R_{\text{fixed}}^{\text{b}} q_{\text{b}} + p_{\text{fixed}}^{\text{b}},$$
$$v_{\text{fixed}} = R_{\text{fixed}}^{\text{b}} v_{\text{b}},$$
$$S_{\text{fixed}} = R_{\text{fixed}}^{\text{b}} S_{\text{b}} + p_{\text{fixed}}^{\text{b}}. \tag{3.2.1}$$

Note that these change-of-frames formulas also hold in the planar case with the corresponding definition of the rotation matrix in SO(2).

Remark 3.13 (Comparison with literature). In our notation, the subscript denotes the frame with respect to which the quantity is expressed. Other references in the literature sometimes adopt the opposite convention, in which the superscript denotes the frame with respect to which the quantity is expressed. •

3.2.2 Physical components

In what follows, we describe our notion of *mobile robots equipped with relative sensors*. We consider a group of n robots moving in an allowable environment $Q \subset \mathbb{R}^d$, for $d \in \{2, 3\}$, and we assume that a reference frame $\Sigma^{[i]}$, for $i \in \{1, \ldots, n\}$, is attached to each robot (see Figure 3.6). Expressed with respect to the fixed frame Σ^{fixed}, the ith frame $\Sigma^{[i]}$ is described by a position $p_{\text{fixed}}^{[i]} \in \mathbb{R}^d$ and an orientation $R_{\text{fixed}}^{[i]} \in \text{SO}(d)$. The continuous-time motion and discrete-time sensing models are described as follows.

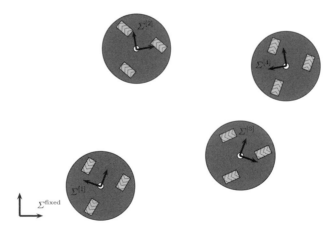

Figure 3.6 A robotic network with relative sensing. A group of four robots moves in $\mathbb{R}^2$. Each robot $i \in \{1, \ldots, 4\}$ has its own reference frame $\Sigma^{[i]}$.

Motion model: We select a simple motion model: for all $t \in \mathbb{R}_{\geq 0}$, the orientation $R^{[i]}_{\text{fixed}}$ is constant in time and robot i translates according to

$$\dot{p}^{[i]}_{\text{fixed}}(t) = R^{[i]}_{\text{fixed}} u^{[i]}_i, \tag{3.2.2}$$

that is, the ith control input $u^{[i]}_i$ is known and applied in the robot frame. Each control input $u^{[i]}_i$, $i \in \{1,\dots,n\}$, takes values in a compact input space U. Clearly, it would be possible to consider a motion model with time-varying orientation and we refer the reader to Exercise E3.1, where we do so.

Sensing model: At each discrete time instant, robot i activates a sensor that detects the presence and returns a measurement about the relative position of any object (robots or environment boundary) inside a given "sensor footprint." We describe the model in two steps. First, each robot measures

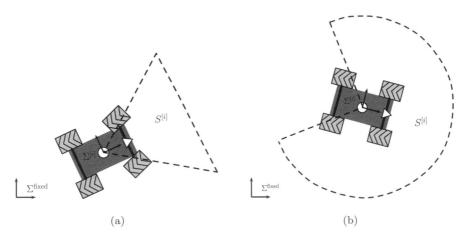

(a)   (b)

Figure 3.7 Examples of sensor footprints. (a) The cone-shaped sensor footprint of a vehicle equipped with a camera. (b) The 270-degree wedge-shaped sensor footprint of a vehicle equipped with a laser scanner.

other robots' positions and the environment as follows.

Sensing other robots' positions. There exists a set $\mathbb{A}_{\text{rbt}}$ containing the null element, called the *sensing alphabet*, and a map rbt-sns : $\mathbb{R}^d \to \mathbb{A}_{\text{rbt}}$, called the *sensing function*, with the interpretation that robot i acquires the symbol rbt-sns$(p^{[j]}_i) \in \mathbb{A}_{\text{rbt}}$ for each robot $j \in \{1,\dots,n\} \setminus \{i\}$.

Sensing the environment. There exists a set $\mathbb{A}_{\text{env}}$ containing the null element, called the *environment sensing alphabet*, and a map env-sns : $\mathbb{P}(\mathbb{R}^d) \to \mathbb{A}_{\text{env}}$, called the *environment sensing function*, with the interpretation that robot i acquires the symbol env-sns$(Q_i) \in \mathbb{A}_{\text{env}}$.

Second, we let $S^{[i]} \subset \mathbb{R}^d$ be the *sensor footprint* of robot i and we let $S_i^{[i]}$ be its expression in the frame $\Sigma^{[i]}$ (see Figure 3.7). For simplicity, we assume that all robot sensors are equal, so that we can write $S_i^{[i]} = S$. We require both sensing functions to provide no information about robots and boundaries that are outside S in the following two meanings: (i) if p is any point outside S, then rbt-sns$(p) = $ null; and (ii) if W is any subset of $\mathbb{R}^d$, env-sns$(W) = $ env-sns$(W \cap S)$.

We summarize this discussion with the following definition.

Definition 3.14 (Network with relative sensing). The physical components of a *network with relative sensing* consist of n mobile robots with identifiers $\{1, \ldots, n\}$, with configurations in $Q \times \mathrm{SO}(d)$, for an allowable environment $Q \subset \mathbb{R}^d$, with dynamics described by equation (3.2.2), and with relative sensors described by the sensor footprint S, sensing alphabets $\mathbb{A}_{\mathrm{rbt}}$ and $\mathbb{A}_{\mathrm{env}}$, and sensing functions rbt-sns and env-sns.   •

To make things concrete, let us present two examples of robotic networks with relative sensing that are analogs of the "communication-based" robotic networks $\mathcal{S}_{\mathrm{disk}}$ and $\mathcal{S}_{\mathrm{vis\text{-}disk}}$ in Examples 3.4 and 3.6.

Example 3.15 (Disk sensor and corresponding relative-sensing network). Given a *sensing range* $r \in \mathbb{R}_{>0}$, the *disk sensor* has sensor footprint $\overline{B}(\mathbf{0}_d, r)$, that is, a disk sensor measures any object (robot and environment boundary) within distance r. Regarding sensing of other robots, we assume that the alphabet is $\mathbb{A}_{\mathrm{rbt}} = \mathbb{R}^d \cup \{$null$\}$ and that the sensing function is rbt-sns$(p_i^{[j]}) = p_i^{[j]}$ for each robot $j \in \{1, \ldots, n\} \setminus \{i\}$, inside the sensor footprint $\overline{B}(\mathbf{0}_d, r)$, and rbt-sns$(p_i^{[j]}) = $ null, otherwise. Regarding sensing of the environment, we assume that the alphabet is $\mathbb{A}_{\mathrm{env}} = \mathbb{P}(\mathbb{R}^d)$ and that the sensing function is env-sns$(Q_i) = Q_i \cap \overline{B}(\mathbf{0}_d, r)$. A group of robots with disk sensors defines the robotic network with relative sensing $\mathcal{S}_{\mathrm{disk}}^{\mathrm{rs}}$.   •

Example 3.16 (Range-limited visibility sensor and corresponding relative-sensing network). Given a *sensing range* $r \in \mathbb{R}_{>0}$, the *range-limited visibility sensor* has sensor footprint $\overline{B}(\mathbf{0}_d, r)$ and performs measurements only of objects within unobstructed line of sight. Regarding sensing of other robots, we assume that the alphabet is $\mathbb{A}_{\mathrm{rbt}} = \mathbb{R}^d \cup \{$null$\}$ and that the sensing function is rbt-sns$(p_i^{[j]}) = p_i^{[j]}$ for each robot $j \in \{1, \ldots, n\} \setminus \{i\}$, inside the range-limited visibility set $\mathrm{Vi}_{\mathrm{disk}}(\mathbf{0}_2; Q_i)$, and rbt-sns$(p_i^{[j]}) = $ null, otherwise. Regarding sensing of the environment, we assume[1] that the alphabet is $\mathbb{A}_{\mathrm{env}} = \mathbb{P}(\mathbb{R}^d)$ and that the environment sensor measures the range-

---

[1] It would be equivalent to assume that the robot can sense every portion of ∂Q that is within distance r and that is visible from the robot's position.

limited visibility set $\mathrm{Vi}_{\mathrm{disk}}(p_{\mathrm{fixed}}^{[i]}; Q)$ expressed with respect to the frame $\Sigma^{[i]}$; for the definition of range-limited visibility set, see Section 2.1.2. In other words, the environment sensing function is $\mathrm{env\text{-}sns}(Q_i) = \mathrm{Vi}_{\mathrm{disk}}(\mathbf{0}_2; Q_i)$. This is illustrated in Figure 3.8. A group of robots with range-limited visi-

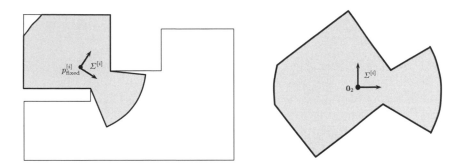

Figure 3.8 The left-hand plot depicts the range-limited visibility set $\mathrm{Vi}_{\mathrm{disk}}(p_{\mathrm{fixed}}^{[i]}; Q)$ expressed with respect to an inertially fixed frame. The right-hand plot depicts the range-limited visibility set expressed with respect to the body-fixed frame $\Sigma^{[i]}$, that is, $\mathrm{Vi}_{\mathrm{disk}}(\mathbf{0}_2; Q_i)$.

bility sensors defines the robotic network with relative sensing $\mathcal{S}_{\mathrm{vis\text{-}disk}}^{\mathrm{rs}}$.   •

Remark 3.17 (Sensing model consequences). The proposed sensing model has the following two consequences:

(i) Robots have no information about the absolute position and orientation of themselves, the other robots or any part of the environment.

(ii) The relative sensing capacity of the robots gives rise to a proximity graph, called the *sensing graph*, whose edges are the collection of robot pairs that are within sensing range. For example, in the network $\mathcal{S}_{\mathrm{disk}}^{\mathrm{rs}}$, the sensing graph is the disk graph $\mathcal{G}_{\mathrm{disk}}(r)$. In general, sensing graphs are directed.   •

3.2.3 Relative-sensing control laws

As we did for robotic networks with interactions based on communication, we define here control laws based on relative sensing and we describe the closed-loop evolution of robotic networks with relative sensing.

First, we consider a robotic network with relative sensing $\mathcal{S}^{\mathrm{rs}}$ characterized by: identifiers $\{1, \dots, n\}$, configurations in $Q \times \mathrm{SO}(d)$, for an allowable environment $Q \subset \mathbb{R}^d$, dynamics described by equation (3.2.2), and relative sensors described by the sensor footprint S, sensing alphabets $\mathbb{A}_{\mathrm{rbt}}$ and

$\mathbb{A}_{\text{env}}$, and sensing functions rbt-sns and env-sns. A *relative-sensing control law* $\mathcal{RSC}$ for the robotic network with relative sensing $\mathcal{S}^{\text{rs}}$ consists of the following tuple:

(i) W, called the *processor state set*, with a corresponding set of *allowable initial values* $W_0 \subseteq W$;

(ii) stf : $W \times \mathbb{A}_{\text{rbt}}^n \times \mathbb{A}_{\text{env}} \to W$, called the *(processor) state-transition function*; and

(iii) ctl : $W \times \mathbb{A}_{\text{rbt}}^n \times \mathbb{A}_{\text{env}} \to U$, called the *(motion) control function*.

As for robotic networks, we say that $\mathcal{RSC}$ is *static* if W is a singleton for all $i \in \{1, \dots, n\}$; in this case, $\mathcal{RSC}$ can be described by a motion control function ctl : $\mathbb{A}_{\text{rbt}}^n \times \mathbb{A}_{\text{env}} \to U$. Additionally, if the environment $Q = \mathbb{R}^d$, then $\mathcal{RSC}$ can be described by a motion control function ctl : $W \times \mathbb{A}_{\text{rbt}}^n \to U$.

Second, the *evolution* of $(\mathcal{S}^{\text{rs}}, \mathcal{RSC})$ from initial conditions $(p_0^{[i]}, R_{\text{fixed}}^{[i]}) \in \mathbb{R}^d \times \text{SO}(d)$ and $w_0^{[i]} \in W_0$, $i \in \{1, \dots, n\}$ is the collection of curves $p_{\text{fixed}}^{[i]} : \mathbb{R}_{\geq 0} \to \mathbb{R}^d$ and $w^{[i]} : \mathbb{Z}_{\geq 0} \to W$, $i \in \{1, \dots, n\}$, defined by

$$\dot{p}_{\text{fixed}}^{[i]}(t) = R_{\text{fixed}}^{[i]} \, \text{ctl}\big(w^{[i]}(\lfloor t \rfloor), y^{[i]}(\lfloor t \rfloor), y_{\text{env}}^{[i]}(\lfloor t \rfloor)\big),$$
$$w^{[i]}(\ell) = \text{stf}(w^{[i]}(\ell - 1), y^{[i]}(\ell), y_{\text{env}}^{[i]}(\ell)),$$

with $p_{\text{fixed}}^{[i]}(0) = p_0^{[i]}$ and $w^{[i]}(-1) = w_0^{[i]}$, $i \in \{1, \dots, n\}$. In the previous equations, $y^{[i]} : \mathbb{Z}_{\geq 0} \to \mathbb{A}_{\text{rbt}}^n$ (describing the robot measurements taken by sensor i) with components $y_j^{[i]}(\ell)$, for $j \in \{1, \dots, n\}$, and $y_{\text{env}}^{[i]} : \mathbb{Z}_{\geq 0} \to \mathbb{A}_{\text{env}}$ (describing the environment measurements taken by sensor i) are defined by

$$y_j^{[i]}(\ell) = \text{rbt-sns}(p_i^{[j]}(\ell)), \quad y_{\text{env}}^{[i]}(\ell) = \text{env-sns}(Q_i(\ell)).$$

In the last equation, $p_i^{[j]}$ and $Q_i(\ell)$ denote the position of the j-th robot and the environment Q as expressed with respect to the moving frame $\Sigma^{[i]}$.

3.2.4 Equivalence between control and communication laws and relative-sensing control laws

Consider a "communication-based" robotic network $\mathcal{S}_1$ with a control and communication law $\mathcal{CC}_1$ with the following properties:

(i) Regarding $\mathcal{S}_1$: the network is uniform, the state space is $X = \mathbb{R}^d$ with states denoted by $x^{[i]} = p^{[i]}$, the communication graph is the r-disk graph, and the robot dynamics are $\dot{p}^{[i]} = u^{[i]}$.

(ii) Regarding $\mathcal{CC}_1$: the control and communication law is uniform and data-sampled, the communication alphabet is $\mathbb{A} = \mathbb{R}^d \cup \{\texttt{null}\}$, and the message-generation function is $\text{msg}(p, w, j) = p$.

Given a network and a law $(\mathcal{S}_1, \mathcal{CC}_1)$ satisfying (i) and (ii), the control and communication law $\mathcal{CC}_1$ is *invariant* if its state transition and control maps satisfy, for all $p \in \mathbb{R}^d$, $w \in W$, $y \in \mathbb{A}^n$, and $R \in \text{SO}(d)$,

$$\text{stf}(p, w, y) = \text{stf}\big(\mathbf{0}_d, w, R(y - p)\big),$$
$$\text{ctl}(p, w, y) = R^T \text{ctl}\big(\mathbf{0}_d, w, R(y - p)\big),$$

where the ith component of $R(y - p) \in \mathbb{A}^n$ is $R(y_i - p)$ if $y_i \in \mathbb{R}^d$, or $\texttt{null}$ if $y_i = \texttt{null}$.

Next, consider a relative-sensing network $\mathcal{S}_2$ with disk sensors as in Example 3.15, that is, assume that the sensing footprint is $\overline{B}(\mathbf{0}_d, r)$, the sensing alphabet is $\mathbb{A}_{\text{rbt}} = \mathbb{R}^d \cup \{\texttt{null}\}$, and the sensing function equals the identity function in $\overline{B}(\mathbf{0}_d, r)$. We assume no environment sensing as we set $Q = \mathbb{R}^d$. The communication and control law $\mathcal{CC}_1$ and the relative-sensing control law $\mathcal{RSC}_2$ for network $\mathcal{S}_2$ are *equivalent* if their processor state sets identical, for example, denoting both by W, and their state transition and control maps satisfy, for all $w \in W$ and $y \in \mathbb{R}^d \cup \{\texttt{null}\} = \mathbb{A}^n = \mathbb{A}_{\text{rbt}}^n$,

$$\text{stf}_1(\mathbf{0}_d, w, y) = \text{stf}_2(w, y), \qquad \text{and} \qquad \text{ctl}_1(\mathbf{0}_d, w, y) = \text{ctl}_2(w, y).$$

Proposition 3.18 (Evolution equivalence). *If $\mathcal{CC}_1$ is invariant and if $\mathcal{CC}_1$ and $\mathcal{RSC}_2$ are equivalent, then the evolutions of the control and communication laws $(\mathcal{S}_1, \mathcal{CC}_1)$ and $(\mathcal{S}_2, \mathcal{RSC}_2)$ from identical initial conditions are identical.*

Proof. Assume that the messages and measurements array $y^{[i]}(t)$ received by the i-th robot at time t in the communication-based network and in the relative-sensing networks are equal to, respectively:

$$p_{\text{fixed}}^{[j_1]}, \dots, p_{\text{fixed}}^{[j_k]}, \qquad \text{and} \qquad p_i^{[j_1]}, \dots, p_i^{[j_k]}.$$

Under this assumption, the evolutions of the communication-based network and of the relative-sensing networks are written, respectively, as,

$$\dot{p}_{\text{fixed}}^{[i]} = \text{ctl}_1(p_{\text{fixed}}^{[i]}, w^{[i]}, p_{\text{fixed}}^{[j_1]}, \dots, p_{\text{fixed}}^{[j_k]}),$$
$$\dot{p}_{\text{fixed}}^{[i]} = R_{\text{fixed}}^{[i]} \text{ctl}_2(w^{[i]}, p_i^{[j_1]}, \dots, p_i^{[j_k]}).$$

From equation (3.2.1), we know that, for all $j \in \{j_1, \dots, j_k\}$,

$$p_{\text{fixed}}^{[j]} = R_{\text{fixed}}^{[i]} p_i^{[j]} + p_{\text{fixed}}^{[i]} \quad \implies \quad p_i^{[j]} = (R_{\text{fixed}}^{[i]})^T (p_{\text{fixed}}^{[j]} - p_{\text{fixed}}^{[i]}).$$

From this equality and from the fact that $\mathcal{CC}_1$ is invariant, we observe that

$$\mathrm{ctl}_1(p^{[i]}_{\mathrm{fixed}}, w^{[i]}, p^{[j_1]}_{\mathrm{fixed}}, \dots, p^{[j_k]}_{\mathrm{fixed}}) = R^{[i]}_{\mathrm{fixed}}\mathrm{ctl}_1(\mathbf{0}_d, w^{[i]}, p^{[j_1]}_i, \dots, p^{[j_k]}_i).$$

Since $\mathcal{CC}_1$ and $\mathcal{RSC}_2$ are equivalent, the two evolution equations coincide. A similar reasoning also shows that the evolutions of the processor states are identical. ∎

Remark 3.19 (Communication-based laws on relative-sensing networks). Proposition 3.18 implies the following fact. An invariant control and communication law for a robotic network satisfying appropriate properties can be implemented on an appropriate relative-sensing network as a relative-sensing control law. •

3.3 COORDINATION TASKS AND COMPLEXITY NOTIONS

In this section, we introduce concepts and tools that are useful analyzing a communication and control law in a robotic network; our treatment is directly generalized to relative-sensing networks. We address the following questions: What is a coordination task for a robotic network? When does a control and communication law achieve a task? And with what time, space, and communication complexity?

3.3.1 Coordination tasks

Our first analysis step is to characterize the correctness properties of a communication and control law. We do so by defining the notions of task and of task achievement by a robotic network.

Definition 3.20 (Coordination task). Let $\mathcal{S}$ be a robotic network and let $\mathcal{W}$ be a set.

(i) A *coordination task* is a map $\mathcal{T}\colon \prod_{i\in I} X^{[i]} \times \mathcal{W}^n \to \{\texttt{true}, \texttt{false}\}$.

(ii) If $\mathcal{W}$ is a singleton, then the coordination task is said to be *static* and can be described by a map $\mathcal{T}\colon \prod_{i\in I} X^{[i]} \to \{\texttt{true}, \texttt{false}\}$.

Additionally, let $\mathcal{CC}$ be a control and communication law for $\mathcal{S}$:

(i) The law $\mathcal{CC}$ is *compatible* with the task $\mathcal{T}\colon \prod_{i\in I} X^{[i]} \times \mathcal{W}^n \to \{\texttt{true}, \texttt{false}\}$ if its processor state takes values in $\mathcal{W}$, that is, if $W^{[i]} = \mathcal{W}$, for all $i \in I$.

(ii) The law $\mathcal{CC}$ *achieves* the task $\mathcal{T}$ if it is compatible with it and if, for all initial conditions $x_0^{[i]} \in X_0^{[i]}$ and $w_0^{[i]} \in W_0^{[i]}$, $i \in I$, there exists

$T \in \mathbb{R}_{>0}$ such that the network evolution $t \mapsto (x(t), w(t))$ has the property that $\mathcal{T}(x(t), w(t)) = \texttt{true}$ for all $t \geq T$. •

Remark 3.21 (Temporal logic). Loosely speaking, the phrase "a law achieves a task" means that the network evolutions reach (and remain at) a specified pattern in the robot physical or processor state. In other words, the task is achieved if *at some time* and *for all subsequent times*, the predicate evaluates to **true** along system trajectories. It is possible to consider more general tasks based on more expressive predicates on trajectories. Such predicates can be defined through various forms of temporal and propositional logic, (see, e.g., Emerson, 1994). In particular, (linear) temporal logic contains certain constructs that allow reasoning in terms of time and is hence appropriate for robotic applications—as argued, for example, by Fainekos et al. (2005). Network tasks such as periodically visiting a desired set of configurations can be encoded with temporal logic statements. •

Example 3.22 (Direction agreement and equidistance tasks). From Example 3.7, consider the uniform network $\mathcal{S}_{\mathrm{circle}}$ of locally connected first-order agents in $\mathbb{S}^1$. From Section 3.1.3, recall the AGREE & PURSUE control and communication law $\mathcal{CC}_{\mathrm{AGREE\ \&\ PURSUE}}$ with processor state taking values in $W = \{\texttt{cc}, \texttt{c}\} \times I$. There are two tasks of interest. First, we define the *direction agreement task* $\mathcal{T}_{\mathtt{dir}} : (\mathbb{S}^1)^n \times W^n \to \{\texttt{true}, \texttt{false}\}$ by

$$\mathcal{T}_{\mathtt{dir}}(\theta, w) = \begin{cases} \texttt{true}, & \text{if } \mathtt{dir}^{[1]} = \cdots = \mathtt{dir}^{[n]}, \\ \texttt{false}, & \text{otherwise}, \end{cases}$$

where $\theta = (\theta^{[1]}, \ldots, \theta^{[n]})$, $w = (w^{[1]}, \ldots, w^{[n]})$, and $w^{[i]} = (\mathtt{dir}^{[i]}, \mathtt{max\text{-}id}^{[i]})$, for $i \in I$. Furthermore, for $\varepsilon > 0$, we define the static *(agent) equidistance task* $\mathcal{T}_{\varepsilon\text{-eqdstnc}} : (\mathbb{S}^1)^n \to \{\texttt{true}, \texttt{false}\}$ to be **true** if and only if

$$\left| \min_{j \neq i} \mathrm{dist}_{\mathtt{c}}(\theta^{[i]}, \theta^{[j]}) - \min_{j \neq i} \mathrm{dist}_{\mathtt{cc}}(\theta^{[i]}, \theta^{[j]}) \right| < \varepsilon, \quad \text{for all } i \in I.$$

In other words, $\mathcal{T}_{\varepsilon\text{-eqdstnc}}$ is true when, for every agent, the distances to the closest clockwise neighbor and to the closest counterclockwise neighbor are approximately equal. •

3.3.2 Complexity notions

We are now ready to define the notions of time, space and communication complexity. These notions describe the cost that a certain control and communication law incurs while completing a certain coordination task. Additionally, the complexity of a task is the infimum of the costs incurred by all laws that achieve that task. We begin by highlighting a difference between what follows and the complexity treatment for synchronous networks.

Remark 3.23 (Termination via task completion). As discussed in Remark 1.44 in Section 1.5, it is possible to consider various algorithm termination notions. Here, we will establish the completion of an algorithm as the instant when a given task is achieved.   •

First, we define the time complexity of an achievable task as the minimum number of communication rounds needed by the agents to achieve the task $\mathcal{T}$.

Definition 3.24 (Time complexity). Let $\mathcal{S}$ be a robotic network and let $\mathcal{T}$ be a coordination task for $\mathcal{S}$. Let $\mathcal{CC}$ be a control and communication law for $\mathcal{S}$ compatible with $\mathcal{T}$:

(i) the *(worst-case) time complexity to achieve $\mathcal{T}$ with $\mathcal{CC}$ from initial conditions* $(x_0, w_0) \in \prod_{i \in I} X_0^{[i]} \times \prod_{i \in I} W_0^{[i]}$ is

$$\mathrm{TC}(\mathcal{T}, \mathcal{CC}, x_0, w_0) = \inf\{\ell \mid \mathcal{T}(x(k), w(k)) = \mathtt{true}, \text{ for all } k \geq \ell\},$$

where $t \mapsto (x(t), w(t))$ is the evolution of $(\mathcal{S}, \mathcal{CC})$ from the initial condition (x_0, w_0);

(ii) the *(worst-case) time complexity to achieve $\mathcal{T}$ with $\mathcal{CC}$* is

$$\mathrm{TC}(\mathcal{T}, \mathcal{CC}) = \sup\left\{ \mathrm{TC}(\mathcal{T}, \mathcal{CC}, x_0, w_0) \mid (x_0, w_0) \in \prod_{i \in I} X_0^{[i]} \times \prod_{i \in I} W_0^{[i]} \right\};$$

(iii) the *(worst-case) time complexity of $\mathcal{T}$* is

$$\mathrm{TC}(\mathcal{T}) = \inf\{\mathrm{TC}(\mathcal{T}, \mathcal{CC}) \mid \mathcal{CC} \text{ compatible with } \mathcal{T}\}.   •$$

Next, we quantify memory and communication requirements of communication and control laws. We assume that elements of the processor state set W or of the alphabet set $\mathbb{A}$ might amount to multiple "basic memory units" or "basic messages." We let $|W|_{\mathrm{basic}}$ and $|\mathbb{A}|_{\mathrm{basic}}$ denote the number of basic memory units and basic messages required to represent elements of W and $\mathbb{A}$, respectively. The $\mathtt{null}$ message has zero cost. To clarify this assumption, we adopt two conventions. First, as in Section 1.5.2, we assume that a "basic memory unit" or a "basic message" contains $\log(n)$ bits. This implies that the $\log(n)$ bits required to store or transmit a robot identifier $i \in \{1, \ldots, n\}$ are equivalent to one "basic memory unit." Second, as mentioned in Remark 3.11, we assume that a processor can store and transmit a (finite number of) integer and real numbers, and we adopt the convention that any such number is quantized and represented by a constant number of basic memory units or basic messages.

We now quantify memory requirements of algorithms and tasks by counting the required number of basic memory units. Let the network $\mathcal{S}$, the task $\mathcal{T}$, and the control and communication law $\mathcal{CC}$ be as in Definition 3.24.

Definition 3.25 (Space complexity).

(i) The *(worst-case) space complexity to achieve $\mathcal{T}$ with $\mathcal{CC}$*, denoted by $\mathrm{SC}(\mathcal{T}, \mathcal{CC})$, is the maximum number of basic memory units required by a robot processor executing the $\mathcal{CC}$ on $\mathcal{S}$ among all robots and among all allowable initial physical and processor states until termination; and

(ii) the space complexity of $\mathcal{T}$ is the infimum among the space complexities of all control and communication laws that achieve $\mathcal{T}$. ●

The set of all non-null messages generated during one communication round from network state (x, w) is denoted by

$$\mathcal{M}(x, w) = \{(i, j) \in E_{\mathrm{cmm}}(x) \mid \mathrm{msg}^{[i]}(x^{[i]}, w^{[i]}, j) \neq \texttt{null}\}.$$

We now quantify the mean and total communication requirements of algorithms and tasks by counting the number of transmitted basic messages.

Definition 3.26 (Mean and Total Communication complexity).

(i) The *(worst-case) mean communication complexity* and the *(worst-case) total communication complexity* to achieve $\mathcal{T}$ with $\mathcal{CC}$ from $(x_0, w_0) \in \prod_{i \in I} X_0^{[i]} \times \prod_{i \in I} W_0^{[i]}$ are, respectively,

$$\mathrm{MCC}(\mathcal{T}, \mathcal{CC}, x_0, w_0) = \frac{|\mathbb{A}|_{\mathrm{basic}}}{\tau} \sum_{\ell=0}^{\tau-1} |\mathcal{M}(x(\ell), w(\ell))|,$$

$$\mathrm{TCC}(\mathcal{T}, \mathcal{CC}, x_0, w_0) = |\mathbb{A}|_{\mathrm{basic}} \sum_{\ell=0}^{\tau-1} |\mathcal{M}(x(\ell), w(\ell))|,$$

where $t \mapsto (x(t), w(t))$ is the evolution of $(\mathcal{S}, \mathcal{CC})$ from the initial condition (x_0, w_0) and where $\tau = \mathrm{TC}(\mathcal{CC}, \mathcal{T}, x_0, w_0)$. Here, MCC is defined only for initial conditions (x_0, w_0) with the property that $\mathcal{T}(x_0, w_0) = \texttt{false}$;

(ii) the *(worst-case) mean communication complexity* (resp. the *(worst-case) total communication complexity*) to achieve $\mathcal{T}$ with $\mathcal{CC}$ is the supremum of $\mathrm{MCC}(\mathcal{T}, \mathcal{CC}, x_0, w_0)$ (resp. $\mathrm{TCC}(\mathcal{T}, \mathcal{CC}, x_0, w_0)$) over all allowable initial states (x_0, w_0); and

(iii) the *(worst-case) mean communication complexity* (resp. the *worst-case total communication complexity*) of $\mathcal{T}$ is the infimum among the mean communication complexity (resp. the total communication complexity) of all control and communication laws achieving $\mathcal{T}$. ●

By construction, one can verify that it always happens that

$$\mathrm{TCC}(\mathcal{T}, \mathcal{CC}) \leq \mathrm{MCC}(\mathcal{T}, \mathcal{CC}) \cdot \mathrm{TC}(\mathcal{T}, \mathcal{CC}). \qquad (3.3.1)$$

We conclude this section with possible variations and extensions of the complexity definitions.

Remark 3.27 (Infinite-horizon mean communication complexity).
The mean communication complexity MCC measures the average cost of the communication rounds required to achieve a task over a finite time horizon; a similar statement holds for the total communication complexity TCC. One might be interested in a notion of mean communication complexity required to maintain the task true for all times. Accordingly, the infinite-horizon mean communication complexity of $\mathcal{CC}$ from initial conditions (x_0, w_0) is

$$\mathrm{IH\text{-}MCC}(\mathcal{CC}, x_0, w_0) = \lim_{\tau \to +\infty} \frac{|\mathbb{A}|_{\mathrm{basic}}}{\tau} \sum_{\ell=0}^{\tau} |\mathcal{M}(x(\ell), w(\ell))|. \qquad \bullet$$

Remark 3.28 (Communication complexity in omnidirectional networks). In omnidirectional wireless networks, the standard operation mode is for all neighbors of a node to receive the signal that it transmits. In other words, the transmission is omnidirectional rather than unidirectional. It is straightforward to require the message-generation function to have the property that the output it generates be independent of the intended receiver. Under such assumptions, it make sense to count as communication complexity not the number of messages transmitted in the network, but the number of transmissions, that is, a unit cost per node rather than a unit cost per edge of the network. $\qquad \bullet$

Remark 3.29 (Energy complexity). Given a model for the energy consumed by the robot to move and to transmit a message, one can easily define a notion of energy complexity for a control and communication law. In modern wireless transmitters, the energy consumption in transmitting a signal at a distance r varies with a power of r. Analogously, energy consumption is an increasing function of distance traveled. We consider this to be a promising avenue for further research. $\qquad \bullet$

3.3.3 Invariance under rescheduling

Here we discuss the invariance properties of time and communication complexity under the *rescheduling* of a control and communication law. The idea behind rescheduling is to "spread" the execution of the law over time without affecting the trajectories described by the robots.

For simplicity we consider the setting of static laws; similar results can be obtained for the general setting. Also, for ease of presentation, we allow our communication and control laws to be time dependent, that is, we consider message-generation functions and motion control functions of the form $\text{msg}^{[i]} : \mathbb{Z}_{\geq 0} \times X^{[i]} \times I \to \mathbb{A}$ and $\text{ctl}^{[i]} : \mathbb{R}_{\geq 0} \times X^{[i]} \times X^{[i]} \times \mathbb{A}^n \to U^{[i]}$, respectively. Definition 3.10 for network evolution can be readily extended to this more general time-dependent setup.

Let $\mathcal{S} = (I, \mathcal{R}, E_{\text{cmm}})$ be a robotic network in which each robot is a driftless control system (see Section 1.3). Let $\mathcal{CC} = (\mathbb{A}, \{\text{msg}^{[i]}\}_{i \in I}, \{\text{ctl}^{[i]}\}_{i \in I})$ be a static control and communication law. In what follows, we define a new control and communication law by modifying $\mathcal{CC}$; to do so, we introduce some notation. Let $s \in \mathbb{N}$, with $s \leq n$, and let $\mathcal{P}_I = \{I_0, \ldots, I_{s-1}\}$ be an *s-partition* of I, that is, $I_0, \ldots, I_{s-1} \subset I$ are disjoint and nonempty and $I = \cup_{k=0}^{s-1} I_k$. For $i \in I$, define the message-generation functions $\text{msg}_{\mathcal{P}_I}^{[i]} : \mathbb{Z}_{\geq 0} \times X^{[i]} \times I \to \mathbb{A}$ by

$$\text{msg}_{\mathcal{P}_I}^{[i]}(\ell, x, j) = \text{msg}^{[i]}(\lfloor \ell/s \rfloor, x, j), \qquad (3.3.2)$$

if $i \in I_k$ and $k = \ell \mod s$, and $\text{msg}_{\mathcal{P}_I}^{[i]}(\ell, x, j) = \texttt{null}$ otherwise. According to this message-generation function, only the agents with a unique identifier in I_k will send messages at time ℓ, where $\ell \in \{k + as\}_{a \in \mathbb{Z}_{\geq 0}}$. Equivalently, this can be stated as follows: according to (3.3.2), the messages originally sent at the time instant ℓ are now rescheduled to be sent at the time instants $F(\ell) - s + 1, \ldots, F(\ell)$, where $F : \mathbb{Z}_{\geq 0} \to \mathbb{Z}_{\geq 0}$ is defined by $F(\ell) = s(\ell + 1) - 1$. Figure 3.9 illustrates this idea. For $i \in I$, define the control functions

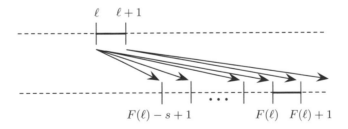

Figure 3.9 Under the rescheduling, the messages that are sent at the time instant ℓ under the control and communication law $\mathcal{CC}$ are rescheduled to be sent over the time instants $F(\ell) - s + 1, \ldots, F(\ell)$ under the control and communication law $\mathcal{CC}_{(s, \mathcal{P}_I)}$.

$\text{ctl}^{[i]} : \mathbb{R}_{\geq 0} \times X^{[i]} \times X^{[i]} \times \mathbb{A}^n \to U^{[i]}$ by

$$\text{ctl}_{\mathcal{P}_I}^{[i]}(t, x, x_{\text{smpld}}, y) = \text{ctl}^{[i]}\left(t - \ell + F^{-1}(\ell), x, x_{\text{smpld}}, y\right), \qquad (3.3.3)$$

if $t \in [\ell, \ell + 1]$ and $\ell = -1 \mod s$, and $\text{ctl}_{\mathcal{P}_I}^{[i]}(t, x, x_{\text{smpld}}, y) = 0$ otherwise.

Here, $F^{-1} : \mathbb{Z}_{\geq 0} \to \mathbb{Z}_{\geq 0}$ is the inverse of F, defined by $F^{-1}(\ell) = \frac{\ell+1}{s} - 1$. Roughly speaking, the control law $\mathrm{ctl}^{[i]}_{\mathcal{P}_I}$ makes the agent i wait for the time intervals $[\ell, \ell+1]$, with $\ell \in \{as-1\}_{a \in \mathbb{N}}$, to execute any motion. Accordingly, the evolution of the robotic network under the original law $\mathcal{CC}$ during the time interval $[\ell, \ell+1]$ now takes place when all the corresponding messages have been transmitted, that is, along the time interval $[F(\ell), F(\ell)+1]$. The following definition summarizes this construction.

Definition 3.30 (Rescheduling of control and communication laws). Let $\mathcal{S} = (I, \mathcal{R}, E_{\mathrm{cmm}})$ be a robotic network with driftless physical agents, and let $\mathcal{CC} = (\mathbb{Z}_{\geq 0}, \mathbb{A}, \{\mathrm{msg}^{[i]}\}_{i \in I}, \{\mathrm{ctl}^{[i]}\}_{i \in I})$ be a static control and communication law. Let $s \in \mathbb{N}$, with $s \leq n$, and let $\mathcal{P}_I$ be an s-partition of I. The control and communication law $\mathcal{CC}_{(s,\mathcal{P}_I)} = (\mathbb{Z}_{\geq 0}, \mathbb{A}, \{\mathrm{msg}^{[i]}_{\mathcal{P}_I}\}_{i \in I}, \{\mathrm{ctl}^{[i]}_{\mathcal{P}_I}\}_{i \in I})$ defined by equations (3.3.2) and (3.3.3) is called a $\mathcal{P}_I$-*rescheduling of* $\mathcal{CC}$. •

The following result, whose proof is presented in Section 3.6.1, shows that the total communication complexity is invariant under rescheduling.

Proposition 3.31 (Complexity of rescheduled laws). *With the assumptions of Definition 3.30, let* $\mathcal{T} : \prod_{i \in I} X^{[i]} \to \{\mathtt{true}, \mathtt{false}\}$ *be a coordination task for* $\mathcal{S}$*. Then, for all* $x_0 \in \prod_{i \in I} X^{[i]}_0$*,*

$$\mathrm{TC}(\mathcal{T}, \mathcal{CC}_{(s,\mathcal{P}_I)}, x_0) = s \cdot \mathrm{TC}(\mathcal{T}, \mathcal{CC}, x_0).$$

Moreover, if $\mathrm{C}_{\mathrm{rnd}}$ *is additive, then, for all* $x_0 \in \prod_{i \in I} X^{[i]}_0$*,*

$$\mathrm{MCC}(\mathcal{T}, \mathcal{CC}_{(s,\mathcal{P}_I)}, x_0) = \frac{1}{s} \cdot \mathrm{MCC}(\mathcal{T}, \mathcal{CC}, x_0),$$

and therefore, $\mathrm{TCC}(\mathcal{T}, \mathcal{CC}_{(s,\mathcal{P}_I)}, x_0) = \mathrm{TCC}(\mathcal{T}, \mathcal{CC}, x_0)$*, that is, the total communication complexity of* $\mathcal{CC}$ *is invariant under rescheduling.*

Remark 3.32 (Appropriate complexity notions for driftless agents). Given the results in the previous theorem, one should be careful in choosing which notion of communication complexity to use in order to evaluate control and communication laws. For driftless physical agents, rather than the *mean* communication complexity MCC, one should really consider the *total* communication complexity TCC, since the latter is invariant with respect to rescheduling. Note that the notion of infinite-horizon mean communication complexity IH-MCC defined in Remark 3.27 satisfies the same relationship as MCC, that is, $\mathrm{IH\text{-}MCC}(\mathcal{CC}_{(s,\mathcal{P}_I)}, x_0) = \frac{1}{s} \mathrm{IH\text{-}MCC}(\mathcal{CC}, x_0)$. •

3.4 COMPLEXITY OF DIRECTION AGREEMENT AND EQUIDISTANCE

From Example 3.7, Section 3.1.3, and Example 3.22, recall the definition of a uniform network $\mathcal{S}_{\text{circle}}$ of locally connected first-order agents in $\mathbb{S}^1$, the AGREE & PURSUE control and communication law $\mathcal{CC}_{\text{AGREE & PURSUE}}$, and the two coordination tasks $\mathcal{T}_{\text{dir}}$ and $\mathcal{T}_{\varepsilon\text{-eqdstnc}}$. In this section, we characterize the complexity to achieve these coordination tasks with $\mathcal{CC}_{\text{AGREE & PURSUE}}$. Because the number of bits required to represent the variable $\texttt{max-id} \in \{1, \ldots, n\}$ is $\log(n)$, note that the space complexity of $\mathcal{CC}_{\text{AGREE & PURSUE}}$ is $\log(n)$ bits, that is, one basic memory unit in our convention discussed in Section 3.3.2.

Motivated by Remark 3.8, we model wireless communication congestion by assuming that the communication range is a monotone non-increasing function $r : \mathbb{N} \to]0, \pi[$ of the number of agents n. Likewise, we assume that the maximum control amplitude $u_{\max}$ is a non-increasing function $u_{\max} : \mathbb{N} \to]0, 1[$; recall that $u_{\max}$ is the maximum robot speed. Finally, it is convenient to define the function $n \mapsto \delta(n) = nr(n) - 2\pi \in \mathbb{R}$ that compares the sum of the communication ranges of all the robots with the length of the unit circle.

We are now ready to state the main result of this section; proofs are postponed to Section 3.6.2.

Theorem 3.33 (Time complexity of agree-and-pursue law). *Given $k_{\text{prop}} \in]0, \frac{1}{2}[$, in the limit as $n \to +\infty$ and $\varepsilon \to 0^+$, the network $\mathcal{S}_{\text{circle}}$ with $u_{\max}(n) \geq k_{\text{prop}}r(n)$, the law $\mathcal{CC}_{\text{AGREE & PURSUE}}$, and the tasks $\mathcal{T}_{\text{dir}}$ and $\mathcal{T}_{\varepsilon\text{-eqdstnc}}$ together satisfy the following properties:*

(i) $\text{TC}(\mathcal{T}_{\text{dir}}, \mathcal{CC}_{\text{AGREE & PURSUE}}) \in \Theta(r(n)^{-1})$.

(ii) If $\delta(n)$ is lower bounded by a positive constant as $n \to +\infty$, then

$$\text{TC}(\mathcal{T}_{\varepsilon\text{-eqdstnc}}, \mathcal{CC}_{\text{AGREE & PURSUE}}) \in \Omega(n^2 \log(n\varepsilon)^{-1}),$$
$$\text{TC}(\mathcal{T}_{\varepsilon\text{-eqdstnc}}, \mathcal{CC}_{\text{AGREE & PURSUE}}) \in O(n^2 \log(n\varepsilon^{-1})).$$

If $\delta(n)$ is upper bounded by a negative constant, then in general the law $\mathcal{CC}_{\text{AGREE & PURSUE}}$ does not achieve $\mathcal{T}_{\varepsilon\text{-eqdstnc}}$.

Next, we study the total communication complexity of the agree-and-pursue control and communication law. First, we note that any message in $\mathbb{A} = \mathbb{S}^1 \times \{\texttt{cc}, \texttt{c}\} \times \{1, \ldots, n\} \cup \{\texttt{null}\}$ requires only a finite number of basic messages to encode, that is, $|\mathbb{A}|_{\text{basic}} \in O(1)$.

Theorem 3.34 (Total communication complexity of agree-and-pur-

sue law). *For $k_{\mathrm{prop}} \in]0, \frac{1}{2}[$, in the limit as $n \to +\infty$ and $\varepsilon \to 0^+$, the network S_{circle} with $u_{\max}(n) \geq k_{\mathrm{prop}} r(n)$, the law $CC_{\mathrm{AGREE\ \&\ PURSUE}}$, and the tasks T_{dir} and $T_{\varepsilon\text{-eqdstnc}}$ together satisfy the following properties:*

(i) If $\delta(n) \geq \pi(1/k_{\mathrm{prop}} - 2)$ as $n \to +\infty$, then

$$\mathrm{TCC}(T_{\mathrm{dir}}, CC_{\mathrm{AGREE\ \&\ PURSUE}}) \in \Theta(n^2 r(n)^{-1});$$

otherwise, if $\delta(n) \leq \pi(1/k_{\mathrm{prop}} - 2)$ as $n \to +\infty$, then

$$\mathrm{TCC}(T_{\mathrm{dir}}, CC_{\mathrm{AGREE\ \&\ PURSUE}}) \in \Omega(n^3 + nr(n)^{-1}),$$
$$\mathrm{TCC}(T_{\mathrm{dir}}, CC_{\mathrm{AGREE\ \&\ PURSUE}}) \in O(n^2 r(n)^{-1}).$$

(ii) If $\delta(n)$ is lower bounded by a positive constant as $n \to +\infty$, then

$$\mathrm{TCC}(T_{\varepsilon\text{-eqdstnc}}, CC_{\mathrm{AGREE\ \&\ PURSUE}}) \in \Omega(n^3 \delta(n) \log(n\varepsilon)^{-1}),$$
$$\mathrm{TCC}(T_{\varepsilon\text{-eqdstnc}}, CC_{\mathrm{AGREE\ \&\ PURSUE}}) \in O(n^4 \log(n\varepsilon^{-1})).$$

Remark 3.35 (Comparison with leader election). Let us compare the agree-and-pursue control and communication law with the classical LE LANN-CHANG-ROBERTS (LCR) ALGORITHM for leader election discussed in Section 1.5.4. The leader election task consists of electing a unique agent among all agents in the network; therefore, it is different from, but closely related to, the coordination task T_{dir}. The LCR ALGORITHM operates on a static network with the ring communication topology, and achieves leader election with time and total communication complexity $\Theta(n)$ and $\Theta(n^2)$, respectively. The agree-and-pursue law operates on a robotic network with the $r(n)$-disk communication topology, and achieves T_{dir} with time and total communication complexity, respectively, $\Theta(r(n)^{-1})$ and $O(n^2 r(n)^{-1})$. If wireless communication congestion is modeled by $r(n)$ of order $1/n$ as in Remark 3.8, then the two algorithms have identical time complexity and the LCR ALGORITHM has better communication complexity. Note that computations on a possibly disconnected, dynamic network are more complex than on a static ring topology. •

3.5 NOTES

The study of multi-robot systems has a long and rich history. Some recent examples include the surveys (Asama, 1992; Cao et al., 1997; Dias et al., 2006), the text by Arkin (1998) on behavior-based robotics, and the special issues (Arai et al., 2002; Abdallah and Tanner, 2007; Bullo et al., 2009). Together with this literature, the starting points for developing the material in this chapter are the standard notions of *synchronous and asynchronous networks* in distributed (Lynch, 1997; Peleg, 2000; Tel, 2001) and parallel (Bertsekas and Tsitsiklis, 1997; Parhami, 1999) computation. The established body of knowledge on synchronous networks is, however, not directly applicable to the robotic network setting because of the agents' mobility and the ensuing dynamic communication topology.

An influential contribution toward a network model of mobile interacting robots is the work by Suzuki and Yamashita (1999). This model consists of a group of identical "distributed anonymous mobile robots" characterized as follows: no explicit communication takes place between them, and at each time instant of an "activation schedule," each robot senses the relative position of all other robots and moves according to a pre-specified algorithm. An artificial intelligence approach to multi-agent behavior in a shared environment is taken in Moses and Tennenholtz (1995). Santoro (2001) provides, with an emphasis on computer science aspects, a brief survey of models, algorithms, and the need for appropriate complexity notions. Recently, a notion of communication complexity for control and communication algorithms in multi-robot systems has been analyzed by Klavins (2003); see also Klavins and Murray (2004). Notions of failures and robustness in robotic networks are discussed by Gupta et al. (2006b). From a broad hybrid networked systems viewpoint, our robotic network model can be regarded as special cases of the general modeling paradigms discussed in Lygeros et al. (2003), Lynch et al. (2003), and Sanfelice et al. (2007).

A key feature of the synchronous robotic network model proposed in this chapter is the adoption of proximity graphs from computational geometry as a basis for our communication model. This design choice is justified by the vast wireless networking literature, where this assumption is made. The simplest communication model, in which two robots communicate only if they are within a fixed communication range, is a common model adopted, for example, in the studies by Gupta and Kumar (2000), Li (2003), Lloyd et al. (2005), and Santi (2005). These works study the proximity graph solutions to various communication optimization problems; this discipline is referred to as *topology control* (cf., Remark 3.8). Although we focus our presentation on the topological aspect of the communication service, more realistic communication models would include randomness, packet losses, coding, quantization, and delays (see, e.g., Toh, 2001; Tse and Viswanath, 2005).

Next, we review some literature on emergent and self-organized swarming behaviors in biological groups. Interesting dynamical systems arise in biological networks at multiple levels of resolution, all the way from interactions among molecules and cells (Miller and Bassler, 2001) to the behavioral ecology of animal groups (Okubo, 1986). Flocks of birds and schools of fish can travel in formation and act as one (see Parrish et al., 2002), allowing these animals to defend themselves against predators and protect their territories. Wildebeest and other animals exhibit complex collective behaviors when migrating, such as obstacle avoidance, leader election, and formation-keeping (see Sinclair, 1977; Gueron and Levin, 1993). Certain foraging behaviors include individual animals partitioning their environment into non-overlapping zones (see Barlow, 1974). Honey bees (Seeley and Buhrman, 1999), gorillas (Stewart and Harcourt, 1994), and whitefaced capuchins (Boinski and Campbell, 1995) exhibit synchronized group activities such as initiation of motion and change of travel direction. These remarkable dynamic capabilities are achieved apparently without following a group leader; see Barlow (1974), Okubo (1986), Gueron and Levin (1993), Stewart and Harcourt (1994), Seeley and Buhrman (1999), Boinski and Campbell (1995), and Parrish et al. (2002) for specific examples of animal species, and Conradt and Roper (2003), and Couzin et al. (2005) for general studies. A comprehensive exposition of bio-inspired optimization and control methods is presented in Passino (2004).

Regarding distributed motion coordination algorithms, much progress has been made on collective pattern formation (Suzuki and Yamashita, 1999; Belta and Kumar, 2004; Justh and Krishnaprasad, 2004; Sepulchre et al., 2007; Paley et al., 2007; Yang et al., 2008), flocking (Olfati-Saber, 2006; Lee and Spong, 2007; Tanner et al., 2007; Moshtagh and Jadbabaie, 2007), motion feasibility of formations (Tabuada et al., 2005), formation control using rigidity and persistence theory (Olfati-Saber and Murray, 2002; Baillieul and Suri, 2003; Hendrickx et al., 2007; Krick, 2007; Yu et al., 2009), formation stability (Tanner et al., 2004; Lafferriere et al., 2005; Kang et al., 2006; Dunbar and Murray, 2006; Smith and Hadaegh, 2007; Zheng et al., 2008), motion camouflage (Justh and Krishnaprasad, 2006), self-assembly (Klavins et al., 2006), swarm aggregation (Gazi and Passino, 2003), gradient climbing (Ögren et al., 2004; Cortés, 2007), cyclic-pursuit (Bruckstein et al., 1991; Marshall et al., 2004; Martínez and Bullo, 2006; Smith et al., 2005; Pavone and Frazzoli, 2007), vehicle routing (Sharma et al., 2007), motion planning with collision avoidance (Lumelsky and Harinarayan, 1997; Hu et al., 2007; Pallottino et al., 2007), and cooperative boundary estimation (Bertozzi et al., 2004; Zhang and Leonard, 2005; Clark and Fierro, 2007; Casbeer et al., 2006; Susca et al., 2008). It is also worth mentioning works on network localization, estimation, and tracking (see, e.g., Aspnes et al., 2006; Barooah and Hespanha, 2007; Oh et al., 2007; and the references therein).

Much research has been devoted to distributed task allocation problems. The work in (Gerkey and Mataric, 2004) proposes a taxonomy of task allocation problems. In papers such as (Godwin et al., 2006; Alighanbari and How, 2006; Schumacher et al., 2003; Moore and Passino, 2007; Tang and Özgüner, 2005), advanced heuristic methods are developed, and their effectiveness is demonstrated through analysis, simulation or real world implementation. Distributed auction algorithms are discussed in (Castañón and Wu, 2003;

Moore and Passino, 2007) building on the classic works in (Bertsekas and Castañón, 1991, 1993). A distributed mixed-integer-linear-programming solver is proposed in (Alighanbari and How, 2006). A spatially distributed receding-horizon scheme is proposed in (Frazzoli and Bullo, 2004; Pavone et al., 2007). There has also been prior work on target assignment problems (Beard et al., 2002; Arslan et al., 2007; Zavlanos and Pappas, 2007a; Smith and Bullo, 2009). Target allocation for vehicles with nonholonomic constraints is studied in (Rathinam et al., 2007; Savla et al., 2008, 2009a).

3.6 PROOFS

This section gathers the proofs of the main results presented in the chapter.

3.6.1 Proof of Proposition 3.31

Proof. Let $t \mapsto x(t)$ and $t \mapsto \tilde{x}(t)$ denote the network evolutions starting from $x_0 \in \prod_{i \in I} X_0^{[i]}$ under $\mathcal{CC}$ and $\mathcal{CC}_{(s,\mathcal{P}_I)}$, respectively. From the definition of rescheduling, one can verify that, for all $k \in \mathbb{Z}_{\geq 0}$,

$$\tilde{x}^{[i]}(t) = \begin{cases} \tilde{x}^{[i]}(F(k-1)+1), & \text{for } t \in \bigcup_{\ell=F(k-1)+1}^{F(k)-1}[\ell, \ell+1], \\ x^{[i]}(t-F(k)+k), & \text{for } t \in [F(k), F(k)+1]. \end{cases} \quad (3.6.1)$$

By the definition of time complexity $\text{TC}(\mathcal{T}, \mathcal{CC}, x_0)$, we have $\mathcal{T}(x(k)) = \texttt{true}$, for all $k \geq \text{TC}(\mathcal{T}, \mathcal{CC}, x_0)$, and $\mathcal{T}(x(\text{TC}(\mathcal{T}, \mathcal{CC}, x_0) - 1)) = \texttt{false}$. We rewrite these equalities in terms of the trajectories of $\mathcal{CC}_{(s,\mathcal{P}_I)}$. From (3.6.1), we write $x^{[i]}(k) = \tilde{x}^{[i]}(F(k))$, for all $i \in I$ and $k \in \mathbb{Z}_{\geq 0}$. Therefore, we have

$$\mathcal{T}(\tilde{x}(F(k))) = \mathcal{T}(x(k)) = \texttt{true}, \qquad \text{for all } F(k) \geq F(\text{TC}(\mathcal{T}, \mathcal{CC}, x_0)),$$
$$\mathcal{T}(\tilde{x}(F(\text{TC}(\mathcal{T}, \mathcal{CC}, x_0) - 1))) = \mathcal{T}(x(\text{TC}(\mathcal{T}, \mathcal{CC}, x_0) - 1)) = \texttt{false},$$

where we have used the rescheduled message-generation function in (3.3.2). Now, note that by equation (3.6.1), $\tilde{x}^{[i]}(\ell) = \tilde{x}^{[i]}(F(\lfloor \ell/s \rfloor - 1) + 1)$, for all $\ell \in \mathbb{Z}_{\geq 0}$ and all $i \in I$. Therefore, $\mathcal{T}(\tilde{x}(F(\text{TC}(\mathcal{T}, \mathcal{CC}, x_0) - 1) + 1)) = \mathcal{T}(\tilde{x}(F(\text{TC}(\mathcal{T}, \mathcal{CC}, x_0))))$ and we can rewrite the previous identities as

$$\mathcal{T}(\tilde{x}(k)) = \texttt{true}, \quad \text{for all } k \geq F(\text{TC}(\mathcal{T}, \mathcal{CC}, x_0) - 1) + 1,$$
$$\mathcal{T}(\tilde{x}(F(\text{TC}(\mathcal{T}, \mathcal{CC}, x_0) - 1))) = \texttt{false},$$

which implies $\text{TC}(\mathcal{T}, \mathcal{CC}_{(s,\mathcal{P}_I)}, x_0) = F(\text{TC}(\mathcal{T}, \mathcal{CC}, x_0) - 1) + 1 = s \, \text{TC}(\mathcal{T}, \mathcal{CC}, x_0)$. As for the mean communication complexity, additivity of C_{rnd} implies

$$\text{C}_{\text{rnd}} \circ \mathcal{M}(\ell, x(\ell))$$
$$= \text{C}_{\text{rnd}} \circ \mathcal{M}(F(\ell) - s + 1, \tilde{x}(F(\ell) - s + 1)) + \cdots + \text{C}_{\text{rnd}} \circ \mathcal{M}(F(\ell), \tilde{x}(F(\ell))),$$

where we have used $F(\ell - 1) + 1 = F(\ell) - s + 1$. We conclude the proof by computing

$$\sum_{\ell=0}^{\mathrm{TC}(\mathcal{T},\mathcal{CC}_{(s,\mathcal{P}_I)},x_0)-1} \mathrm{C}_{\mathrm{rnd}} \circ \mathcal{M}(\ell, \tilde{x}(\ell)) = \sum_{\ell=0}^{F(\mathrm{TC}(\mathcal{T},\mathcal{CC},x_0)-1)} \mathrm{C}_{\mathrm{rnd}} \circ \mathcal{M}(\ell, \tilde{x}(\ell))$$

$$= \sum_{\ell=0}^{\mathrm{TC}(\mathcal{T},\mathcal{CC},x_0)-1} \sum_{k=F(\ell)-s+1}^{F(\ell)} \mathrm{C}_{\mathrm{rnd}} \circ \mathcal{M}(k, \tilde{x}(k))$$

$$= \sum_{\ell=0}^{\mathrm{TC}(\mathcal{T},\mathcal{CC},x_0)-1} \mathrm{C}_{\mathrm{rnd}} \circ \mathcal{M}(\ell, x(\ell)).$$

∎

3.6.2 Proof of Theorem 3.33

Proof. In the following four *STEPS*, we prove the two upper bounds and the two lower bounds.

STEP 1: We start by proving the upper bound in statement (i). We claim that $\mathrm{TC}(\mathcal{T}_{\mathtt{dir}}, \mathcal{CC}_{\mathrm{AGREE\ \&\ PURSUE}}) \leq 2\pi/(k_{\mathrm{prop}}r(n))$, and we reason by contradiction, that is, we assume that there exists an initial condition which gives rise to an execution with time complexity strictly larger than $2\pi/(k_{\mathrm{prop}}r(n))$. Without loss of generality, assume $\mathtt{dir}^{[n]}(0) = \mathtt{c}$. For $\ell \leq 2\pi/(k_{\mathrm{prop}}r(n))$, define

$$k(\ell) = \mathrm{argmin}\{\mathrm{dist}_{\mathtt{cc}}(\theta^{[n]}(0), \theta^{[i]}(\ell)) \mid \mathtt{dir}^{[i]}(\ell) = \mathtt{cc}, i \in \{1,\ldots,n\}\}.$$

In other words, agent $k(\ell)$ is the agent moving counterclockwise that has smallest counterclockwise distance from the initial position of agent n. Note that $k(\ell)$ is well-defined since, by hypothesis of contradiction, $\mathcal{T}_{\mathtt{dir}}$ is $\mathtt{false}$ for $\ell \leq 2\pi/(k_{\mathrm{prop}}r(n))$. According to the state-transition function of the law $\mathcal{CC}_{\mathrm{AGREE\ \&\ PURSUE}}$ (cf., Section 3.1.3), messages with $\mathtt{dir} = \mathtt{cc}$ can only travel counterclockwise, while messages with $\mathtt{dir} = \mathtt{c}$ can only travel clockwise. Therefore, the position of agent $k(\ell)$ at time ℓ can only belong to the counterclockwise interval from the position of agent $k(0)$ at time 0 to the position of agent n at time 0.

Let us examine how fast the message from agent n travels clockwise. To this end, for $\ell \leq 2\pi/(k_{\mathrm{prop}}r(n))$, define

$$j(\ell) = \mathrm{argmax}\{\mathrm{dist}_{\mathtt{c}}(\theta^{[n]}(0), \theta^{[i]}(\ell)) \mid \mathtt{max\text{-}id}^{[i]}(\ell) = n, i \in \{1,\ldots,n\}\}.$$

In other words, agent $j(\ell)$ has $\mathtt{max\text{-}id}$ equal to n, is moving clockwise, and is the agent furthest from the initial position of agent n in the clockwise direction with these two properties. Initially, $j(0) = n$. Additionally, for $\ell \leq 2\pi/(k_{\mathrm{prop}}r(n))$, we claim that

$$\mathrm{dist}_{\mathrm{c}}(\theta^{[j(\ell)]}(\ell), \theta^{[j(\ell+1)]}(\ell+1)) \geq k_{\mathrm{prop}}r(n).$$

This happens because either (1) there is no agent clockwise-ahead of $\theta^{[j(\ell)]}(\ell)$ within clockwise distance $r(n)$, and therefore, the claim is obvious, or (2) there are such agents. In case (2), let m denote the agent whose clockwise distance to agent $j(\ell)$ is maximal within the set of agents with clockwise distance $r(n)$ from $\theta^{[j(\ell)]}(\ell)$. Then,

$$
\begin{aligned}
\mathrm{dist}_{\mathrm{c}}&(\theta^{[j(\ell)]}(\ell), \theta^{[j(\ell+1)]}(\ell+1)) \\
&= \mathrm{dist}_{\mathrm{c}}(\theta^{[j(\ell)]}(\ell), \theta^{[m]}(\ell+1)) \\
&= \mathrm{dist}_{\mathrm{c}}(\theta^{[j(\ell)]}(\ell), \theta^{[m]}(\ell)) + \mathrm{dist}_{\mathrm{c}}(\theta^{[m]}(\ell), \theta^{[m]}(\ell+1)) \\
&\geq \mathrm{dist}_{\mathrm{c}}(\theta^{[j(\ell)]}(\ell), \theta^{[m]}(\ell)) + k_{\mathrm{prop}}\big(r(n) - \mathrm{dist}_{\mathrm{c}}(\theta^{[j(\ell)]}(\ell), \theta^{[m]}(\ell))\big) \\
&= k_{\mathrm{prop}}r(n) + (1 - k_{\mathrm{prop}})\,\mathrm{dist}_{\mathrm{c}}(\theta^{[j(\ell)]}(\ell), \theta^{[m]}(\ell)) \geq k_{\mathrm{prop}}r(n),
\end{aligned}
$$

where the first inequality follows from the fact that at time ℓ there can be no agent whose clockwise distance to agent m is less than $(r(n) - \mathrm{dist}_{\mathrm{c}}(\theta^{[j(\ell)]}(\ell), \theta^{[m]}(\ell)))$. Therefore, after a number of communication rounds larger than $2\pi/(k_{\mathrm{prop}}r(n))$, the message with `max-id` $= n$ has traveled the whole circle in the clockwise direction, and must therefore have reached agent $k(\ell)$. This is a contradiction.

STEP 2: We prove the lower bound in statement (i). If $r(n) > \pi$ for all n, then $1/r(n) < 1/\pi$, and the upper bound is $\mathrm{TC}(\mathcal{T}_{\mathtt{dir}}, \mathcal{CC}_{\mathrm{AGREE\ \&\ PURSUE}}) \in O(1)$. Obviously, the time complexity of any evolution with an initial configuration where $\mathtt{dir}^{[i]}(0) = \mathtt{cc}$ for $i \in \{1, \ldots, n-1\}$, $\mathtt{dir}^{[n]}(0) = \mathtt{c}$, and $\mathcal{E}_{\mathcal{G}_{\mathrm{disk}}(r)}(\theta^{[1]}(0), \ldots, \theta^{[n]}(0))$ is the complete graph, is lower bounded by 1. Therefore, $\mathrm{TC}(\mathcal{T}_{\mathtt{dir}}, \mathcal{CC}_{\mathrm{AGREE\ \&\ PURSUE}}) \in \Omega(1)$. If $r(n) > \pi$ for all n, then we conclude $\mathrm{TC}(\mathcal{T}_{\mathtt{dir}}, \mathcal{CC}_{\mathrm{AGREE\ \&\ PURSUE}}) \in \Theta(r(n)^{-1})$. Assume now that $r(n) \leq \pi$ for sufficiently large n. Consider an initial configuration where $\mathtt{dir}^{[i]}(0) = \mathtt{cc}$ for $i \in \{1, \ldots, n-1\}$, $\mathtt{dir}^{[n]}(0) = \mathtt{c}$, and the agents are placed as depicted in Figure 3.10. Note that, after each communication round, agent 1 has moved $k_{\mathrm{prop}}r(n)$ in the counterclockwise direction, while agent n has moved $k_{\mathrm{prop}}r(n)$ in the clockwise direction. These two agents keep moving at full speed toward each other until they become neighbors at a time lower bounded by

$$\frac{2\pi - r(n)}{2k_{\mathrm{prop}}r(n)} > \frac{\pi}{k_{\mathrm{prop}}r(n)} - 1.$$

We conclude that $\mathrm{TC}(\mathcal{T}_{\mathtt{dir}}, \mathcal{CC}_{\mathrm{AGREE\ \&\ PURSUE}}) \in \Omega(r(n)^{-1})$.

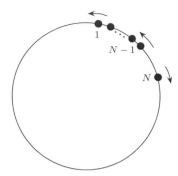

Figure 3.10 Initial condition for the lower bound of $TC(\mathcal{T}_{\mathtt{dir}}, \mathcal{CC}_{\text{AGREE \& PURSUE}})$, with $0 < \text{dist}_c(\theta^{[n-1]}(0), \theta^{[n]}(0)) - r(n) < \varepsilon$ and $\text{dist}_c(\theta^{[1]}(0), \theta^{[n-1]}(0)) \leq r(n) - \varepsilon$, for some fixed $\varepsilon > 0$.

STEP 3: We now prove the upper bound in (ii). We begin by noting that the lower bound on δ implies $r(n)^{-1} \in O(n)$. Therefore, we know that $TC(\mathcal{T}_{\mathtt{dir}}, \mathcal{CC}_{\text{AGREE \& PURSUE}})$ belongs to $O(n)$ and is negligible as compared with the claimed upper bound estimates for $TC(\mathcal{T}_{\varepsilon\text{-eqdstnc}}, \mathcal{CC}_{\text{AGREE \& PURSUE}})$. In what follows, we therefore assume that $\mathcal{T}_{\mathtt{dir}}$ has been achieved and that, without loss of generality, all agents are moving clockwise. We now prove a fact regarding connectivity. At time $\ell \in \mathbb{Z}_{\geq 0}$, let $H(\ell)$ be the union of all the empty "circular segments" of length at least $r(n)$, that is, let

$$H(\ell) = \{x \in \mathbb{S}^1 \mid \min_{i \in \{1,\dots,n\}} \text{dist}_c(x, \theta^{[i]}(\ell)) + \min_{j \in \{1,\dots,n\}} \text{dist}_{cc}(x, \theta^{[j]}(\ell)) > r(n)\}.$$

In other words, $H(\ell)$ does not contain any point between two agents separated by a distance less than $r(n)$, and each connected component of $H(\ell)$ has length at least $r(n)$. Let $n_H(\ell)$ be the number of connected components of $H(\ell)$; if $H(\ell)$ is empty, then we take the convention that $n_H(\ell) = 0$. Clearly, $n_H(\ell) \leq n$. We claim that if $n_H(\ell) > 0$, then $\tau \mapsto n_H(\ell + \tau)$ is non-increasing. Let $d(\ell) < r(n)$ be the distance between any two consecutive agents at time ℓ. Because both agents move in the same direction, a simple calculation shows that

$$d(\ell + 1) \leq d(\ell) + k_{\text{prop}}(r - d(\ell)) = (1 - k_{\text{prop}})d(\ell) + k_{\text{prop}}r(n)$$
$$< (1 - k_{\text{prop}})r + k_{\text{prop}}r(n) = r(n).$$

This means that the two agents remain within distance $r(n)$ and, therefore, connected at the following time instant. Because the number of connected components of $\mathcal{E}_{\mathcal{G}_{\text{disk}}(r)}(\theta^{[1]}, \dots, \theta^{[n]})$ does not increase, it follows that the number of connected components of H cannot increase. Next, we claim that if $n_H(\ell) > 0$, then there exists $\tau > \ell$ such that $n_H(\tau) < n_H(\ell)$. By contradiction, assume that $n_H(\ell) = n_H(\tau)$ for all $\tau \geq \ell$. Without loss of generality, let $\{1, \dots, m\}$ be a set of agents with the properties that $\text{dist}_{cc}\left(\theta^{[i]}(\ell), \theta^{[i+1]}(\ell)\right) \leq r(n)$, for $i \in \{1, \dots, m\}$, that $\theta^{[1]}(\ell)$ and $\theta^{[m]}(\ell)$

belong to the boundary of $H(\ell)$, and that there is no other set with the same properties and more agents. (Note that this implies that the agents $1, \ldots, m$ are in counterclockwise order.) One can show that, for $\tau \geq \ell$,

$$\theta^{[1]}(\tau + 1) = \theta^{[1]}(\tau) - k_{\mathrm{prop}} r(n),$$
$$\theta^{[i]}(\tau + 1) = \theta^{[i]}(\tau) - k_{\mathrm{prop}} \mathrm{dist}_{\mathrm{c}}(\theta^{[i]}(\tau), \theta^{[i-1]}(\tau)),$$

for $i \in \{2, \ldots, m\}$. If we consider the inter-agent distances

$$d(\tau) = \left(\mathrm{dist}_{\mathrm{cc}}(\theta^{[1]}(\tau), \theta^{[2]}(\tau)), \ldots, \mathrm{dist}_{\mathrm{cc}}(\theta^{[m-1]}(\tau), \theta^{[m]}(\tau)) \right) \in \mathbb{R}_{>0}^{m-1},$$

then the previous equations can be rewritten as

$$d(\tau + 1) = \mathrm{Trid}_{m-1}(k_{\mathrm{prop}}, 1 - k_{\mathrm{prop}}, 0)\, d(\tau) + r(n) k_{\mathrm{prop}} \boldsymbol{e}_1,$$

where the linear map $(a, b, c) \mapsto \mathrm{Trid}_{m-1}(a, b, c) \in \mathbb{R}^{(m-1) \times (m-1)}$ is defined in Section 1.6.4. This is a discrete-time affine time-invariant dynamical system with unique equilibrium point $r(n)\mathbf{1}_{m-1}$. By construction, the initial condition of this system satisfies $\|d(0) - r(n)\mathbf{1}_{m-1}\|_2 \leq r(n)\sqrt{m-1}$. By Theorem 1.79(ii) in Section 1.6.4, for $\eta_1 \in\,]0, 1[$, the solution $\tau \mapsto d(\tau)$ to this system reaches a ball of radius η_1 centered at the equilibrium point in time $O(m \log m + \log \eta_1^{-1})$. (Here we have used the fact that the initial condition of this system is bounded.) In turn, this implies that $\tau \mapsto \sum_{i=1}^{m} d_i(\tau)$ is larger than $(m-1)(r(n) - \eta_1)$ in time $O(m \log m + \log \eta_1^{-1})$. We are now ready to find the contradiction, and show that $n_H(\tau)$ cannot remain equal to $n_H(\ell)$ for all time τ. After time $O(m \log m + \log \eta_1^{-1}) = O(n \log n + \log \eta_1^{-1})$, we have

$$2\pi \geq n_H(\ell) r(n) + \sum_{j=1}^{n_H(\ell)} (r(n) - \eta_1)(m_j - 1)$$
$$= n_H(\ell) r(n) + (n - n_H(\ell))(r(n) - \eta_1) = n_H(\ell)\eta_1 + n(r(n) - \eta_1).$$

Here, $m_1, \ldots, m_{n_H(\ell)}$ are the numbers of agents in each isolated group, and each connected component of $H(\ell)$ has length at least $r(n)$. Now, take $\eta_1 = (nr(n) - 2\pi)n^{-1} = \delta(n)n^{-1}$, and the contradiction follows from

$$2\pi \geq n_H(\ell)\eta_1 + nr(n) - n\eta_1$$
$$= n_H(\ell)\eta_1 + nr(n) + 2\pi - nr(n) = n_H(\ell)\eta_1 + 2\pi.$$

In summary, this shows that the number of connected components of $H(\ell)$ decreases by one in time $O(n \log n + \log \eta_1^{-1}) = O(n \log n + \log(n\delta(n)^{-1}))$. Note that δ being lower bounded implies that $n\delta(n)^{-1} = O(n)$ and, therefore, $O(n \log n + \log(n\delta(n)^{-1})) = O(n \log n)$. Iterating this argument n times, in time $O(n^2 \log n)$ the set H will become empty. At that time, the resulting network will obey the discrete-time linear time-invariant dynamical system

$$d(\tau + 1) = \mathrm{Circ}_n(k_{\mathrm{prop}}, 1 - k_{\mathrm{prop}}, 0)\, d(\tau), \tag{3.6.2}$$

where the linear map $(a, b, c) \mapsto \mathrm{Circ}_n(a, b, c) \in \mathbb{R}^{n \times n}$ is defined in Section 1.6.4 and where $d : \mathbb{Z}_{\geq 0} \to \mathbb{R}^n_{>0}$ is defined by

$$d(\tau) = \big(\mathrm{dist}_{\mathrm{cc}}(\theta^{[1]}(\tau), \theta^{[2]}(\tau)), \ldots, \mathrm{dist}_{\mathrm{cc}}(\theta^{[n]}(\tau), \theta^{[n+1]}(\tau)) \big),$$

with the convention $\theta^{[n+1]} = \theta^{[1]}$. By Theorem 1.79(iii) in Section 1.6.4, in time $O\big(n^2 \log \varepsilon^{-1}\big)$, the error 2-norm satisfies the contraction inequality $\|d(\tau) - d_*\|_2 \leq \varepsilon \|d(0) - d_*\|_2$, for $d_* = \frac{2\pi}{n} \mathbf{1}_n$. We convert this inequality on 2-norms into an appropriate inequality on ∞-norms as follows. Note that $\|d(0) - d_*\|_\infty = \max_{i \in \{1, \ldots, n\}} |d^{[i]}(0) - d_*^{[i]}| \leq 2\pi$. For $\eta_2 \in {]0, 1[}$ and for τ of order $n^2 \log \eta_2^{-1}$,

$$\begin{aligned} \|d(\tau) - d_*\|_\infty &\leq \|d(\tau) - d_*\|_2 \leq \eta_2 \|d(0) - d_*\|_2 \\ &\leq \eta_2 \sqrt{n} \|d(0) - d_*\|_\infty \leq \eta_2 2\pi \sqrt{n}. \end{aligned}$$

This means that the desired configuration is achieved for $\eta_2 2\pi \sqrt{n} = \varepsilon$, that is, in time $O(n^2 \log \eta_2^{-1}) = O(n^2 \log(n\varepsilon^{-1}))$. In summary, the equidistance task is achieved in time $O(n^2 \log(n\varepsilon^{-1}))$.

STEP 4: Finally, we prove the lower bound in (ii). As we reasoned before, $\mathrm{TC}(\mathcal{T}_{\mathrm{dir}}, \mathcal{CC}_{\mathrm{AGREE~\&~PURSUE}})$ is negligible as compared with the claimed lower bound estimate for $\mathrm{TC}(\mathcal{T}_{\varepsilon\text{-eqdstnc}}, \mathcal{CC}_{\mathrm{AGREE~\&~PURSUE}})$ and, therefore, we assume that $\mathcal{T}_{\mathrm{dir}}$ has been achieved. We consider an initial configuration with the properties that: (i) agents are counterclockwise-ordered according to their unique identifier; (ii) the set $H(0)$ is empty; and (iii) the inter-agent distances $d(0) = \big(\mathrm{dist}_{\mathrm{cc}}(\theta^{[1]}(0), \theta^{[2]}(0)), \ldots, \mathrm{dist}_{\mathrm{cc}}(\theta^{[n]}(0), \theta^{[1]}(0)) \big)$ are

$$d(0) = \frac{2\pi}{n} \mathbf{1}_n + \frac{\pi - \varepsilon'}{n} (\mathbf{v}_n + \overline{\mathbf{v}}_n),$$

where $\varepsilon' \in {]\pi, 0[}$ and where $\mathbf{v}_n$ is the eigenvector of $\mathrm{Circ}_n(k_{\mathrm{prop}}, 1 - k_{\mathrm{prop}}, 0)$ corresponding to the eigenvalue $1 - k_{\mathrm{prop}} + k_{\mathrm{prop}} \cos\left(\frac{2\pi}{n}\right) - k_{\mathrm{prop}} \sqrt{-1} \sin\left(\frac{2\pi}{n}\right)$ (see Section 1.6.4). Straightforward calculations show the equality $\mathbf{v}_n + \overline{\mathbf{v}}_n = 2(1, \cos(2\pi/n), \ldots, \cos((n-1)2\pi/n))$ and that $\|\mathbf{v}_n + \overline{\mathbf{v}}_n\|_2 = \sqrt{2n}$. In turn, this implies that $d(0) \in \mathbb{R}^n_{>0}$ and that $\|d(0) - \frac{2\pi}{n} \mathbf{1}_n\|_2 \in O(1/\sqrt{n})$. Take $\eta_3 \in {]0, 1[}$. The argument described in the proof of Theorem 1.79(iii) leads to the following statement: the 2-norm of the difference between $\ell \mapsto d(\ell)$ and the desired configuration $\frac{2\pi}{n} \mathbf{1}_n$ decreases by a factor η_3 in time of order $n^2 \log \eta_3^{-1}$. Given an initial error of order $O(1/\sqrt{n})$ and a final desired error of order ε, we set $\eta_3 = \varepsilon \sqrt{n}$ and obtain the desired result that it takes time of order $n^2 \log(n\varepsilon)^{-1}$ to reduce the 2-norm error and, therefore, the ∞-norm error to size ε. This concludes the proof. $\blacksquare$

3.6.3 Proof of Theorem 3.34

Proof. Note that the number of edges in $\mathcal{S}_{\text{circle}}$ is at most $O(n^2)$, as it is possible that all robots are within distance $r(n)$ of each other. The upper bounds in (i) and (ii) then follow from inequality (3.3.1) and Theorem 3.33. To prove the lower bounds, we follow the steps and notation in the proof of Theorem 3.33. Regarding the lower bounds in (i), we examine the evolution of the initial configuration depicted in Figure 3.10. From *STEP 2:* in the proof of Theorem 3.33, recall that the time it takes agent 1 to receive the message with `max-id` $= n$ is lower bounded by $\pi/(k_{\text{prop}}r(n)) - 1$. Our proof strategy is to lower bound the number of edges in the graph until this event happens. Note that, at initial time, there are $(n-1)^2$ edges in the communication graph of the network and, therefore, $(n-1)^2$ messages get transmitted. At the next communication round, agent 1 has moved $k_{\text{prop}}r(n)$ counterclockwise and, therefore, the number of edges is lower bounded by $(n-2)^2$. Iterating this reasoning, we see that after $i < \pi/(k_{\text{prop}}r(n))$ communication rounds, the number of edges is lower bounded by $(n-i)^2$. Now, if $\delta(n) > \pi(1/k_{\text{prop}} - 2)$, then $n > \pi/k_{\text{prop}}r(n))$, and therefore, the total communication complexity is lower bounded by

$$\sum_{i=1}^{\frac{\pi}{k_{\text{prop}}r(n)}} (n-i)^2 \in \Omega(n^2 r(n)^{-1}).$$

On the other hand, if $\delta(n) < \pi(1/k_{\text{prop}}-2)$, then $n < \pi/k_{\text{prop}}r(n))$, and after n time steps, we lower bound the number of edges in the communication graph by the number of edges in a chain of length n, that is, $n-1$. Therefore, the total communication complexity is lower bounded by

$$\sum_{i=1}^{n}(n-i)^2 + (n-1)\left(\frac{\pi}{k_{\text{prop}}r(n)} - n\right) \in \Omega(n^3 + nr(n)^{-1}).$$

The two lower bounds match when $\delta(n) = \pi(1/k_{\text{prop}} - 2)$.

Regarding the lower bound in (ii), we consider first the case when $n_H(0) = 0$. In this case, the network obeys the discrete-time linear time-invariant dynamical system (3.6.2). Consider the initial condition $d(0)$ that we adopted for *STEP 4:*. We know it takes time of order $n^2 \log(n\varepsilon)^{-1}$ for the appropriate contraction property to hold. At $d(0)$, the maximal inter-agent distance is $(4\pi - \varepsilon')/n$ and it decreases during the evolution. Because each robot can communicate with any other robot within a distance $r(n)$, the number of agents within communication range of a given agent is of order $r(n)n/(4\pi - \varepsilon')$, that is, of order $\delta(n)$. From here, we deduce that the total communication complexity belongs to $\Omega(n^3\delta(n)\log(n\varepsilon)^{-1})$. ∎

3.7 EXERCISES

E3.1   **(Orientation dynamics).** We review some basic kinematic concepts about orientation dynamics, (see, e.g., Bullo and Lewis, 2004; Spong et al., 2006. Define the set of skew-symmetric matrices in $\mathbb{R}^{d \times d}$ as

$$\mathfrak{so}(d) = \{S \in \mathbb{R}^{d \times d} \mid S = -S^T\}.$$

Let $\times$ denote the cross-product on $\mathbb{R}^3$ and define the linear map $\widehat{\cdot} : \mathbb{R}^3 \to \mathfrak{so}(3)$ by $\widehat{x}y = x \times y$ for all $y \in \mathbb{R}^3$.

(i) Show that, if $x = (x_1, x_2, x_3)$, then:

$$\widehat{x} = \begin{bmatrix} 0 & -x_3 & x_2 \\ x_3 & 0 & -x_1 \\ -x_2 & x_1 & 0 \end{bmatrix}.$$

(ii) Given a differentiable curve $R : [0, T] \to \mathrm{SO}(3)$, show that there exists a curve $\omega : [0, T] \to \mathbb{R}^3$ such that

$$\dot{R}(t) = R(t)\widehat{\omega}(t).$$

These two results lead to a motion model of a relative sensing network with time-varying orientation. Generalizing the constant-orientation model in equation (3.2.2), the complete position and orientation dynamics may be written as

$$\dot{p}^{[i]}_{\text{fixed}}(t) = R^{[i]}_{\text{fixed}}(t) \, u_i^{[i]},$$
$$\dot{R}^{[i]}_{\text{fixed}}(t) = R^{[i]}_{\text{fixed}}(t) \, \widehat{\omega}_i^{[i]},$$

where, for $i \in \{1, \dots, n\}$, $u_i^{[i]}$ and $\omega_i^{[i]}$ are the linear and the body angular velocities of robot i, respectively.

E3.2   **(Variation of the agree & pursue control and communication law).** Consider the AGREE & PURSUE control and communication law defined in Section 3.1.3, with the state transition function replaced by the following:

```
function stf(θ, w, y)
1: for each non-null message (θrcvd, (dirrcvd, max-idrcvd)) in y do
2:     if (max-idrcvd > max-id) then
3:         new-dir := dirrcvd
4:         new-id := max-idrcvd
5: return (new-dir, new-id)
```

The only difference between this law and the AGREE & PURSUE law in Section 3.1.3 is that, in each communication round, each agent picks the message with the largest value of max-id among all messages received (instead of among the messages received only from agents moving towards its position). We refer to this law as MOD-AGREE & PURSUE.

Consider the direction agreement task $\mathcal{T}_{\text{dir}} : (\mathbb{S}^1)^n \times W^n \to \{\texttt{true}, \texttt{false}\}$ defined in Example 3.22. Assume that $\text{dir}^{[n]}(0) = \texttt{c}$, and let $k \in \{1, \dots, n-1\}$ be the largest identity such that $\text{dir}^{[k]}(0) = \texttt{cc}$. Do the following tasks:

(i) Show that if the message from agent k gets delivered to agents clockwise-placed with respect to agent k along two consecutive communication rounds, then the message from agent k has traveled at least $(1 - k_{\text{prop}})r(n)$ along the circle in the clockwise direction.

(ii) Show that, if $\text{dist}_{\text{cc}}(\theta^{[n]}(0), \theta^{[k]}(0)) < 2\,r(n)$, then

$$\text{TC}(\mathcal{T}_{\text{dir}}, \mathcal{CC}_{\text{ MOD-AGREE \& PURSUE}}, x_0, w_0) = \Theta(r(n)^{-1}).$$

(iii) Implement the algorithm in your favorite simulation software (for example, Mathematica© Matlab© or Maple©), and compute the time complexity of multiple executions of the algorithm starting from different initial conditions. Does your simulation analysis support the conjecture that

$$\text{TC}(\mathcal{T}_{\text{dir}}, \mathcal{CC}_{\text{ MOD-AGREE \& PURSUE}}) = \Theta(r(n)^{-1})?$$

For the simulation analysis to be relevant, you should use a large number of randomly generated initial physical positions and processor states.

E3.3 **(Leader-following flocking).** Consider a group of robots moving in $\mathbb{R}^2$ according to the following discrete-time version of the planar vehicle dynamics introduced in Example 3.1:

$$x(\ell+1) = x(\ell) + v\cos(\theta(\ell)),$$
$$y(\ell+1) = y(\ell) + v\sin(\theta(\ell)),$$
$$\theta(\ell+1) = \theta(\ell) + \omega.$$

We let $\{(p^{[1]}, \theta^{[1]}), \ldots, (p^{[n]}, \theta^{[n]})\}$ denote the robot physical states, where $p^{[i]} = (x^{[i]}, y^{[i]}) \in \mathbb{R}^2$ corresponds to the position and $\theta^{[i]} \in [0, 2\pi)$ corresponds to the orientation of the robot $i \in I$. As communication graph, we adopt the r-disk graph $\mathcal{G}_{\text{disk}}(r)$ introduced in Section 2.2.

Assume that all agents move at unit speed, $v = 1$, and update their heading according to the leader-following version of Vicsek's model (see equation (1.6.5)):

$$\theta^{[1]}(\ell+1) = \theta^{[1]}(\ell), \qquad\qquad\qquad\qquad\qquad\qquad (\text{E3.1})$$
$$\theta^{[i]}(\ell+1) = \text{avrg}\Big(\{\theta^{[i]}(\ell)\} \cup \{\theta^{[j]}(\ell) \mid j \text{ s.t. } \|p^{[j]}(\ell) - p^{[i]}(\ell)\|_2 \le r\}\Big),$$

for $i \in \{2, \ldots, n\}$. Do the following tasks:

(i) Write the algorithm formally as a control and communication law as defined in Section 3.1.2.

(ii) Given initial conditions for the position and orientation of the robots, express (E3.1) as the time-dependent linear iteration associated to a sequence of matrices $\{F(\ell) \mid \ell \in \mathbb{Z}_{\geq 0}\}$. Are these matrices stochastic? Are they symmetric? Is the sequence non-degenerate?

(iii) We loosely define the flocking task as achieving agreement on the heading of the agents. Using Theorem 1.63, identify connectivity conditions on the sequence of graphs determined by the evolution of the network that guarantee that agents achieve flocking. What is the final orientation in which the network flocks?

Chapter Four

Connectivity maintenance and rendezvous

The aims of this chapter are twofold. First, we introduce the rendezvous problem and analyze various coordination algorithms that achieve it, providing upper and lower bounds on their time complexity. Second, we introduce the problem of maintaining connectivity among a group of mobile robots and use geometric approaches to preserve this topological property of the network.

Loosely speaking, the *rendezvous objective* is to achieve agreement over the physical location of as many robots as possible, that is, to steer the robots to a common location. This objective is to be achieved with the limited information flow described in the model of the network. Typically, it will be impossible to solve the rendezvous problem for all robots if the robots are placed in such a way that they do not form a connected communication graph. Therefore, it is reasonable to assume that the network is connected at initial time, and that a good property of any rendezvous algorithm is that of maintaining some form of connectivity among robots. This discussion motivates the *connectivity maintenance problem*. Once a model for when two robots can acquire each other's relative position is adopted, this problem is of particular relevance, as the inter-robot topology depends on the physical states of the robots. Our exposition here is mainly based on Ando et al. (1999), Cortés et al. (2006), and Ganguli et al. (2009).

The chapter is organized as follows. In the first section, we formally introduce the two coordination problems. In the second section, we define various connectivity constraint sets to limit the motion of robots in order to maintain network connectivity. These notions of constraint sets allows us to study in the next section various rendezvous algorithms with connectivity maintenance properties. We study numerous variations of the circumcenter algorithm for the rendezvous objective and we characterize its complexity. Additionally, we introduce the perimeter-minimizing algorithm for nonconvex environments. The fourth section presents various simulations of the proposed motion-coordination algorithms. We end the chapter with three sections on, respectively, bibliographic notes, proofs of the results presented in the chapter, and exercises. Our technical treatment is based on the LaSalle Invariance Principle, on linear distributed algorithms, and on geometric tools such as proximity graphs and robust visibility.

4.1 PROBLEM STATEMENT

We begin this section by reviewing the classes of networks and the types of problems that will be considered in the chapter.

4.1.1 Networks with discrete-time motion

In the course of the chapter, we will consider the robotic networks $\mathcal{S}_{\text{disk}}$, $\mathcal{S}_{\text{LD}}$, and $\mathcal{S}_{\infty\text{-disk}}$, and the relative-sensing networks $\mathcal{S}_{\text{disk}}^{\text{rs}}$ and $\mathcal{S}_{\text{vis-disk}}^{\text{rs}}$ presented in Example 3.4 and in Section 3.2.2.

For the robotic networks $\mathcal{S}_{\text{disk}}$, $\mathcal{S}_{\text{LD}}$, and $\mathcal{S}_{\infty\text{-disk}}$, we will, however, assume that the robots move in discrete time, that is, we adopt the discrete-time motion model

$$p^{[i]}(\ell + 1) = p^{[i]}(\ell) + u^{[i]}(\ell), \quad i \in \{1, \dots, n\}. \tag{4.1.1}$$

Similarly, for the relative-sensing networks $\mathcal{S}_{\text{disk}}^{\text{rs}}$ and $\mathcal{S}_{\text{vis-disk}}^{\text{rs}}$, we adopt the discrete-time motion model

$$p_{\text{fixed}}^{[i]}(\ell + 1) = p_{\text{fixed}}^{[i]}(\ell) + R_{\text{fixed}}^{[i]} u_i^{[i]}(\ell), \quad i \in \{1, \dots, n\}. \tag{4.1.2}$$

As an aside, if we express the previous equation with respect to frame i at time t, then equation (4.1.2) reads

$$p_{(\text{frame } i \text{ at time } \ell)}^{[i]}(\ell + 1) = u_{(\text{frame } i \text{ at time } \ell)}^{[i]}(\ell), \quad i \in \{1, \dots, n\}.$$

We present the treatment in discrete time for simplicity. It is easy to show that any control law for the discrete-time motion model can be implemented in the continuous-time networks. In what follows, we begin our discussion by assuming no bound on the control magnitude and we later introduce an upper bound denoted by $u_{\max}$.

4.1.2 The rendezvous task

Next, we discuss the rendezvous problem. There are different ways of formulating this objective in terms of task maps. Let $\mathcal{S} = (\{1, \dots, n\}, \mathcal{R}, E_{\text{cmm}})$ be a uniform robotic network. The *(exact) rendezvous task* $\mathcal{T}_{\text{rndzvs}} : X^n \to$

$\{\texttt{true}, \texttt{false}\}$ for $\mathcal{S}$ is the coordination task defined by

$$\mathcal{T}_{\text{rndzvs}}(x^{[1]}, \dots, x^{[n]})$$
$$= \begin{cases} \texttt{true}, & \text{if } x^{[i]} = x^{[j]}, \text{ for all } (i, j) \in E_{\text{cmm}}(x^{[1]}, \dots, x^{[n]}), \\ \texttt{false}, & \text{otherwise}. \end{cases}$$

Next, assume that, for the same network $\mathcal{S} = (\{1, \dots, n\}, \mathcal{R}, E_{\text{cmm}})$, the robots' physical state space is $X \subset \mathbb{R}^d$. It is convenient to review some basic notation consistent with what we adopted in Chapter 2. We let $\mathcal{P} = \{p^{[1]}, \dots, p^{[n]}\}$ denote the set of agents' location in $X \subset \mathbb{R}^d$ and we let P be an array of n points in $\mathbb{R}^d$. Furthermore, we let avrg denote the average of a finite point set in $\mathbb{R}^d$, that is,

$$\text{avrg}(\{q_1, \dots, q_k\}) = \frac{1}{k}(q_1 + \dots + q_k).$$

For $\varepsilon \in \mathbb{R}_{>0}$, the ε-*rendezvous task* $\mathcal{T}_{\varepsilon\text{-rndzvs}} : (\mathbb{R}^d)^n \to \{\texttt{true}, \texttt{false}\}$ for $\mathcal{S}$ is defined as follows: $\mathcal{T}_{\varepsilon\text{-rndzvs}}$ is $\texttt{true}$ at P if and only if each robot position $p^{[i]}$, for $i \in \{1, \dots, n\}$, is at distance less than ε from the average position of its E_{cmm}-neighbors. Formally,

$$\mathcal{T}_{\varepsilon\text{-rndzvs}}(P) = \texttt{true}$$
$$\iff \|p^{[i]} - \text{avrg}\left(\{p^{[j]} \mid (i, j) \in E_{\text{cmm}}(P)\}\right)\|_2 < \varepsilon, \quad i \in \{1, \dots, n\}.$$

4.1.3 The connectivity maintenance problem

Assume that the communication graph, computed as a function of the robot positions, is connected: How should the robots move in such a way that their communication graph is again connected? Clearly, the problem depends upon: (1) how the robots move; and (2) what proximity graph describes the communication graph or, in the case of relative-sensing networks, what sensor model is available on each robot.

The key idea is to restrict the allowable motion of each agent. Different motion constraint sets correspond to different communication or sensing graphs. We have three objectives in doing so. First, we aim to achieve this objective only based on local measurements or 1-hop communication, that is, without introducing processor states explicitly dedicated to this task. Second, the constraint sets should depend continuously on the position of the robots. Third, we have the somehow informal objective to design the constraint sets as "large" as possible so as to minimally constrain the motion of the robots.

4.2 CONNECTIVITY MAINTENANCE ALGORITHMS

In this section, we present some algorithms that might be used by a robotic network to maintain communication connectivity. The results presented in this section start with the original idea introduced by Ando et al. (1999) for first-order robots communicating along the edges of a disk graph, that is, for the network described in Example 3.4. This idea is then generalized to a number of useful settings. The properties of proximity graphs presented in Section 2.2 play a key role in formulating and solving the connectivity problem.

4.2.1 Enforcing range-limited links

First, we aim to constrain the motion of two first-order agents in order to maintain a communication link between them. We assume that the communication takes place over the disk graph $\mathcal{G}_{\mathrm{disk}}(r)$ with communication range $r > 0$.

Loosely stated, the *pairwise connectivity maintenance problem* is as follows: given two neighbors in the proximity graph $\mathcal{G}_{\mathrm{disk}}(r)$, find a rich set of control inputs for both agents with the property that, after moving, both agents are again within distance r. We provide a solution to this problem as follows.

Definition 4.1 (Pairwise connectivity constraint set). Consider two agents i and j at positions $p^{[i]} \in \mathbb{R}^d$ and $p^{[j]} \in \mathbb{R}^d$ such that $\|p^{[i]} - p^{[j]}\|_2 \leq r$. The *connectivity constraint set* of agent i with respect to agent j is

$$\mathcal{X}_{\mathrm{disk}}\big(p^{[i]}, p^{[j]}\big) = \overline{B}\Big(\frac{p^{[j]} + p^{[i]}}{2}, \frac{r}{2}\Big). \qquad \bullet$$

Note that both robots, i and j, can independently compute their respective connectivity constraint sets. The proof of the following result is straightforward.

Lemma 4.2 (Maintaining pairwise connectivity). *Assume that the distance between agents $p^{[i]}$ and $p^{[j]}$ is no more than r, at some time ℓ. If the control $u^{[i]}(\ell)$ takes value in*

$$u^{[i]}(\ell) \in \mathcal{X}_{\mathrm{disk}}\big(p^{[i]}(\ell), p^{[j]}(\ell)\big) - p^{[i]}(\ell) = \overline{B}\Big(\frac{p^{[j]}(\ell) - p^{[i]}(\ell)}{2}, \frac{r}{2}\Big),$$

and, similarly, $u^{[j]}(\ell) \in \mathcal{X}_{\mathrm{disk}}\big(p^{[j]}(\ell), p^{[i]}(\ell)\big) - p^{[j]}(\ell)$, then, according to the discrete-time motion model (4.1.1):

(i) *the positions of both agents at time $\ell + 1$ are inside the connectivity constraint set $\mathcal{X}_{\text{disk}}\big(p^{[i]}(\ell), p^{[j]}(\ell)\big)$; and*

(ii) *the distance between the agents' positions at time $\ell + 1$ is no more than r.*

We illustrate these pairwise connectivity maintenance concepts in Figure 4.1.

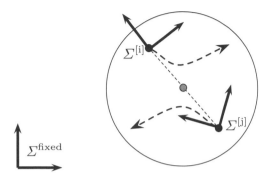

Figure 4.1 An illustration of the connectivity maintenance constraint. Starting from positions $p^{[i]}$ and $p^{[j]}$, the robots are restricted to moving inside the disk centered at $\mathcal{X}_{\text{disk}}(p^{[i]}, p^{[j]}) = \frac{1}{2}\big(p^{[i]} + p^{[j]}\big)$ with radius $\frac{r}{2}$.

Remark 4.3 (Constraints for relative-sensing networks). Let us consider a relative-sensing network with a disk sensor of radius r (see Example 3.15). Recall the following facts about this model. First, agent i measures the position of robot j in its frame $\Sigma^{[i]}$, that is, robot i measures $p_i^{[j]}$. Second, $p_i^{[i]} = \mathbf{0}_d$. Third, if $W \subset \mathbb{R}^d$, then W_i denotes its expression in the frame $\Sigma^{[i]}$. Combining these notions and assuming that the interagent distance is no more than r, the pairwise connectivity constraint set in Definition 4.1 satisfies

$$\Big(\mathcal{X}_{\text{disk}}(p^{[i]}, p^{[j]})\Big)_i = \mathcal{X}_{\text{disk}}\big(\mathbf{0}_d, p_i^{[j]}\big) = \overline{B}\Big(\frac{p_i^{[j]}}{2}, \frac{r}{2}\Big). \qquad \bullet$$

4.2.2 Enforcing network connectivity

Here, we focus on how to constrain the mobility of multiple agents in order to maintain connectivity for the entire network that they form. We again consider the case of first-order agents moving according to the discrete-time equation (4.1.1) and communicating over $\mathcal{G}_{\text{disk}}(r)$.

Loosely stated, the *network connectivity maintenance problem* is as follows: Given n agents at positions $\mathcal{P}(\ell) = \{p^{[1]}(\ell), \ldots, p^{[n]}(\ell)\}$ in which they form a connected r-disk graph $\mathcal{G}_{\text{disk}}(r)$, the objective is to find a rich set of control inputs for all agents with the property that, at time $\ell + 1$, the agents' new positions $\mathcal{P}(\ell + 1)$ again form a connected r-disk graph $\mathcal{G}_{\text{disk}}(r)$. We provide a simple, but potentially conservative, solution to this problem as follows.

Definition 4.4 (Connectivity constraint set for groups of agents). Consider a group of agents at positions $\mathcal{P} = \{p^{[1]}, \ldots, p^{[n]}\} \subset \mathbb{R}^d$. The *connectivity constraint set* of agent i with respect to $\mathcal{P}$ is

$$\mathcal{X}_{\text{disk}}(p^{[i]}, \mathcal{P}) = \left\{ x \in \mathcal{X}_{\text{disk}}(p^{[i]}, q) \mid q \in \mathcal{P} \setminus \{p^{[i]}\} \text{ s.t. } \|q - p^{[i]}\|_2 \leq r \right\}. \quad \bullet$$

In other words, if $q_1, \ldots, q_l$ are agents' positions whose distance from $p^{[i]}$ is no more than r, then the connectivity constraint set for agent i is the intersection of the constraint sets $\overline{B}\big(\frac{1}{2}(q_k + p^{[i]}), \frac{r}{2}\big)$ for $k \in \{1, \ldots, l\}$ (see Figure 4.2).

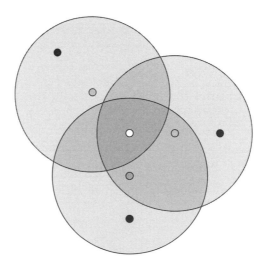

Figure 4.2 An illustration of network connectivity maintenance. The connectivity $\mathcal{X}_{\text{disk}}$-constraint set of the white-colored agent is the intersection of the individual constraint sets determined by its neighbors.

The following result is a consequence of Lemma 4.2.

Lemma 4.5 (Maintaining network connectivity). *Consider a group of agents at positions $\mathcal{P}(\ell) = \{p^{[1]}(\ell), \ldots, p^{[n]}(\ell)\} \subset \mathbb{R}^d$ at time ℓ. If each* *agent's control $u^{[i]}(\ell)$ takes value in*

$$u^{[i]}(\ell) \in \mathcal{X}_{\mathrm{disk}}\big(p^{[i]}(\ell), \mathcal{P}(\ell)\big) - p^{[i]}(\ell), \quad i \in \{1, \ldots, n\},$$

then, according to the discrete-time motion model (4.1.1):

(i) each agent remains in its connectivity constraint set, that is, $p^{[i]}(\ell + 1) \in \mathcal{X}_{\mathrm{disk}}(p^{[i]}(\ell), \mathcal{P}(\ell))$;

(ii) each edge of $\mathcal{G}_{\mathrm{disk}}(r)$ at $\mathcal{P}(\ell)$ is maintained after the motion step, that is, if $\|p^{[i]}(\ell) - p^{[j]}(\ell)\|_2 \leq r$, then also $\|p^{[i]}(\ell+1) - p^{[j]}(\ell+1)\|_2 \leq r$;

(iii) if $\mathcal{G}_{\mathrm{disk}}(r)$ at time ℓ is connected, then $\mathcal{G}_{\mathrm{disk}}(r)$ at time $\ell + 1$ is connected; and

(iv) the number of connected components of the graph $\mathcal{G}_{\mathrm{disk}}(r)$ at time $\ell+1$ is equal to or smaller than the number of connected components of the graph $\mathcal{G}_{\mathrm{disk}}(r)$ at time ℓ.

Remark 4.6 (Constraints for relative-sensing networks: cont'd).
Following up on Remark 4.3, the connectivity constraint set in Definition 4.4, written in the frame $\Sigma^{[i]}$, is

$$\mathcal{X}_{\mathrm{disk}}(\mathbf{0}_d, \{p_i^{[1]}, \ldots, p_i^{[n]}\})$$
$$= \Big\{ x \in \overline{B}\Big(\frac{p_i^{[j]}}{2}, \frac{r}{2}\Big) \mid j \neq i \text{ such that } \|p^{[j]} - p^{[i]}\|_2 \leq r \Big\}. \quad \bullet$$

Next, we relax the constraints in Definition 4.4 to provide the network nodes with larger, and therefore less conservative, motion-constraint sets. Recall from Section 2.2 the relative neighborhood graph $\mathcal{G}_{\mathrm{RN}}$, the Gabriel graph $\mathcal{G}_{\mathrm{G}}$, and the r-limited Delaunay graph $\mathcal{G}_{\mathrm{LD}}(r)$. These proximity graphs are illustrated in Figure 2.8. From Theorem 2.8 and Proposition 2.9, respectively, recall that the proximity graphs $\mathcal{G}_{\mathrm{RN}} \cap \mathcal{G}_{\mathrm{disk}}(r)$, $\mathcal{G}_{\mathrm{G}} \cap \mathcal{G}_{\mathrm{disk}}(r)$, and $\mathcal{G}_{\mathrm{LD}}(r)$ have the following properties:

(i) they have the same connected components as $\mathcal{G}_{\mathrm{disk}}(r)$, that is, for all point sets $\mathcal{P} \subset \mathbb{R}^d$, all graphs have the same number of connected components consisting of the same vertices; and

(ii) they are spatially distributed over $\mathcal{G}_{\mathrm{disk}}(r)$.

These mathematical facts have two implications. First, to maintain or decrease the number of connected components of a disk graph, it is sufficient to maintain or decrease the number of connected components of any of the three proximity graphs $\mathcal{G}_{\mathrm{RN}} \cap \mathcal{G}_{\mathrm{disk}}(r)$, $\mathcal{G}_{\mathrm{G}} \cap \mathcal{G}_{\mathrm{disk}}(r)$, and $\mathcal{G}_{\mathrm{LD}}(r)$. Because each of these graphs is more sparse than the disk graph, that is, they are subgraphs of $\mathcal{G}_{\mathrm{disk}}(r)$, fewer connectivity constraints need to be imposed.

Second, because these proximity graphs are spatially distributed over the disk graph, it is possible for each agent to determine which of its neighbors in $\mathcal{G}_{\text{disk}}(r)$ are also its neighbors in these subgraphs. We formalize this discussion as follows.

Definition 4.7 ($\mathcal{G}$-connectivity constraint set). Let $\mathcal{G}$ be a proximity graph that is spatially distributed over $\mathcal{G}_{\text{disk}}(r)$ and that has the same connected components as $\mathcal{G}_{\text{disk}}(r)$. Consider a group of agents at positions $\mathcal{P} = \{p^{[1]}, \ldots, p^{[n]}\} \subset \mathbb{R}^d$. The $\mathcal{G}$-connectivity constraint set of agent i with respect to $\mathcal{P}$ is

$$\mathcal{X}_{\text{disk},\mathcal{G}}(p^{[i]}, \mathcal{P})$$
$$= \left\{ x \in \mathcal{X}_{\text{disk}}(p^{[i]}, q) \mid q \in \mathcal{P} \text{ s.t. } (q, p^{[i]}) \text{ is an edge of } \mathcal{G}(\mathcal{P}) \right\}. \quad \bullet$$

Lemma 4.8 (Maintaining connectivity of sparser networks). *Let $\mathcal{G}$ be a proximity graph that is spatially distributed over $\mathcal{G}_{\text{disk}}(r)$ and that has the same connected components as $\mathcal{G}_{\text{disk}}(r)$. Consider a group of agents at positions $\mathcal{P}(\ell) = \{p^{[1]}(\ell), \ldots, p^{[n]}(\ell)\} \subset \mathbb{R}^d$ at time ℓ. If each agent's control $u^{[i]}(\ell)$ takes value in*

$$u^{[i]}(\ell) \in \mathcal{X}_{\text{disk},\mathcal{G}}\big(p^{[i]}(\ell), \mathcal{P}(\ell)\big) - p^{[i]}(\ell), \quad i \in \{1, \ldots, n\},$$

then, according to the discrete-time motion model (4.1.1):

(i) each agent remains in its $\mathcal{G}$-connectivity constraint set;

(ii) two agents that are in the same connected component of $\mathcal{G}$ remain at the same connected component after the motion step; and

(iii) the number of connected components of the graph $\mathcal{G}$ at $\mathcal{P}(\ell+1)$ is equal to or smaller than the number of connected components of the graph $\mathcal{G}$ at $\mathcal{P}(\ell)$.

The reader is asked to provide a proof of this result in Exercise E4.1.

4.2.3 Enforcing range-limited line-of-sight links and network connectivity

Here, we consider the connectivity maintenance problem for a group of agents with range-limited line-of-sight communication, as described in Example 3.6. It is convenient to treat directly and only the case of a compact allowable nonconvex environment $Q \subset \mathbb{R}^2$ contracted into $Q_\delta = \{q \in Q \mid \text{dist}(q, \partial Q) \geq \delta\}$ for a small positive δ. We present a solution based on designing constraint sets that guarantee that every edge of the range-limited visibility graph $\mathcal{G}_{\text{vis-disk},Q_\delta}$ is preserved.

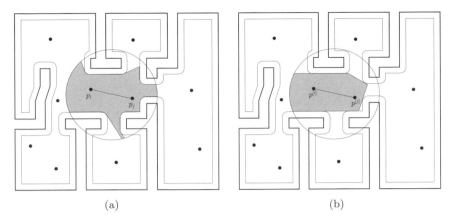

(a)   (b)

Figure 4.3 Image (a) shows the set $\mathrm{Vi}_{\mathrm{disk}}(p^{[i]}; Q_\delta) \cap \overline{B}(\frac{1}{2}(p^{[i]} + p^{[j]}), \frac{r}{2})$. Image (b) illustrates the execution of the ITERATED TRUNCATION ALGORITHM. Robots i and j are constrained to remain inside the shaded region in (b), which is a convex subset of Q_δ and of the closed ball with center $\frac{1}{2}(p^{[i]} + p^{[j]})$ and radius $\frac{r}{2}$.

We begin with a useful observation and a corresponding geometric algorithm. Assume that, at time ℓ, robot j is inside the range-limited visibility set from $p^{[i]}$ in Q_δ, that is, with the notation of Section 2.1.2,

$$p^{[j]}(\ell) \in \mathrm{Vi}_{\mathrm{disk}}(p^{[i]}(\ell); Q_\delta) = \mathrm{Vi}(p^{[i]}(\ell); Q_\delta) \cap \overline{B}(p^{[i]}(\ell), r).$$

This property holds also at time $\ell + 1$ if $\|p^{[i]}(\ell+1) - p^{[j]}(\ell+1)\|_2 \le r$ and $[p^{[i]}(\ell+1), p^{[j]}(\ell+1)] \subset Q_\delta$. A sufficient condition is therefore that

$$p^{[i]}(\ell + 1), \; p^{[j]}(\ell + 1) \in \mathcal{X},$$

for some convex subset $\mathcal{X}$ of $Q_\delta \cap \overline{B}(\frac{1}{2}(p^{[i]}(\ell) + p^{[j]}(\ell)), \frac{r}{2})$. Intuitively speaking, $\mathcal{X}$ plays the role of $\mathcal{X}$-constraint set for the proximity graph $\mathcal{G}_{\mathrm{vis\text{-}disk}, Q_\delta}$. The following geometric algorithm, given the positions $p^{[i]}$ and $p^{[j]}$ in an environment Q_δ, computes precisely one such convex subset:

> **function** ITERATED TRUNCATION$(p^{[i]}, p^{[j]}; Q_\delta)$
> % *Executed by robot i at position $p^{[i]}$ assuming that robot j is at position $p^{[j]}$ within range-limited line of sight of $p^{[i]}$*
> 1: $\mathcal{X}_{\mathrm{temp}} := \mathrm{Vi}_{\mathrm{disk}}(p^{[i]}; Q_\delta) \cap \overline{B}(\frac{1}{2}(p^{[i]} + p^{[j]}), \frac{r}{2})$
> 2:   **while** $\partial\mathcal{X}_{\mathrm{temp}}$ contains a concavity **do**
> 3: $v :=$ a strictly concave point of $\partial\mathcal{X}_{\mathrm{temp}}$ closest to $[p^{[i]}, p^{[j]}]$
> 4: $\mathcal{X}_{\mathrm{temp}} := \mathcal{X}_{\mathrm{temp}} \cap H_{Q_\delta}(v)$
> 5:   **return** $\mathcal{X}_{\mathrm{temp}}$

Note: in step 3: multiple points belonging to distinct concavities may satisfy the required property. If so, v may be chosen as any of them.

Figure 4.3 illustrates an example convex constraint set computed by the

ITERATED TRUNCATION ALGORITHM. Figure 4.4 illustrates the step-by-step execution required to generate Figure 4.3(b).

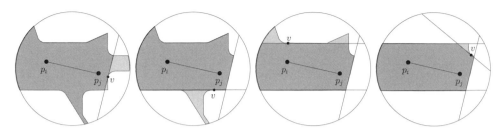

Figure 4.4 From left to right, a sample run of the ITERATED TRUNCATION ALGORITHM. The set $\mathcal{X}_{\text{temp}} := \text{Vi}_{\text{disk}}(p^{[i]}; Q_\delta) \cap \overline{B}(\frac{1}{2}(p^{[i]} + p^{[j]}), \frac{r}{2})$ is shown in Figure 4.3(a). The lightly and darkly shaded regions together represent $\mathcal{X}_{\text{temp}}$ at the current iteration. The darkly shaded region represents $\mathcal{X}_{\text{temp}} \cap H_{Q_\delta}(v)$, where v is as described in step 3:. The outcome of the execution is shown in Figure 4.3(b).

Next, we characterize the main properties of the ITERATED TRUNCATION ALGORITHM. It is convenient to define the set

$$J = \{(p,q) \in Q_\delta \times Q_\delta \mid [p,q] \in Q_\delta \text{ and } \|p - q\|_2 \leq r\}.$$

Proposition 4.9 (Properties of the iterated truncation algorithm). *Consider the δ-contraction of a compact allowable environment Q_δ with κ strict concavities, and let $(p^{[i]}, p^{[j]}) \in J$. The following statements hold:*

(i) *The* ITERATED TRUNCATION ALGORITHM, *invoked with arguments* $(p^{[i]}, p^{[j]}; Q_\delta)$, *terminates in at most κ steps; denote its output by* $\mathcal{X}_{\text{vis-disk}}(p^{[i]}, p^{[j]}; Q_\delta)$.

(ii) $\mathcal{X}_{\text{vis-disk}}(p^{[i]}, p^{[j]}; Q_\delta)$ *is nonempty, compact and convex.*

(iii) $\mathcal{X}_{\text{vis-disk}}(p^{[i]}, p^{[j]}; Q_\delta) = \mathcal{X}_{\text{vis-disk}}(p^{[j]}, p^{[i]}; Q_\delta)$.

(iv) *The set-valued map* $(p, q) \mapsto \mathcal{X}_{\text{vis-disk}}(p, q; Q_\delta)$ *is closed at J.*

In the interest of brevity, we do not include the proof here and instead refer the reader to Ganguli et al. (2009). We just mention that fact (iii) is a consequence of the fact that all relevant concavities in the computation of $\mathcal{X}_{\text{vis-disk}}(p^{[i]}, p^{[j]}; Q_\delta)$ are visible from both agents $p^{[i]}$ and $p^{[j]}$. We are finally ready to provide analogs of Definition 4.4 and Lemma 4.5.

Definition 4.10 (Line-of-sight connectivity constraint set). Consider a nonconvex allowable environment Q_δ and two agents i and j within range-limited line of sight. We call $\mathcal{X}_{\text{vis-disk}}(p^{[i]}, p^{[j]}; Q_\delta)$ the *pairwise line-of-sight connectivity constraint set* of agent i with respect to agent j. Furthermore, given agents at positions $\mathcal{P} = \{p^{[1]}, \ldots, p^{[n]}\} \subset Q_\delta$ that are all within range-limited line of sight of agent i, the *line-of-sight connectivity constraint sets* of agent i with respect to $\mathcal{P}$ is

$$\mathcal{X}_{\text{vis-disk}}(p^{[i]}, \mathcal{P}; Q_\delta) = \{x \in \mathcal{X}_{\text{vis-disk}}(p^{[i]}, q; Q_\delta) \mid q \in \mathcal{P} \setminus \{p^{[i]}\}\}. \qquad \bullet$$

The following result is a consequence of Proposition 4.9.

Lemma 4.11 (Maintaining network line-of-sight connectivity). *Consider a group of agents at positions $\mathcal{P}(\ell) = \{p^{[1]}(\ell), \ldots, p^{[n]}(\ell)\} \subset Q_\delta$ at time ℓ. If each agent's control $u^{[i]}(\ell)$ takes value in*

$$u^{[i]}(\ell) \in \mathcal{X}_{\text{vis-disk}}\big(p^{[i]}(\ell), \mathcal{P}(\ell); Q_\delta\big) - p^{[i]}(\ell), \quad i \in \{1, \ldots, n\},$$

then, according to the discrete-time motion model (4.1.1):

(i) *each agent remains in its constraint set, that is,*

$$p^{[i]}(\ell + 1) \in \mathcal{X}_{\text{vis-disk}}(p^{[i]}(\ell), \mathcal{P}(\ell); Q_\delta);$$

(ii) *each edge of $\mathcal{G}_{\text{vis-disk},Q_\delta}$ at $\mathcal{P}(\ell)$ is maintained after the motion step, that is, if $p^{[i]}$ and $p^{[j]}$ are within range-limited line of sight at time ℓ, then they are within range-limited line of sight also at time $\ell + 1$;*

(iii) *if $\mathcal{G}_{\text{vis-disk},Q_\delta}$ at $\mathcal{P}(\ell)$ is connected, then $\mathcal{G}_{\text{vis-disk},Q_\delta}$ at $\mathcal{P}(\ell + 1)$ is connected; and*

(iv) *the number of connected components of the graph $\mathcal{G}_{\text{vis-disk},Q_\delta}$ at $\mathcal{P}(\ell + 1)$ is equal to or smaller than the number of connected components of the graph $\mathcal{G}_{\text{vis-disk},Q_\delta}$ at $\mathcal{P}(\ell)$.*

Remark 4.12 (Constraints for relative-sensing networks: cont'd). Following up on Remarks 4.3 and 4.6, we consider a relative-sensing network with range-limited visibility sensors (see Example 3.16). To compute the connectivity constraint set for this network, it suffices to provide a relative sensing version of the ITERATED TRUNCATION ALGORITHM:

> **function** RELATIVE-SENSING ITERATED TRUNCATION$(y; y_{\text{env}})$
> % *Executed by robot i with range-limited visibility sensor:*
> *robot measurement is $y = p_i^{[j]} \in \text{Vi}_{\text{disk}}(\mathbf{0}_2; (Q_\delta)_i)$ for $j \neq i$*
> *environment measurement is $y_{\text{env}} = \text{Vi}_{\text{disk}}(\mathbf{0}_2; (Q_\delta)_i)$*
> 1: $\mathcal{X}_{\text{temp}} := y_{\text{env}} \cap \overline{B}\big(\frac{p_i^{[j]}}{2}, \frac{r}{2}\big)$
> 2: **while** $\partial \mathcal{X}_{\text{temp}}$ contains a concavity **do**
> 3: $\quad v :=$ a strictly concave point of $\partial \mathcal{X}_{\text{temp}}$ closest to $[\mathbf{0}_2, y]$
> 4: $\quad \mathcal{X}_{\text{temp}} := \mathcal{X}_{\text{temp}} \cap H_{y_{\text{env}}}(v)$
> 5: **return** $\mathcal{X}_{\text{temp}}$

The algorithm output is $\mathcal{X}_{\text{vis-disk}}(\mathbf{0}_d, y)$, for $y = p_i^{[j]} \in \text{Vi}_{\text{disk}}(\mathbf{0}_2; (Q_\delta)_i)$. $\quad \bullet$

Next, we relax the constraints in Definition 4.10 to provide the network nodes with larger, and therefore less conservative, motion constraint sets. Similarly to Section 4.2.2, we seek to enforce the preservation of a smaller number of range-limited line-of-sight links, while still making sure that the overall network connectivity is preserved. To do this, we recall from Section 2.2 the notion of locally cliqueless graph $\mathcal{G}_{\text{lc},\mathcal{G}}$ of a proximity graph $\mathcal{G}$. This proximity graph is illustrated in Figure 2.12. Let us use the shorthand notation $\mathcal{G}_{\text{lc-vis-disk},Q_\delta} \equiv \mathcal{G}_{\text{lc},\mathcal{G}_{\text{vis-disk},Q_\delta}}$. From Theorems 2.11(ii) and (iii), respectively, recall that $\mathcal{G}_{\text{lc-vis-disk},Q_\delta}$ has the following properties:

(i) it has the same connected components as $\mathcal{G}_{\text{vis-disk},Q_\delta}$, that is, for all point sets $\mathcal{P} \subset \mathbb{R}^d$, the graph has the same number of connected components consisting of the same vertices; and

(ii) it is spatially distributed over $\mathcal{G}_{\text{vis-disk},Q_\delta}$.

Because of (i), to maintain or decrease the number of connected components of a range-limited visibility graph, it is sufficient to maintain or decrease the number of connected components of the sparser graph $\mathcal{G}_{\text{lc-vis-disk},Q_\delta}$. Because of (ii), it is possible for each agent to determine which of its neighbors in $\mathcal{G}_{\text{vis-disk},Q_\delta}$ are its neighbors also in $\mathcal{G}_{\text{lc-vis-disk},Q_\delta}$. We formalize this discussion as follows.

Definition 4.13 (Locally cliqueless line-of-sight connectivity constraint set). Consider a nonconvex allowable environment $Q_\delta \subset \mathbb{R}^2$ and a group of agents at positions $\mathcal{P} = \{p^{[1]}, \ldots, p^{[n]}\} \subset Q$. The *locally cliqueless line-of-sight connectivity constraint set* of agent i with respect to $\mathcal{P}$ is

$$\mathcal{X}_{\text{lc-vis-disk}}(p^{[i]}, \mathcal{P}; Q_\delta) = \big\{ x \in \mathcal{X}_{\text{vis-disk}}(p^{[i]}, q; Q_\delta) \mid$$
$$q \in \mathcal{P} \text{ s.t. } (q, p^{[i]}) \text{ is an edge of } \mathcal{G}_{\text{lc-vis-disk},Q_\delta}(\mathcal{P})\big\}. \quad \bullet$$

The following result is a direct consequence of the previous arguments.

Lemma 4.14 (Maintaining connectivity of sparser networks). *Consider a group of agents at positions $\mathcal{P}(\ell) = \{p^{[1]}(\ell), \ldots, p^{[n]}(\ell)\} \subset Q_\delta$ at time ℓ. If each agent's control $u^{[i]}(\ell)$ takes value in*

$$u^{[i]}(\ell) \in \mathcal{X}_{\text{lc-vis-disk}}\big(p^{[i]}(\ell), \mathcal{P}(\ell); Q_\delta\big) - p^{[i]}(\ell), \quad i \in \{1, \ldots, n\},$$

then, according to the discrete-time motion model (4.1.1):

(i) each agent remains in its locally cliqueless line-of-sight connectivity constraint set;

(ii) two agents that are in the same connected component of $\mathcal{G}_{\text{lc-vis-disk},Q_\delta}$ remain at the same connected component after the motion step; and

(iii) the number of connected components of the graph $\mathcal{G}_{\text{lc-vis-disk},Q_\delta}$ at

$\mathcal{P}(\ell+1)$ *is equal to or smaller than the number of connected components of the graph* $\mathcal{G}_{\text{lc-vis-disk},Q_\delta}$ *at* $\mathcal{P}(\ell)$.

4.3 RENDEZVOUS ALGORITHMS

In this section, we present some algorithms that might be used by a robotic network to achieve rendezvous. Throughout the section, we mainly focus on the uniform network $\mathcal{S}_{\text{disk}}$ of locally connected first-order agents in $\mathbb{R}^d$; this robotic network was introduced in Example 3.4.

4.3.1 Averaging control and communication law

We first study a behavior in which agents move toward a position computed as the average of the received messages. This law is related to the distributed linear algorithms discussed in Section 1.6 and, in particular, to adjacency-based agreement algorithms and Vicsek's model. This algorithm has also been studied in the context of "opinion dynamics under bounded confidence" and is known in the literature as the Krause model.

We loosely describe the AVERAGING law, which we denote by $\mathcal{CC}_{\text{AVERAGING}}$, as follows:

> *[Informal description]* In each communication round each agent performs the following tasks: (i) it transmits its position and receives its neighbors' positions; (ii) it computes the average of the point set comprised of its neighbors and of itself. Between communication rounds, each robot moves toward the average point that it computed.

We next formulate the algorithm, using the description model of Chapter 3. The law is uniform, static, and data-sampled, with standard message-generation function. (Recall from Definition 3.9 and Remark 3.11 that a control and coordination law (1) is uniform if processor state set, message-generation, state-transition and control functions are the same for each agent; (2) is static if the processor state set is a singleton, i.e., the law requires no memory; (3) is data-sampled if if the control functions are independent of the current position of the robot and depend only upon the robots position at the last sample time.)

`Robotic Network:` $\mathcal{S}_{\text{disk}}$ with motion model (4.1.1) in $\mathbb{R}^d$,
with absolute sensing of own position, and
with communication range r

Distributed Algorithm: AVERAGING

Alphabet: $\mathbb{A} = \mathbb{R}^d \cup \{\text{null}\}$

function $\text{msg}(p, i)$
1: **return** p

function $\text{ctl}(p, y)$
1: **return** $\text{avrg}(\{p\} \cup \{p_{\text{rcvd}} \mid p_{\text{rcvd}} \text{ is a non-null message in } y\}) - p$

An implementation of this control and communication law is shown in Figure 4.5 for $d = 1$. Note that, along the evolution, (1) several robots *rendezvous*, that is, agree upon a common location, and (2) some robots are connected at the simulation's beginning and not connected at the simulation's end (e.g., robots number 8 and 9, counting from the left). Our analysis

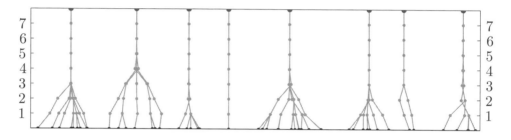

Figure 4.5 The evolution of a robotic network $\mathcal{S}_{\text{disk}}$, with $r = 1.5$, under the AVERAGING control and communication law. The vertical axis corresponds to the elapsed time, and the horizontal axis to the positions of the agents in the real line. The 51 agents are initially randomly deployed over the interval $[-15, 15]$.

of the performance of this law is contained in the following theorem, whose proof is postponed to Section 4.6.1.

Theorem 4.15 (Correctness and time complexity of averaging law). *For $d = 1$, the network $\mathcal{S}_{\text{disk}}$, the law $CC_{\text{AVERAGING}}$ achieves the task $\mathcal{T}_{\text{rndzvs}}$ with time complexity*

$$\text{TC}(\mathcal{T}_{\text{rndzvs}}, CC_{\text{AVERAGING}}) \in O(n^5),$$
$$\text{TC}(\mathcal{T}_{\text{rndzvs}}, CC_{\text{AVERAGING}}) \in \Omega(n).$$

4.3.2 Circumcenter control and communication laws

Here, we define the CRCMCNTR control and communication law for the network $\mathcal{S}_{\text{disk}}$. The law solves the rendezvous problem while keeping the network connected. This law was introduced by Ando et al. (1999) and later studied by Lin et al. (2007a) and Cortés et al. (2006).

We begin by recalling two useful geometric concepts: (i) given a bounded set S, its circumcenter $\mathrm{CC}(S)$ is the center of the closed ball of minimum radius containing S (see Section 2.1.3); (ii) given a point p in a convex set Q and a second point q, the from-to-inside map $\mathrm{fti}(p, q, S)$ is the point in the closed segment $[p, q]$ which is at the same time closest to q and inside S (see Section 2.1.1). Finally, recall also the connectivity constraint set introduced in Definition 4.4.

We loosely describe the CRCMCNTR law, denoted by $\mathcal{CC}_{\mathrm{CRCMCNTR}}$, as follows:

> *[Informal description]* In each communication round each agent performs the following tasks: (i) it transmits its position and receives its neighbors' positions; (ii) it computes the circumcenter of the point set comprised of its neighbors and of itself. Between communication rounds, each robot moves toward this circumcenter point while maintaining connectivity with its neighbors using appropriate connectivity constraint sets.

We next formulate the algorithm, using the description model of Chapter 3. The law is uniform, static, and data-sampled, with standard message-generation function:

Robotic Network: $\mathcal{S}_{\mathrm{disk}}$ with discrete-time motion model (4.1.1), with absolute sensing of own position, and with communication range r, in $\mathbb{R}^d$

Distributed Algorithm: CRCMCNTR

Alphabet: $\mathbb{A} = \mathbb{R}^d \cup \{\texttt{null}\}$

function $\mathrm{msg}(p, i)$
  1: **return** p

function $\mathrm{ctl}(p, y)$
  1: $p_{\mathrm{goal}} := \mathrm{CC}(\{p\} \cup \{p_{\mathrm{rcvd}} \mid \text{for all non-null } p_{\mathrm{rcvd}} \in y\})$
  2: $\mathcal{X} := \mathcal{X}_{\mathrm{disk}}(p, \{p_{\mathrm{rcvd}} \mid \text{for all non-null } p_{\mathrm{rcvd}} \in y\})$
  3: **return** $\mathrm{fti}(p, p_{\mathrm{goal}}, \mathcal{X}) - p$

This algorithm is illustrated in Figure 4.6.

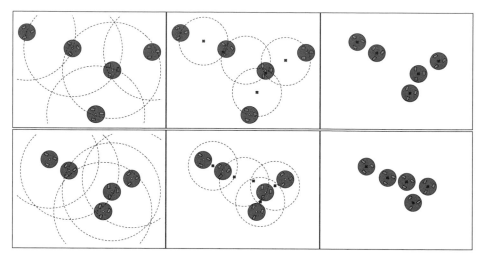

Figure 4.6 An illustration of the execution of the CRCMCNTR ALGORITHM. Each row of
plots represents an iteration of the law. In each round, each agent computes
its goal point and its constraint set, and then moves toward the goal while
remaining in the constraint set.

Next, let us note that it is possible and straightforward to implement
the circumcenter law as a static relative-sensing control law on the relative-
sensing network with disk sensors $\mathcal{S}_{\mathrm{disk}}^{\mathrm{rs}}$ introduced in Example 3.15:

Relative Sensing Network: $\mathcal{S}_{\mathrm{disk}}^{\mathrm{rs}}$ with motion model (4.1.2),
no communication, relative sensing for robot i given by:
robot measurements y contains $p_i^{[j]} \in \overline{B}(\mathbf{0}_2, r)$ for all $j \neq i$

Distributed Algorithm: RELATIVE-SENSING CRCMCNTR

function $\mathrm{ctl}(y)$

  1: $p_{\mathrm{goal}} := \mathrm{CC}(\{\mathbf{0}_d\} \cup \{p_{\mathrm{snsd}} \mid \text{for all non-null } p_{\mathrm{snsd}} \in y\})$
  2: $\mathcal{X} := \mathcal{X}_{\mathrm{disk}}(\mathbf{0}_d, \{p_{\mathrm{snsd}} \mid \text{for all non-null } p_{\mathrm{snsd}} \in y\})$
  3: **return** $\mathrm{fti}(\mathbf{0}_d, p_{\mathrm{goal}}, \mathcal{X})$

In the remainder of this section, we generalize the circumcenter law in a
number of ways: (i) we modify the constraint set by imposing bounds on
the control inputs and by relaxing the connectivity constraint as much as
possible, while maintaining connectivity guarantees; and (ii) we implement
the circumcenter law on two distinct communication graphs. Let us note
that many of these generalized circumcenter laws can also be implemented
as relative-sensing control laws; in the interest of brevity, we do not present
the details.

4.3.2.1 Circumcenter law with control bounds and relaxed connectivity constraints

First, assume that the agents have a compact input space $U = \overline{B}(\mathbf{0}_d, u_{\max})$, with $u_{\max} \in \mathbb{R}_{>0}$. Additionally, we adopt the relaxed $\mathcal{G}$-connectivity constraint sets as follows. Let $\mathcal{G}$ be a proximity graph that is spatially distributed over $\mathcal{G}_{\text{disk}}(r)$ and that has the same connected components as $\mathcal{G}_{\text{disk}}(r)$; examples include $\mathcal{G}_{\text{RN}} \cap \mathcal{G}_{\text{disk}}(r)$, $\mathcal{G}_{\text{G}} \cap \mathcal{G}_{\text{disk}}(r)$, and $\mathcal{G}_{\text{LD}}(r)$. Recall the $\mathcal{G}$-connectivity constraint set from Definition 4.7. Combining the relaxed connectivity constraint and the control magnitude bound, we redefine the control function in the CRCMCNTR law to be:

> **function** $\text{ctl}(p, y)$
> *% Includes control bound and relaxed $\mathcal{G}$-connectivity constraint*
> 1: $p_{\text{goal}} := \text{CC}(\{p\} \cup \{p_{\text{rcvd}} \mid \text{for all non-null } p_{\text{rcvd}} \in y\})$
> 2: $\mathcal{X} := \mathcal{X}_{\text{disk},\mathcal{G}}(p, \{p_{\text{rcvd}} \mid \text{for all non-null } p_{\text{rcvd}} \in y\}) \cap \overline{B}(p, u_{\max})$
> 3: **return** $\text{fti}(p, p_{\text{goal}}, \mathcal{X}) - p$

Second, the circumcenter law can be implemented also on robotic networks with different proximity graphs. For example, we can implement the circumcenter algorithm without any change on the following network.

4.3.2.2 Circumcenter law on the limited Delaunay graph

We consider the same set of physical agents as in $\mathcal{S}_{\text{disk}}$. For $r \in \mathbb{R}_{>0}$, we adopt as communication graph the *r-limited Delaunay graph* $\mathcal{G}_{\text{LD}}(r)$, described in Section 2.2. These data define the uniform robotic network $\mathcal{S}_{\text{LD}} = (\{1, \dots, n\}, \mathcal{R}, \mathcal{E}_{\text{LD}})$, as described in Example 3.4. On this network, we implement the CRCMCNTR law without any change, that is, with the same message-generation and control function as we did for the implementation on the network $\mathcal{S}_{\text{disk}}$.

4.3.2.3 Parallel circumcenter law on the ∞-disk graph

We consider the network $\mathcal{S}_{\infty\text{-disk}}$ of first-order robots in $\mathbb{R}^d$, connected according to the $\mathcal{G}_{\infty\text{-disk}}(r)$ graph (see Example 3.4). For this network, we define the PLL-CRCMCNTR law, which we denoted by $\mathcal{CC}_{\text{PLL-CRCMCNTR}}$, by designing d decoupled circumcenter laws running in parallel on each coordinate axis of $\mathbb{R}^d$. As before, this law is uniform and static. What is remarkable, however, is that no constraint is required to maintain connectivity (see Exercise E4.4).

The parallel circumcenter of the set S, denoted by $\mathrm{PCC}(S)$, is the center of the smallest axis-aligned rectangle containing S. In other words, $\mathrm{PCC}(S)$ is the component-wise circumcenter of S (see Figure 4.7). We state the

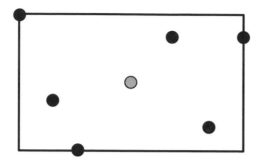

Figure 4.7 The gray point is the parallel circumcenter of the collection of black points.

parallel circumcenter law as follows:

Robotic Network: $\mathcal{S}_{\infty\text{-disk}}$ with discrete-time motion model (4.1.1) in $\mathbb{R}^d$,
with absolute sensing of own position, and
with communication range r in L^∞-metric

Distributed Algorithm: PLL-CRCMCNTR

Alphabet: $\mathbb{A} = \mathbb{R}^d \cup \{\texttt{null}\}$

function $\mathrm{msg}(p, i)$
  1: **return** p

function $\mathrm{ctl}(p, y)$
  1: $p_{\mathrm{goal}} := \mathrm{PCC}(\{p\} \cup \{p_{\mathrm{rcvd}} \mid \text{for all non-null } p_{\mathrm{rcvd}} \in y\})$
  2: **return** $p_{\mathrm{goal}} - p$

4.3.3 Correctness and complexity of circumcenter laws

In this section, we characterize the convergence and complexity properties of the circumcenter law and of its variations. The following theorem summarizes the results known in the literature about the asymptotic properties of the circumcenter law.

Theorem 4.16 (Correctness of the circumcenter laws). *For $d \in \mathbb{N}$, $r \in \mathbb{R}_{>0}$, and $\varepsilon \in \mathbb{R}_{>0}$, the following statements hold:*

(i) *on the network* $\mathcal{S}_{\mathrm{disk}}$, *the law* $\mathcal{CC}_{\mathrm{CRCMCNTR}}$ *(with control magnitude bounds and relaxed* $\mathcal{G}$-*connectivity constraints) achieves the exact rendezvous task* $\mathcal{T}_{\mathrm{rndzvs}}$;

(ii) *on the network* $\mathcal{S}_{\mathrm{LD}}$, *the law* $\mathcal{CC}_{\mathrm{CRCMCNTR}}$ *achieves the* ε-*rendezvous task* $\mathcal{T}_{\varepsilon\text{-rndzvs}}$; *and*

(iii) *on the network* $\mathcal{S}_{\infty\text{-disk}}$, *the law* $\mathcal{CC}_{\mathrm{PLL\text{-}CRCMCNTR}}$ *achieves the exact rendezvous task* $\mathcal{T}_{\mathrm{rndzvs}}$.

Furthermore, the evolutions of $(\mathcal{S}_{\mathrm{disk}}, \mathcal{CC}_{\mathrm{CRCMCNTR}})$, $(\mathcal{S}_{\mathrm{LD}}, \mathcal{CC}_{\mathrm{CRCMCNTR}})$, *and* $(\mathcal{S}_{\infty\text{-disk}}, \mathcal{CC}_{\mathrm{PLL\text{-}CRCMCNTR}})$ *have the following properties:*

(iv) *If any two agents belong to the same connected component of the respective communication graph at* $\ell \in \mathbb{Z}_{\geq 0}$, *then they continue to belong to the same connected component for all subsequent times* $k \geq \ell$.

(v) *For each evolution, there exists* $P^* = (p_1^*, \ldots, p_n^*) \in (\mathbb{R}^d)^n$ *such that:*

    (a) *the evolution asymptotically approaches* P^*; *and*

    (b) *for each* $i, j \in \{1, \ldots, n\}$, *either* $p_i^* = p_j^*$, *or* $\|p_i^* - p_j^*\|_2 > r$ *(for the networks* $\mathcal{S}_{\mathrm{disk}}$ *and* $\mathcal{S}_{\mathrm{LD}}$*) or* $\|p_i^* - p_j^*\|_\infty > r$ *(for the network* $\mathcal{S}_{\infty\text{-disk}}$*).*

The proof of this theorem is given in Section 4.6.2. The robustness of the circumcenter control and communication laws can be characterized with respect to link failures (see Cortés et al., 2006).

Next, we analyze the time complexity of $\mathcal{CC}_{\mathrm{CRCMCNTR}}$. As we will see, next, the complexity of $\mathcal{CC}_{\mathrm{CRCMCNTR}}$ differs dramatically when applied to robotic networks with different communication graphs. We provide complete results for the networks $\mathcal{S}_{\mathrm{disk}}$ and $\mathcal{S}_{\mathrm{LD}}$ only for the case $d = 1$.

Theorem 4.17 (Time complexity of circumcenter laws). *For* $r \in \mathbb{R}_{>0}$ *and* $\varepsilon \in\,]0, 1[$, *the following statements hold:*

(i) *on the network* $\mathcal{S}_{\mathrm{disk}}$, *evolving on the real line* $\mathbb{R}$ *(i.e., with* $d = 1$*),* $\mathrm{TC}(\mathcal{T}_{\mathrm{rndzvs}}, \mathcal{CC}_{\mathrm{CRCMCNTR}}) \in \Theta(n)$;

(ii) *on the network* $\mathcal{S}_{\mathrm{LD}}$, *evolving on the real line* $\mathbb{R}$ *(i.e., with* $d = 1$*),* $\mathrm{TC}(\mathcal{T}_{(r\varepsilon)\text{-rndzvs}}, \mathcal{CC}_{\mathrm{CRCMCNTR}}) \in \Theta(n^2 \log(n\varepsilon^{-1}))$; *and*

(iii) *on the network* $\mathcal{S}_{\infty\text{-disk}}$, *evolving on Euclidean space (i.e., with* $d \in \mathbb{N}$*),* $\mathrm{TC}(\mathcal{T}_{\mathrm{rndzvs}}, \mathcal{CC}_{\mathrm{PLL\text{-}CRCMCNTR}}) \in \Theta(n)$.

The proof of this result is contained in Martínez et al. (2007b).

Remark 4.18 (Analysis in higher dimensions). The results in Theorems 4.17(i) and (ii) induce lower bounds on the time complexity of the circumcenter law in higher dimensions. Indeed, for arbitrary $d \geq 1$, we have the following:

(i) on the network $\mathcal{S}_{\mathrm{disk}}$, $\mathrm{TC}(\mathcal{T}_{\mathrm{rndzvs}}, \mathcal{CC}_{\mathrm{CRCMCNTR}}) \in \Omega(n)$;

(ii) on the network $\mathcal{S}_{\mathrm{LD}}$, $\mathrm{TC}(\mathcal{T}_{(r\varepsilon)\text{-rndzvs}}, \mathcal{CC}_{\mathrm{CRCMCNTR}}) \in \Omega(n^2 \log(n\varepsilon^{-1}))$.

We have performed extensive numerical simulations for the case $d = 2$ and the network $\mathcal{S}_{\mathrm{disk}}$. We run the algorithm starting from generic initial configurations (where, in particular, the robots' positions are not aligned) contained in a bounded region of $\mathbb{R}^2$. We have consistently obtained that the time complexity to achieve $\mathcal{T}_{\mathrm{rndzvs}}$ with $\mathcal{CC}_{\mathrm{CRCMCNTR}}$ starting from these initial configurations is independent of the number of robots. This leads us to conjecture that initial configurations where all robots are aligned (equivalently, the 1-dimensional case) give rise to the worst possible performance of the algorithm. In other words, we conjecture that, for $d \geq 2$, $\mathrm{TC}(\mathcal{T}_{\mathrm{rndzvs}}, \mathcal{CC}_{\mathrm{CRCMCNTR}}) = \Theta(n)$. •

Remark 4.19 (Congestion effects). As discussed in Remark 3.8, one way of incorporating congestion effects into the network operation is to assume that the parameters of the physical components of the network depend upon the number of robots—for instance, by assuming that the communication range decreases with the number of robots. Theorem 4.17 presents an alternative, equivalent, way of looking at congestion: the results hold under the assumption that the communication range is constant, but allow for the diameter of the initial network configuration (the maximum inter-agent distance) to grow unbounded with the number of robots. •

4.3.4 The circumcenter law in nonconvex environments

In this section, we adapt the circumcenter algorithm to work on networks in planar nonconvex allowable environments. Throughout the section, we only consider the case of a compact allowable nonconvex environment Q contracted into Q_δ for a small positive δ. We present the algorithm in two formats: for the communication-based network $\mathcal{S}_{\mathrm{vis\text{-}disk}}$ described in Example 3.6, and for the relative-sensing network $\mathcal{S}^{\mathrm{rs}}_{\mathrm{vis\text{-}disk}}$ described in Example 3.16.

We modify the circumcenter algorithm in three ways: first, we adopt the connectivity constraints described in the previous section for range-limited line-of-sight links; second, we further restrict the robot motion to remain inside the relative convex hull of the sensed robot positions; and third, we move towards the circumcenter of the constraint set, instead of the circumcenter of the neighbors positions. The details of the algorithm are as follows:

Robotic Network: $\mathcal{S}_{\text{vis-disk}}$ with discrete-time motion model (4.1.1), absolute sensing of own position and of Q_δ, and communication range r within line of sight ($\mathcal{G}_{\text{vis-disk},Q_\delta}$)

Distributed Algorithm: NONCONVEX CRCMCNTR

Alphabet: $\mathbb{A} = \mathbb{R}^2 \cup \{\texttt{null}\}$

function $\text{msg}(p, i)$

  1: **return** p

function $\text{ctl}(p, y)$

  1: $\mathcal{X}_1 := \mathcal{X}_{\text{vis-disk}}(p, \{p_{\text{rcvd}} \mid \text{for all non-null } p_{\text{rcvd}} \in y\}; Q_\delta)$
  2: $\mathcal{X}_2 := \text{rco}(\{p\} \cup \{p_{\text{rcvd}} \mid \text{for all non-null } p_{\text{rcvd}} \in y\}; \text{Vi}(p; Q_\delta))$
  3: $p_{\text{goal}} := \text{CC}(\mathcal{X}_1 \cap \mathcal{X}_2)$
  4: **return** $\text{fti}(p, p_{\text{goal}}, \overline{B}(p, u_{\max})) - p$

Next, we present the relative sensing version; recall that $p_i^{[i]} = \mathbf{0}_2$ and that, as discussed in Section 3.2.3 in the context of the evolution of a relative sensing network with environment sensors, y_{env} denotes the environment measurement provided by the range-limited visibility sensor:

Relative Sensing Network: $\mathcal{S}_{\text{vis-disk}}^{\text{rs}}$ with motion model (4.1.2) in Q_δ, no communication, relative sensing for robot i given by:

robot measurements y contains $p_i^{[j]} \in \text{Vi}_{\text{disk}}(\mathbf{0}_2; (Q_\delta)_i)$ for $j \neq i$
environment sensing is $y_{\text{env}} = \text{Vi}_{\text{disk}}(\mathbf{0}_2; (Q_\delta)_i)$

Distributed Algorithm: NONCONVEX RELATIVE-SENSING CRCMCNTR

function $\text{ctl}(y, y_{\text{env}})$

  1: $\mathcal{X}_1 := \mathcal{X}_{\text{vis-disk}}(\mathbf{0}_2, \{p_{\text{snsd}} \mid \text{for all non-null } p_{\text{snsd}} \in y\}; y_{\text{env}})$
  2: $\mathcal{X}_2 := \text{rco}(\{\mathbf{0}_2\} \cup \{p_{\text{snsd}} \mid \text{for all non-null } p_{\text{snsd}} \in y\}; y_{\text{env}})$
  3: $p_{\text{goal}} := \text{CC}(\mathcal{X}_1 \cap \mathcal{X}_2)$
  4: **return** $\text{fti}(\mathbf{0}_2, p_{\text{goal}}, \overline{B}(\mathbf{0}_2, u_{\max}))$

Theorem 4.20 (Correctness of the circumcenter law in nonconvex environments). *For $\delta > 0$, let Q_δ be a contraction of a compact allowable nonconvex environment Q. For $r \in \mathbb{R}_{>0}$ and $\varepsilon \in \mathbb{R}_{>0}$, on the network $\mathcal{S}_{\text{vis-disk}}$, the law $\mathcal{CC}_{\text{NONCONVEX CRCMCNTR}}$ (with control magnitude bounds) achieves the ε-rendezvous task $\mathcal{T}_{\varepsilon\text{-rndzvs}}$. Furthermore, the evolution has the following properties:*

(i) If any two agents belong to the same connected component of the graph $\mathcal{G}_{\text{vis-disk},Q_\delta}$ at $\ell \in \mathbb{Z}_{\geq 0}$, then they continue to belong to the same connected component for all subsequent times $k \geq \ell$.

(ii) There exists $P^* = (p_1^*, \ldots, p_n^*) \in Q_\delta^n$ such that:

    (a) the evolution asymptotically approaches P^*; and

    (b) for each $i, j \in \{1, \ldots, n\}$, either $p_i^* = p_j^*$, or p_i^* and p_j^* are not within range-limited line of sight.

The proof of this result can be found in Ganguli et al. (2009). A brief sketch of the proof steps is presented in Section 4.6.4. The complexity of the NONCONVEX CRCMCNTR law has not been characterized. However, note that the evolution from any initial configuration such that $\mathcal{G}_{\text{vis},Q_\delta}$ is complete is also an evolution of the CRCMCNTR law discussed in Section 4.3.2, and hence Theorem 4.17(i) induces a lower bound on the time complexity.

4.4 SIMULATION RESULTS

In this section, we illustrate the execution of some circumcenter control and communication laws introduced in this chapter. The CRCMCNTR law is implemented on the networks $\mathcal{S}_{\text{disk}}$, $\mathcal{S}_{\text{LD}}$, and $\mathcal{S}_{\infty\text{-disk}}$ in Mathematica® as a library of routines and a main program running the simulation. The packages `PlanGeom.m` and `SpatialGeom.m` contain routines for the computation of geometric objects in $\mathbb{R}^2$ and $\mathbb{R}^3$, respectively. These routines are freely available at the book webpage http://coordinationbook.info

First, we show evolutions of $(\mathcal{S}_{\text{disk}}, \text{CRCMCNTR})$ in two and three dimensions in Figures 4.8 and 4.9, respectively. Measuring displacements in meters, we consider random initial positions over the square $[-7, 7] \times [-7, 7]$ and the cube $[-7, 7] \times [-7, 7] \times [-7, 7]$. The 25 robotic agents have a communication radius $r = 4$ and a compact input space $U = \overline{B}(\mathbf{0}_d, u_{\max})$, with $u_{\max} = 0.15$. As the simulations show, the task $\mathcal{T}_{\text{rndzvs}}$ is achieved, as guaranteed by Theorem 4.16(i).

Second, within the same setup, we show an evolution of $(\mathcal{S}_{\text{LD}}, \text{CRCMCNTR})$ in two dimensions in Figure 4.10. As the simulation shows, the task $\mathcal{T}_{\varepsilon\text{-rndzvs}}$ is achieved, as guaranteed by Theorem 4.16(ii).

Third, we show an evolution of $(\mathcal{S}_{\infty\text{-disk}}, \text{PLL-CRCMCNTR})$ in two dimensions in Figure 4.11. As the simulations show, the task $\mathcal{T}_{\text{rndzvs}}$ is achieved, as guaranteed by Theorem 4.16(iii).

Finally, we refer the interested reader to Ganguli et al. (2009) for simulation results for the NONCONVEX CRCMCNTR algorithm.

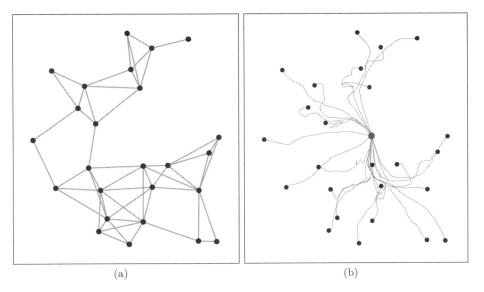

(a)   (b)

Figure 4.8 The evolution of $(\mathcal{S}_{\mathrm{disk}}, \textsc{crcmcntr})$ with $n = 25$ robots in 2 dimensions: (a)
shows the initial connected network configuration; (b) shows the evolution of
the individual agents until rendezvous is achieved.

4.5 NOTES

The rendezvous problem and the circumcenter algorithm were originally
introduced by Ando et al. (1999). The circumcenter algorithm has been
extended to other control policies, including asynchronous implementations,
in Lin et al. (2007a,b). The circumcenter algorithm has been extended
beyond planar problems to arbitrary dimensions in Cortés et al. (2006),
where its robustness properties are also characterized. Regarding Theo-
rem 4.16, the results on $\mathcal{S}_{\mathrm{disk}}$ appeared originally in Ando et al. (1999); the
results on $\mathcal{S}_{\mathrm{LD}}$ and on $\mathcal{S}_{\infty\text{-disk}}$ appeared originally in Cortés et al. (2006) and
in Martínez et al. (2007b), respectively. Variations of the circumcenter law
in the presence of noise and sensor errors are studied in Martínez (2009b).
The continuous-time version of the circumcenter law, with no connectivity
constraints, is analyzed in Lin et al. (2007c). Continuous-time control laws
for groups of robots with simple first-order dynamics and unicycle dynamics
are proposed in Lin et al. (2004, 2005) and Dimarogonas and Kyriakopoulos (2007). In these works, the inter-robot topology is time dependent and
assumed *a priori* to be connected at all times. Rendezvous under communication quantization is studied in Fagnani et al. (2004) and Carli and Bullo
(2009). Rendezvous for unicycle robots with minimal sensing capabilities
is studied by Yu et al. (2008). Relationships with classic curve-shortening
flows are studied by Smith et al. (2007).

Rendezvous has also been studied within the computer science literature,

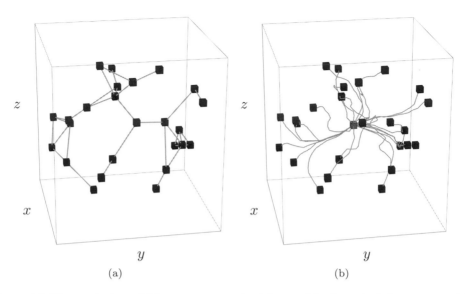

Figure 4.9 The evolution of $(\mathcal{S}_{\mathrm{disk}}, \mathrm{CRCMCNTR})$ with $n = 25$ robots in 3 dimensions: (a) shows the initial connected network configuration; (b) shows the evolution of the individual agents until rendezvous is achieved.

where the problem is referred to as the "gathering," or point formation, problem. Flocchini et al. (1999) and Suzuki and Yamashita (1999) study the point formation problem under the assumption that each robot is capable of sensing all other robots. Flocchini et al. (2005) propose asynchronous algorithms to solve the gathering problem, and Agmon and Peleg (2006) study the solvability of the problem in the presence of faulty robots.

Multi-robot rendezvous with line-of-sight sensors is considered in Roy and Dudek (2001), where solutions are proposed based on the exploration of the unknown environment and the selection of appropriate rendezvous points at pre-specified times. Hayes et al. (2003) also consider rendezvous at a specified location for visually guided agents, but the proposed solution requires each agent to have knowledge of the location of all other agents. The problem of computing a multi-robot rendezvous point in polyhedral surfaces made of triangular faces is considered in Lanthier et al. (2005). The perimeter-minimizing algorithm presented by Ganguli et al. (2009) solves the rendezvous problem for sensor-based networks with line-of-sight range-limited sensors in nonconvex environments.

Regarding the connectivity maintenance problem, a number of works have addressed the problem of designing a coordination algorithm that achieves a general, non-specified task while preserving connectivity. The centralized solution proposed in Zavlanos and Pappas (2005) allows for a general range of agent motions. The distributed solution presented by Savla et al. (2009b)

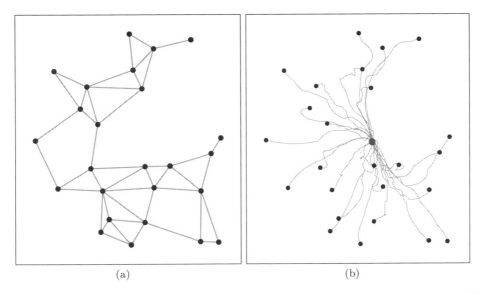

(a)   (b)

Figure 4.10 The evolution of $(\mathcal{S}_{\mathrm{LD}}, \mathrm{CRCMCNTR})$ with $n = 25$ robots in 2 dimensions: (a) shows the initial connected network configuration; (b) shows the evolution of the individual agents until rendezvous is achieved.

gives connectivity maintaining constraints for second-order control systems with input magnitude bounds. A distributed algorithm to perform graph rearrangements that preserve the connectivity is presented in Schuresko and Cortés (2007). Connectivity problems have been studied also in other contexts. Langbort and Gupta (2009) study the impact of the connectivity of the interconnection topology in a class of network optimization problems. Spanos and Murray (2005) generate connectivity-preserving motions between pairs of formations. Ji and Egerstedt (2007) design Laplacian-based control laws to solve formation control problems while preserving connectivity. Various works have focused on designing the network motion so that some desired measure of connectivity (e.g., algebraic connectivity) is maximized under position constraints. Boyd (2006) and de Gennaro and Jadbabaie (2006) consider convex constraints, while Kim and Mesbahi (2006) deal with a class of nonconvex constraints. Zavlanos and Pappas (2007b) use potential fields to maximize algebraic connectivity.

A continuous-time version of the averaging control and communication law is also known as the Hegselmann-Krause model for "opinion dynamics under bounded confidence" (see Hegselmann and Krause, 2002; Lorenz, 2007). In this model, each agent may change its opinion by averaging it with that of neighbors who are in an ε-confidence area. In other words, the difference between the agent's opinion and those of its neighbors' should be bounded by ε. A similar model where the communication between agents is random is the Deffuant-Weisbuch model, inspired by a model of dissemination of

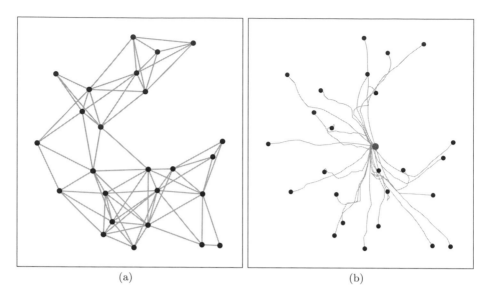

Figure 4.11 The evolution of $(\mathcal{S}_{\infty\text{-disk}}, \text{PLL-CRCMCNTR})$ with $n = 25$ robots in 2 dimen-
        sions: (a) shows the initial connected network configuration; (b) shows the
        evolution of the individual agents until rendezvous is achieved.

culture (see Deffuant et al., 2000; Axelrod, 1997).

4.6 PROOFS

This section gathers the proofs of the main results presented in the chapter.

4.6.1 Proof of Theorem 4.15

Proof. One can easily prove that, along the evolution of the network, the ordering of the agents is preserved, that is, the inequality $p^{[i]} \leq p^{[j]}$ is preserved
at the next time step. However, links between agents are not necessarily preserved (see, e.g., Figure 4.8). Indeed, connected components may split along
the evolution. However, merging events do not occur. Consider two contigu-
ous connected components C_1 and C_2 of $\mathcal{G}_{\text{disk}}(r)$, with C_1 to the left of C_2.
By definition, the rightmost agent in the component C_1 and the leftmost
agent in the component C_2 are at a distance strictly larger than r. Now,
by executing the algorithm, they can only but increase that distance, since
the rightmost agent in C_1 will move to the left, and the leftmost agent in
C_2 will move to the right. Therefore, connected components do not merge.

Consider first the case of an initial network configuration for which the
communication graph remains connected throughout the evolution. Without

loss of generality, assume that the agents are ordered from left to right according to their identifier, that is, $p^{[1]}(0) \leq \cdots \leq p^{[n]}(0)$. Let $\alpha \in \{3,\ldots,n\}$ have the property that agents $\{2,\ldots,\alpha-1\}$ are neighbors of agent 1, and agent α is not. (If, instead, all agents are within an interval of length r, then rendezvous is achieved in 1 time instant, and the statement in theorem is easily seen to be true.) Note that we can assume that agents $\{2,\ldots,\alpha-1\}$ are also neighbors of agent α. If this is not the case, then those agents that are neighbors of agent 1 and not of agent α rendezvous with agent 1 at the next time instant. At the time instant $\ell = 1$, the new updated positions satisfy

$$p^{[1]}(1) = \frac{1}{\alpha-1} \sum_{k=1}^{\alpha-1} p^{[k]}(0),$$

$$p^{[\gamma]}(1) \in \Big[\frac{1}{\alpha} \sum_{k=1}^{\alpha} p^{[k]}(0), *\Big], \quad \gamma \in \{2,\ldots,\alpha-1\},$$

where $*$ denotes a certain unimportant point.

Now, we show that

$$p^{[1]}(\alpha-1) - p^{[1]}(0) \geq \frac{r}{\alpha(\alpha-1)}. \tag{4.6.1}$$

Let us first show the inequality for $\alpha = 3$. Because of the assumption that the communication graph remains connected, agent 2 is still a neighbor of agent 1 at the time instant $\ell = 1$. Therefore, $p^{[1]}(2) \geq \frac{1}{2}(p^{[1]}(1) + p^{[2]}(1))$, and from here we deduce

$$p^{[1]}(2) - p^{[1]}(0) \geq \frac{1}{2}\big(p^{[2]}(1) - p^{[1]}(0)\big)$$

$$\geq \frac{1}{2}\Big(\frac{1}{3}\big(p^{[1]}(0) + p^{[2]}(0) + p^{[3]}(0)\big) - p^{[1]}(0)\Big) \geq \frac{1}{6}\big(p^{[3]}(0) - p^{[1]}(0)\big) \geq \frac{r}{6}.$$

Let us now proceed by induction. Assume that inequality (4.6.1) is valid for $\alpha - 1$, and let us prove it for α. Consider first the possibility, when at the time instant $\ell = 1$, that the agent $\alpha - 1$ is still a neighbor of agent 1. In this case, $p^{[1]}(2) \geq \frac{1}{\alpha-1} \sum_{k=1}^{\alpha-1} p^{[k]}(1)$, and from here we deduce

$$p^{[1]}(2) - p^{[1]}(0) \geq \frac{1}{\alpha-1}\Big(p^{[\alpha-1]}(1) - p^{[1]}(0)\Big)$$

$$\geq \frac{1}{\alpha-1}\Big(\frac{1}{\alpha} \sum_{k=1}^{\alpha} p^{[k]}(0) - p^{[1]}(0)\Big)$$

$$\geq \frac{1}{\alpha(\alpha-1)}\Big(p^{[\alpha]}(0) - p^{[1]}(0)\Big) \geq \frac{r}{\alpha(\alpha-1)},$$

which, in particular, implies (4.6.1). Consider then the case when agent $\alpha - 1$ is not a neighbor of agent 1 at the time instant $\ell = 1$. Let $\beta < \alpha$

such that agent $\beta - 1$ is a neighbor of agent 1 at $\ell = 1$, but agent β is not. Since $\beta < \alpha$, we have by induction $p^{[1]}(\beta) - p^{[1]}(1) \geq \frac{r}{\beta(\beta-1)}$. From here, we deduce that $p^{[1]}(\alpha - 1) - p^{[1]}(0) \geq \frac{r}{\alpha(\alpha-1)}$.

It is clear that after $\ell_1 = \alpha - 1$, we could again consider two complementary cases (either agent 1 has all others as neighbors or not) and repeat the same argument once again. In that way, we would find ℓ_2 such that the distance traveled by agent 1 after ℓ_2 rounds would be lower bounded by $\frac{2r}{n(n-1)}$. Repeating this argument iteratively, the worst possible case is one in which agent 1 keeps moving to the right and, at each time step, there is always another agent which is not a neighbor. Since the diameter of the initial condition P_0 is upper bounded by $(n - 1)r$, in the worst possible situation, there exists some time ℓ_k such that $\frac{kr}{(n-1)n} = O(r(n - 1))$. This implies that $k = O((n - 1)^2 n)$. Now we can upper bound the total convergence time ℓ_k by $\ell_k = \sum_{i=1}^{k} \alpha_i - k \leq k(n - 1)$, where we have used that $\alpha_i \leq n$ for all $i \in \{1, \ldots, n\}$. From here, we see that $\ell_k = O((n - 1)^3 n)$, and hence we deduce that in $O(n(n - 1)^3)$ time instants there cannot be any agent which is not a neighbor of the agent 1. Hence, all agents rendezvous at the next time instant. Consequently,

$$\mathrm{TC}(\mathcal{T}_{\mathrm{rndzvs}}, \mathcal{CC}_{\mathrm{AVERAGING}}, P_0) = O(n(n - 1)^3).$$

Finally, for a general initial configuration P_0, because there are a finite number of agents, only a finite number of splittings (at most $n - 1$) of the connected components of the communication graph can take place along the evolution. Therefore, we conclude that $\mathrm{TC}(\mathcal{T}_{\mathrm{rndzvs}}, \mathcal{CC}_{\mathrm{AVERAGING}}) = O(n^5)$.

Let us now prove the lower bound. Consider an initial configuration $P_0 \in \mathbb{R}^n$ where all agents are positioned in increasing order according to their identity, and exactly at a distance r apart—say, $p^{[i+1]}(0) - p^{[i]}(0) = r$, $i \in \{1, \ldots, n - 1\}$. Assume for simplicity that n is odd—when n is even, one can reason in an analogous way. Because of the symmetry of the initial condition, in the first time step, only agents 1 and n move. All the remaining agents remain in their position, because it coincides with the average of its neighbors' position and its own. At the second time step, only agents 1, 2, $n - 1$, and n move, and the others remain static because of the symmetry. Applying this idea iteratively, one deduces that the time step when agents $\frac{n-1}{2}$ and $\frac{n+3}{2}$ move for the first time is lower bounded by $\frac{n-1}{2}$. Since both agents have still at least a neighbor (agent $\frac{n+1}{2}$), the task $\mathcal{T}_{\mathrm{rndzvs}}$ has not been achieved yet at this time step. Therefore, $\mathrm{TC}(\mathcal{T}_{\mathrm{rndzvs}}, \mathcal{CC}_{\mathrm{AVERAGING}}, P_0) \geq \frac{n-1}{2}$, and the result follows. ∎

4.6.2 Proof of Theorem 4.16

Proof. We divide the proof of the theorem into three groups, one per network.

STEP 1: Facts on $(\mathcal{S}_{\text{disk}}, \mathcal{CC}_{\text{CRCMCNTR}})$. Fact (iv) for $(\mathcal{S}_{\text{disk}}, \mathcal{CC}_{\text{CRCMCNTR}})$ is a direct consequence of the control function definition of the CRCMCNTR law and Lemma 4.8.

Let us show fact (i). Because $\mathcal{G}$ has the same connected components as $\mathcal{G}_{\text{disk}}(r)$, fact (iv) implies that the number of connected components of $\mathcal{G}_{\text{disk}}(r)$ can only but decrease. In other words, the number of agents in each of the connected components of $\mathcal{G}_{\text{disk}}(r)$ is non-decreasing. Since there is a finite number of agents, there must exist ℓ_0 such that the identity of the agents in each connected component of $\mathcal{G}_{\text{disk}}(r)$ is fixed for all $\ell \geq \ell_0$ (that is, no more agents are added to the connected component afterwards). In what follows, without loss of generality, we assume that there is only one connected component after ℓ_0, i.e., the graph is connected (if this is not the case, then the same argument follows through for each connected component).

We prove that the law $\mathcal{CC}_{\text{CRCMCNTR}}$ (with control magnitude bounds and relaxed $\mathcal{G}$-connectivity constraints) achieves the exact rendezvous task $\mathcal{T}_{\text{rndzvs}}$ in the following two steps:

(a) We first define a set-valued dynamical system $((\mathbb{R}^d)^n, (\mathbb{R}^d)^n, T)$ such that the evolutions of $(\mathcal{S}_{\text{disk}}, \mathcal{CC}_{\text{CRCMCNTR}})$, starting from an initial configuration where $\mathcal{G}_{\text{disk}}(r)$ is connected, are contained in the set of evolutions of the set-valued dynamical system.

(b) We then establish that any evolution of $((\mathbb{R}^d)^n, (\mathbb{R}^d)^n, T)$ converges to a point in $\text{diag}((\mathbb{R}^d)^n)$ (the point might be different for different evolutions).

This strategy is analogous to the discussion regarding the Overapproximation Lemma for time-dependent systems in Section 1.3.5.

Let as perform (a). Given a connected graph G with vertices $\{1, \ldots, n\}$, let us consider the constraint sets and goal points defined with respect to G. In other words, given $P = (p_1, \ldots, p_n) \in (\mathbb{R}^d)^n$, define for each $i \in \{1, \ldots, n\}$,

$$(p_{\text{goal}})_i := \text{CC}(\{p_i\} \cup \{p_j \mid j \in \mathcal{N}_G(i)\}),$$
$$\mathcal{X}_i := \bigcap \{\overline{B}(\tfrac{p_i + p_j}{2}, \tfrac{r_i(P)}{2}) \mid j \in \mathcal{N}_G(i)\} \cap \overline{B}(p_i, u_{\max}),$$

where $r_i(P) = \max\{r, \max\{\|p_i - p_j\|_2 \mid j \in \mathcal{N}_G(i)\}\}$. Since two neighbors according to G can be arbitrarily far from each other in $\mathbb{R}^d$, we need to modify the definition of the constraint set with the radius $r_i(P)$ to prevent $\mathcal{X}_i$ from becoming empty. Note that if $\|p_i - p_j\|_2 \leq r$ for all $j \in \mathcal{N}_G(i)$, then $r_i(P) = r$ and, therefore, $\mathcal{X}_i = \mathcal{X}_{\mathrm{disk},G}(p_i, P) \cap \overline{B}(p_i, u_{\max})$. It is also worth observing that both $(p_{\mathrm{goal}})_i$ and $\mathcal{X}_i$ change continuously with $(p_1, \ldots, p_n)$.

Define the map $\mathrm{fti}_G : (\mathbb{R}^d)^n \to (\mathbb{R}^d)^n$ by

$$\mathrm{fti}_G(p_1, \ldots, p_n) = (\mathrm{fti}(p_1, (p_{\mathrm{goal}})_1, \mathcal{X}_1), \ldots, \mathrm{fti}(p_n, (p_{\mathrm{goal}})_n, \mathcal{X}_n)).$$

One can think of fti_G as a circumcenter law where the neighboring relationships among the agents never change. Because fti is continuous, and $(p_{\mathrm{goal}})_i$ and $\mathcal{X}_i$, $i \in \{1, \ldots, n\}$, change continuously with $(p_1, \ldots, p_n)$, we deduce that fti_G is continuous.

We now define a set-valued dynamical system $((\mathbb{R}^d)^n, (\mathbb{R}^d)^n, T)$ through the set-valued map $T : (\mathbb{R}^d)^n \rightrightarrows (\mathbb{R}^d)^n$ given by

$$T(p_1, \ldots, p_n) = \{\mathrm{fti}_G(p_1, \ldots, p_n) \mid G \text{ is a strongly connected digraph}\}.$$

Note that the evolution of the CRCMCNTR law using a proximity graph such as $\mathcal{G}_{\mathrm{disk}}(r)$ is just one of the multiple evolutions described by this set-valued map. This concludes (a).

Let us now perform (b). To characterize the convergence properties of the set-valued dynamical system, we use the LaSalle Invariance Principle in Theorem 1.21. With the notation of this result, we select $W = (\mathbb{R}^d)^n$. This set is clearly strongly positively invariant for $((\mathbb{R}^d)^n, (\mathbb{R}^d)^n, T)$.

Closedness of the set-valued map. Since fti_G is continuous for each digraph G and there is a finite number of strongly connected digraphs on the vertices $\{1, \ldots, n\}$, Exercise E1.9 implies that T is closed.

Common Lyapunov function. Define the function $V_{\mathrm{diam}} : (\mathbb{R}^d)^n \to \mathbb{R}_{\geq 0}$ by

$$V_{\mathrm{diam}}(P) = \max\{\|p_i - p_j\| \mid i, j \in \{1, \ldots, n\}\}.$$

With a slight abuse of notation, we denote by $\mathrm{co}(P)$ the convex hull of $\{p_1, \ldots, p_n\} \subset \mathbb{R}^d$. Note that $V_{\mathrm{diam}}(P) = \mathrm{diam}(\mathrm{co}(P))$. The function V_{diam} has the following properties:

(i) V_{diam} is continuous and invariant under permutations of its arguments.

(ii) $V_{\mathrm{diam}}(P) = 0$ if and only if $P \in \mathrm{diag}((\mathbb{R}^d)^n)$, where we recall that $\mathrm{diag}((\mathbb{R}^d)^n) = \{(p_1, \ldots, p_n) \in (\mathbb{R}^d)^n \mid p^{[i]} = \cdots = p^{[n]} \in \mathbb{R}^d\}$ denotes the diagonal set of $(\mathbb{R}^d)^n$. This fact is an immediate consequence of the fact that, given a set $S \subset (\mathbb{R}^d)^n$, $\mathrm{diam}(\mathrm{co}(S)) = 0$ if and only if S is a singleton.

(iii) V_{diam} is non-increasing along T on $(\mathbb{R}^d)^n$. Consider a finite set of points $S \in \mathbb{F}((\mathbb{R}^d)^n)$ and let $CC(S)$ be its circumcenter. From Lemma 2.2(i), we have $CC(S) \in \text{co}(S)$. Therefore, for any strongly connected digraph G, we have that $\text{co}(\text{fti}_G(P)) \subset \text{co}(P)$ for any $P \in (\mathbb{R}^d)^n$. Since for any two sets $S_1, S_2 \subset (\mathbb{R}^d)^n$ such that $\text{co}(S_2) \subset \text{co}(S_2)$ it holds that $V_{\text{diam}}(S_2) \le V_{\text{diam}}(S_1)$, then $V_{\text{diam}}(\text{fti}_G(P)) \le V_{\text{diam}}(P)$ for any strongly connected digraph G, which implies that V_{diam} is non-increasing along T on $(\mathbb{R}^d)^n$.

Bounded evolutions. Consider any initial condition $(p_1(0), \dots, p_n(0)) \in (\mathbb{R}^d)^n$. For any strongly connected digraph, G, we have

$$\text{fti}_G(p_1(\ell), \dots, p_n(\ell)) \in \text{co}(p_1(0), \dots, p_n(0)),$$

for all $\ell \in \mathbb{Z}_{\ge 0}$. Therefore, any evolution of the set-valued dynamical system $((\mathbb{R}^d)^n, (\mathbb{R}^d)^n, T)$ is bounded.

Characterization of the invariant set. By the LaSalle Invariance for set-valued dynamical systems in Theorem 1.21, any evolution with initial condition in $W = (\mathbb{R}^d)^n$ approaches the largest weakly positively invariant set M contained in

$$\{P \in (\mathbb{R}^d)^n \mid \exists P' \in T(P) \text{ such that } V_{\text{diam}}(P') = V_{\text{diam}}(P)\}.$$

We show that $M = \text{diag}((\mathbb{R}^d)^n)$. Clearly, $\text{diag}((\mathbb{R}^d)^n) \subset M$. To prove the other inclusion, we reason by contradiction. Assume that $P \in M \setminus \text{diag}((\mathbb{R}^d)^n)$ and, therefore, $V_{\text{diam}}(P) > 0$. Let G be a strongly connected digraph and consider $\text{fti}_G(P)$. For each $i \in \{1, \dots, n\}$, we distinguish two cases depending on whether p_i is or is not a vertex of $\text{co}(P)$. If $p_i \notin \text{Ve}(\text{co}(P))$, then Lemma 2.2(i) implies that $\text{fti}(p_i, (p_{\text{goal}})_i, \mathcal{X}_i) \in \text{co}(P) \setminus \text{Ve}(\text{co}(P))$.

If $p_i \in \text{Ve}(\text{co}(P))$, then we must take into consideration the possibility of having more than one agent located at the same point. If the location of all the neighbors of i in the digraph G coincides with p_i, then agent i will not move, and hence $\text{fti}(p_i, (p_{\text{goal}})_i, \mathcal{X}_i) \in \text{Ve}(\text{co}(P))$. However, we can show that the application of fti_G strictly decreases the number of agents located at p_i. Let us denote this number by N_i, that is,

$$N_i = |\{j \in \{1, \dots, n\} \mid p_j = p_i \text{ and } p_j \in \{p_1, \dots, p_n\}\}|.$$

Since the digraph G is strongly connected, there must exist at least an agent located at p_i with a neighbor which is not located at p_i (otherwise, all agents would be at p_i, which is a contradiction). In other words, there exist $i_*, j \in \{1, \dots, n\}$ such that $p_{i_*} = p_i$, $p_j \ne p_i$, and $j \in \mathcal{N}_G(i_*)$. By Lemma 2.2(i), we have that $(p_{\text{goal}})_{i_*} \in \text{co}(P) \setminus \text{Ve}(\text{co}(P))$ and, therefore,

$(p_{\text{goal}})_{i_*} \neq p_{i_*}$. Combining this with the fact that

$$\{p_i\} \cup \{p_j \mid j \in \mathcal{N}_G(i)\} \subset \overline{B}(p_{i_*}, r_{i_*}(P)),$$

we can apply Lemma 2.2(ii) to ensure that $]p_{i_*}, (p_{\text{goal}})_{i_*}[$ has nonempty intersection with $\mathcal{X}_{i_*}$. Therefore, $\text{fti}(p_{i_*}, (p_{\text{goal}})_{i_*}, \mathcal{X}_{i_*}) \in \text{co}(P) \setminus \text{Ve}(\text{co}(P))$, and the number N_i of agents located at p_i decreases at least by one with the application of fti_G.

Next, we show that, after a finite number of steps, no agents remain at the location p_i. Define $N = \max\{N_i \mid p_i \in \text{Ve}(\text{co}(P))\} < n - 1$. Then all agents in the configuration $\text{fti}_{G_1}(\text{fti}_{G_2}(\ldots \text{fti}_{G_N}(P)))$ are contained in $\text{co}(P) \setminus \text{Ve}(\text{co}(P))$, for any collection of strongly connected directed graphs $G_1, \ldots, G_N$. Therefore, $\text{diam}(\text{co}(\text{fti}_{G_1}(\text{fti}_{G_2}(\ldots \text{fti}_{G_N}(P))))) < \text{diam}(\text{co}(P))$, which contradicts the fact that M is weakly invariant.

Point convergence. We have proved that any evolution of $((\mathbb{R}^d)^n, (\mathbb{R}^d)^n, T)$ approaches the set $\text{diag}((\mathbb{R}^d)^n)$. To conclude the proof, let us show that the convergence of each trajectory is to a point, rather than to the diagonal set. Let $\{P(\ell) \mid \ell \in \mathbb{Z}_{\geq 0}\}$ be an evolution of the set-valued dynamical system. Since the sequence is contained in the compact set $\text{co}(P(0))$, there exists a convergent subsequence $\{P(\ell_k) \mid k \in \mathbb{Z}_{\geq 0}\}$, that is, there exists $p \in \mathbb{R}^d$ such that

$$\lim_{k \to +\infty} P(\ell_k) = (p, \ldots, p). \tag{4.6.2}$$

Let us show that the whole sequence $\{P(\ell) \mid \ell \in \mathbb{Z}_{\geq 0}\}$ converges to $(p, \ldots, p)$. Because of (4.6.2), for any $\varepsilon > 0$, there exists k_0 such that for $k \geq k_0$ one has $\text{co}(P(\ell_k)) \subset \overline{B}(p, \varepsilon/\sqrt{n})$. From this, we deduce that $\text{co}(P(\ell)) \subset \overline{B}(p, \varepsilon/\sqrt{n})$ for all $\ell \geq \ell_{k_0}$, which in turn implies that $\|P(\ell) - (p, \ldots, p)\|_2 \leq \varepsilon$ for all $\ell \geq \ell_{k_0}$, as claimed. This concludes (b).

The steps (a) and (b) imply that any evolution of $(\mathcal{S}_{\text{disk}}, \mathcal{CC}_{\text{CRCMCNTR}})$ starting from an initial configuration where $\mathcal{G}_{\text{disk}}(r)$ is connected converges to a point in $\text{diag}((\mathbb{R}^d)^n)$. To conclude the proof of fact (i), we only need to establish that this convergence is in finite time. This last fact is a consequence of Exercise E4.5.

Fact (v) for $(\mathcal{S}_{\text{disk}}, \mathcal{CC}_{\text{CRCMCNTR}})$ is a consequence of facts (i) and (iv).

STEP 2: Facts on $(\mathcal{S}_{\text{LD}}, \mathcal{CC}_{\text{CRCMCNTR}})$. The proof of facts (i), (iv), and (v) for $(\mathcal{S}_{\text{LD}}, \mathcal{CC}_{\text{CRCMCNTR}})$ is analogous to the proof of these facts for the pair $(\mathcal{S}_{\text{disk}}, \mathcal{CC}_{\text{CRCMCNTR}})$, and we leave it to the reader.

STEP 3: Facts on $(\mathcal{S}_{\infty\text{-disk}}, \mathcal{CC}_{\text{PLL-CRCMCNTR}})$. From the expression for the control function of $\mathcal{CC}_{\text{PLL-CRCMCNTR}}$, we deduce that the evolution under $\mathcal{CC}_{\text{PLL-CRCMCNTR}}$ of the robotic network $\mathcal{S}_{\infty\text{-disk}}$ (in d dimensions) can be alternatively described as the evolution under $\mathcal{CC}_{\text{CRCMCNTR}}$ of d robotic networks $\mathcal{S}_{\text{disk}}$ in $\mathbb{R}$ (see Exercise E4.4). Therefore, facts (i), (iv), and (v) for the pair $(\mathcal{S}_{\infty\text{-disk}}, \mathcal{CC}_{\text{PLL-CRCMCNTR}})$ follow from facts (i), (iv), and (v) for the pair $(\mathcal{S}_{\text{disk}}, \mathcal{CC}_{\text{CRCMCNTR}})$. ■

4.6.3 Proof of Theorem 4.17

Proof. Let $P_0 = (p^{[1]}(0), \ldots, p^{[n]}(0)) \in \mathbb{R}^n$ denote the initial condition.

Fact (i). For $d = 1$, the connectivity constraints on each agent $i \in \{1, \ldots, n\}$ imposed by the constraint set

$$\mathcal{X}_{\text{disk}}(p^{[i]}, \{p_{\text{rcvd}} \mid \text{for all non-null } p_{\text{rcvd}} \in y^{[i]}\}) \qquad (4.6.3)$$

are superfluous. In other words, the goal configuration resulting from the evaluation by agent i of the control function of the CRCMCNTR law belongs to the constraint set in (4.6.3). Moreover, the order of the robots on the real line is preserved from one time step to the next. Both observations are a consequence of Exercise E4.3.

Let us first establish the upper bound in fact (i). Consider the case when $\mathcal{G}_{\text{disk}}(r)$ is connected at P_0. Without loss of generality, assume that the agents are ordered from left to right according to their identifier, that is, $p^{[1]}(0) \leq \cdots \leq p^{[n]}(0)$. Let $\alpha \in \{3, \ldots, n\}$ have the property that agents $\{2, \ldots, \alpha - 1\}$ are neighbors of agent 1, and agent α is not. (If, instead, all agents are within an interval of length r, then rendezvous is achieved after one time step, and the upper bound in fact (i) is easily seen to be true.) Figure 4.12 presents an illustration of the definition of α. Note that we can

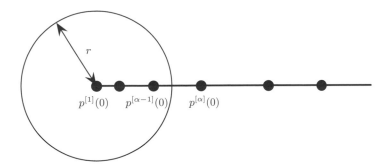

Figure 4.12 The definition of $\alpha \in \{3, \ldots, n\}$ for an initial network configuration.

assume that agents $\{2, \ldots, \alpha - 1\}$ are also neighbors of agent α. If this is not the case, then those agents that are neighbors of agent 1 and not of agent

α, rendezvous with agent 1 after one time step. At the time instant $\ell = 1$, the new updated positions satisfy

$$p^{[1]}(1) = \frac{p^{[1]}(0) + p^{[\alpha-1]}(0)}{2},$$

$$p^{[\gamma]}(1) \in \left[\frac{p^{[1]}(0) + p^{[\alpha]}(0)}{2}, \frac{p^{[1]}(0) + p^{[\gamma]}(0) + r}{2} \right],$$

for $\gamma \in \{2, \ldots, \alpha - 1\}$. These equalities imply that $p^{[1]}(1) - p^{[1]}(0) = \frac{1}{2}\left(p^{[\alpha-1]}(0) - p^{[1]}(0)\right) \leq \frac{1}{2}r$. Analogously, we deduce $p^{[1]}(2) - p^{[1]}(1) \leq \frac{1}{2}r$ and, therefore,

$$p^{[1]}(2) - p^{[1]}(0) \leq r. \qquad (4.6.4)$$

On the other hand, from $p^{[1]}(2) \in \left[\frac{1}{2}\left(p^{[1]}(1) + p^{[\alpha-1]}(1)\right), *\right]$ (where the symbol $*$ represents a certain unimportant point in $\mathbb{R}$), we deduce

$$p^{[1]}(2) - p^{[1]}(0) \geq \frac{1}{2}\left(p^{[1]}(1) + p^{[\alpha-1]}(1)\right) - p^{[1]}(0)$$

$$\geq \frac{1}{2}\left(p^{[\alpha-1]}(1) - p^{[1]}(0)\right) \geq \frac{1}{2}\left(\frac{p^{[1]}(0) + p^{[\alpha]}(0)}{2} - p^{[1]}(0)\right)$$

$$= \frac{1}{4}\left(p^{[\alpha]}(0) - p^{[1]}(0)\right) \geq \frac{1}{4}r. \qquad (4.6.5)$$

Inequalities (4.6.4) and (4.6.5) mean that, after at most two time steps, agent 1 has traveled a distance greater than $r/4$. In turn, this implies that

$$\frac{1}{r}\operatorname{diam}(\operatorname{co}(P_0)) \leq \operatorname{TC}(\mathcal{T}_{\mathrm{rndzvs}}, \mathcal{CC}_{\mathrm{CRCMCNTR}}, P_0) \leq \frac{4}{r}\operatorname{diam}(\operatorname{co}(P_0)).$$

If $\mathcal{G}_{\mathrm{disk}}(r)$ is not connected at P_0, note that along the network evolution, the connected components of the r-disk graph do not change. Using the previous characterization on the distance traveled by the leftmost agent of each connected component in at most two time steps, we deduce that

$$\operatorname{TC}(\mathcal{T}_{\mathrm{rndzvs}}, \mathcal{CC}_{\mathrm{CRCMCNTR}}, P_0) \leq \frac{4}{r} \max_{C \in \mathcal{C}(P_0)} \operatorname{diam}(\operatorname{co}(C)),$$

where $\mathcal{C}(P_0)$ denotes the collection of connected components of $\mathcal{G}_{\mathrm{disk}}(r)$ at P_0. The connectedness of each $C \in \mathcal{C}(P_0)$ implies that $\operatorname{diam}(\operatorname{co}(C)) \leq (n-1)r$, and therefore, $\operatorname{TC}(\mathcal{T}_{\mathrm{rndzvs}}, \mathcal{CC}_{\mathrm{CRCMCNTR}}) \in O(n)$.

The lower bound in fact (i) is established by considering $P_0 \in \mathbb{R}^n$ such that $p^{[i+1]}(0) - p^{[i]}(0) = r$, $i \in \{1, \ldots, n-1\}$. For this configuration, we have $\operatorname{diam}(\operatorname{co}(P_0)) = (n-1)r$ and, therefore, $\operatorname{TC}(\mathcal{T}_{\mathrm{rndzvs}}, \mathcal{CC}_{\mathrm{CRCMCNTR}}, P_0) \geq n-1$.

Fact (ii). In the r-limited Delaunay graph, two agents on the line that are at most at a distance r from each other are neighbors if and only if there are no other agents between them. Also, note that the r-limited Delaunay graph and the r-disk graph have the same connected components (cf., Theorem 2.8). An argument similar to the one used in the proof of fact (i) above guarantees that the connectivity constraints imposed by the constraint sets $\mathcal{X}_{\text{disk}}(p^{[i]}, \{p_{\text{rcvd}} \mid \text{for all non-null } p_{\text{rcvd}} \in y^{[i]}\})$ are again superfluous.

Consider first the case when $\mathcal{G}_{\text{LD}}(r)$ is connected at P_0. Note that this is equivalent to $\mathcal{G}_{\text{disk}}(r)$ being connected at P_0. Without loss of generality, assume that the agents are ordered from left to right according to their identifier, that is, $p^{[1]}(0) \leq \cdots \leq p^{[n]}(0)$. The evolution of the network under $\mathcal{CC}_{\text{CRCMCNTR}}$ can then be described as the discrete-time dynamical system

$$p^{[1]}(\ell + 1) = \frac{1}{2}(p^{[1]}(\ell) + p^{[2]}(\ell)),$$

$$p^{[2]}(\ell + 1) = \frac{1}{2}(p^{[1]}(\ell) + p^{[3]}(\ell)),$$

$$\vdots$$

$$p^{[n-1]}(\ell + 1) = \frac{1}{2}(p^{[n-2]}(\ell) + p^{[n]}(\ell)),$$

$$p^{[n]}(\ell + 1) = \frac{1}{2}(p^{[n-1]}(\ell) + p^{[n]}(\ell)).$$

Note that this evolution respects the ordering of the agents. Equivalently, we can write $P(\ell + 1) = A P(\ell)$, where $A \in \mathbb{R}^{n \times n}$ is the matrix given by

$$A = \begin{bmatrix} \frac{1}{2} & \frac{1}{2} & 0 & \cdots & \cdots & 0 \\ \frac{1}{2} & 0 & \frac{1}{2} & \cdots & \cdots & 0 \\ 0 & \frac{1}{2} & 0 & \frac{1}{2} & \cdots & 0 \\ \vdots & & \ddots & \ddots & \ddots & \vdots \\ 0 & \cdots & \cdots & \frac{1}{2} & 0 & \frac{1}{2} \\ 0 & \cdots & \cdots & 0 & \frac{1}{2} & \frac{1}{2} \end{bmatrix}.$$

Note that $A = \text{ATrid}_n^+\left(\frac{1}{2}, 0\right)$, as defined in Section 1.6.4. Theorem 1.80(i) implies that, for $P_{\text{ave}} = \frac{1}{n}\mathbf{1}_n^T P_0$, we have that $\lim_{\ell \to +\infty} P(\ell) = P_{\text{ave}}\mathbf{1}_n$, and that the maximum time required for $\|P(\ell) - P_{\text{ave}}\mathbf{1}_n\|_2 \leq \eta\|P_0 - P_{\text{ave}}\mathbf{1}_n\|_2$ (over all initial conditions in $\mathbb{R}^n$) is $\Theta(n^2 \log \eta^{-1})$. (Note that this also implies that agents rendezvous at the location given by the average of their initial positions. In other words, the asymptotic rendezvous position for this case can be expressed in closed form, as opposed to the case with the r-disk graph.)

Next, let us convert the contraction inequality on 2-norms into an appropriate inequality on ∞-norms. Note that $\text{diam}(\text{co}(P_0)) \leq (n-1)r$ because

$\mathcal{G}_{\mathrm{LD}}(r)$ is connected at P_0. Therefore,

$$\|P_0 - P_{\mathrm{ave}}\mathbf{1}\|_\infty = \max_{i\in\{1,\ldots,n\}} |p^{[i]}(0) - P_{\mathrm{ave}}| \le |p^{[1]}(0) - p^{[n]}(0)| \le (n-1)r.$$

For ℓ of order $n^2\log\eta^{-1}$, we use this bound on $\|P_0 - P_{\mathrm{ave}}\mathbf{1}\|_\infty$ and the basic inequalities $\|v\|_\infty \le \|v\|_2 \le \sqrt{n}\|v\|_\infty$ for all $v \in \mathbb{R}^n$, to obtain

$$\|P(\ell) - P_{\mathrm{ave}}\mathbf{1}\|_\infty \le \|P(\ell) - P_{\mathrm{ave}}\mathbf{1}\|_2 \le \eta\|P_0 - P_{\mathrm{ave}}\mathbf{1}\|_2$$
$$\le \eta\sqrt{n}\|P_0 - P_{\mathrm{ave}}\mathbf{1}\|_\infty \le \eta\sqrt{n}(n-1)r.$$

This means that $(r\varepsilon)$-rendezvous is achieved for $\eta\sqrt{n}(n-1)r = r\varepsilon$, that is, in time $O(n^2\log\eta^{-1}) = O(n^2\log(n\varepsilon^{-1}))$.

Next, we show the lower bound. Consider the unit-length eigenvector $\mathbf{v}_n = \sqrt{\frac{2}{n+1}}(\sin\frac{\pi}{n+1},\ldots,\sin\frac{n\pi}{n+1})^T \in \mathbb{R}^n$ of $\mathrm{Trid}_{n-1}(\frac{1}{2},0,\frac{1}{2})$ corresponding to the largest singular value $\cos(\frac{\pi}{n})$. For $\mu = \frac{-1}{10\sqrt{2}}rn^{5/2}$, we then define the initial condition

$$P_0 = \mu P_+ \begin{bmatrix} 0 \\ \mathbf{v}_{n-1} \end{bmatrix} \in \mathbb{R}^n.$$

One can show that $p^{[i]}(0) < p^{[i+1]}(0)$ for $i \in \{1,\ldots,n-1\}$, that $P_{\mathrm{ave}} = 0$, and that $\max\{p^{[i+1]}(0) - p^{[i]}(0) \mid i \in \{1,\ldots,n-1\}\} \le r$. Using Lemma 1.82 and because $\|w\|_\infty \le \|w\|_2 \le \sqrt{n}\|w\|_\infty$ for all $w \in \mathbb{R}^n$, we compute

$$\|P_0\|_\infty = \frac{rn^{5/2}}{10\sqrt{2}}\left\|P_+\begin{bmatrix}0\\\mathbf{v}_{n-1}\end{bmatrix}\right\|_\infty \ge \frac{rn^2}{10\sqrt{2}}\left\|P_+\begin{bmatrix}0\\\mathbf{v}_{n-1}\end{bmatrix}\right\|_2$$
$$\ge \frac{rn}{10\sqrt{2}}\|\mathbf{v}_{n-1}\|_2 = \frac{rn}{10\sqrt{2}}.$$

The trajectory $P(\ell) = (\cos(\frac{\pi}{n}))^\ell P_0$ therefore satisfies

$$\|P(\ell)\|_\infty = \left(\cos\left(\frac{\pi}{n}\right)\right)^\ell\|P_0\|_\infty \ge \frac{rn}{10\sqrt{2}}\left(\cos\left(\frac{\pi}{n}\right)\right)^\ell.$$

Therefore, $\|P(\ell)\|_\infty$ is larger than $\frac{1}{2}r\varepsilon$ so long as $\frac{1}{10\sqrt{2}}n(\cos(\frac{\pi}{n}))^\ell > \frac{1}{2}\varepsilon$, that is, so long as

$$\ell < \frac{\log(\varepsilon^{-1}n) - \log(5\sqrt{2})}{-\log\left(\cos(\frac{\pi}{n})\right)}.$$

In exercise E4.7, the reader is asked to show that the asymptotics of this bound correspond to the lower bound in fact (i).

Now consider the case when $\mathcal{G}_{\mathrm{LD}}(r)$ is not connected at P_0. Note that the connected components do not change along the network evolution. Therefore, the previous reasoning can be applied to each connected component.

Since the number of agents in each connected component is strictly less than n, the time complexity can only but improve. Therefore, we conclude that

$$\mathrm{TC}(\mathcal{T}_{\mathrm{rndzvs}}, \mathcal{CC}_{\mathrm{CRCMCNTR}}) \in \Theta(n^2 \log(n\varepsilon^{-1})).$$

Fact (iii). Recall from the proof of Theorem 4.6.2 that the evolution under $\mathcal{CC}_{\mathrm{PLL\text{-}CRCMCNTR}}$ of the robotic network $\mathcal{S}_{\infty\text{-}\mathrm{disk}}$ (in d dimensions) can be alternatively described as the evolution under $\mathcal{CC}_{\mathrm{CRCMCNTR}}$ of d robotic networks $\mathcal{S}_{\mathrm{disk}}$ in $\mathbb{R}$ (see Exercise E4.4). Fact (iii) now follows from fact (i). ■

4.6.4 Proof sketch of Theorem 4.20

Here, we only provide a sketch of the proof of Theorem 4.20. Fact (i) is a consequence of the control function definition of the NONCONVEX CRCMCNTR law in Section 4.3.4 and Lemma 4.11. Fact (ii) follows from the fact that the law $\mathcal{CC}_{\mathrm{NONCONVEX\ CRCMCNTR}}$ (with control magnitude bounds) achieves the ε-rendezvous task $\mathcal{T}_{\varepsilon\text{-}\mathrm{rndzvs}}$ and fact (i).

To show that, on the network $\mathcal{S}_{\mathrm{vis\text{-}disk}}$, the law $\mathcal{CC}_{\mathrm{NONCONVEX\ CRCMCNTR}}$ (with control magnitude bounds) achieves the ε-rendezvous task $\mathcal{T}_{\varepsilon\text{-}\mathrm{rndzvs}}$, one can follow the same overapproximation strategy that we used in the proof of Theorem 4.16, *STEP 1:*, that is,

(a) define a set-valued dynamical system $(Q_\delta^n, Q_\delta^n, T)$ such that the evolutions of $(\mathcal{S}_{\mathrm{vis\text{-}disk}}, \mathcal{CC}_{\mathrm{NONCONVEX\ CRCMCNTR}})$ starting from an initial configuration where $\mathcal{G}_{\mathrm{vis\text{-}disk},Q_\delta}$ is connected are contained in the set of evolutions of the set-valued dynamical system; and

(b) establish that any evolution of $(Q_\delta^n, Q_\delta^n, T)$ converges to a point in $\mathrm{diag}(Q_\delta^n)$ (note that the point might be different for different evolutions).

We refer to Ganguli et al. (2009) for a detailed development of this proof strategy. Here, we only remark that in order to carry out (b), the proof uses the LaSalle Invariance Principle in Theorem 1.21, with the perimeter of the relative convex hull of a set of points as Lyapunov function.

4.7 EXERCISES

E4.1 **(Maintaining connectivity of sparser networks).** Prove Lemma 4.8.
     Hint: Use Lemma 4.2 and the fact that $\mathcal{G}$ and $\mathcal{G}_{\mathrm{disk}}(r)$ have the same connected components.

E4.2 **(Maintaining network line-of-sight connectivity).** Prove Lemma 4.11.
     Hint: Use Proposition 4.9.

E4.3   **(Enforcing range-limited links is unnecessary for the crcmcntr law on $\mathbb{R}$).** Let $\mathcal{P} = \{p_1, \ldots, p_n\} \in \mathbb{F}(\mathbb{R})$. For $r \in \mathbb{R}_{>0}$, we work with the r-disk proximity graph $\mathcal{G}_{\text{disk}}(r)$ evaluated at $\mathcal{P}$. Let $i \in \{1, \ldots, n\}$ and consider the circumcenter of the set comprised of p_i and of its neighbors:

$$(p_{\text{goal}})_i = \text{CC}(\{p_i\} \cup \mathcal{N}_{\mathcal{G}_{\text{disk}}(r), p_i}(\mathcal{P})).$$

Show that the following hold:

  (i) if p_i and p_j are neighbors in $\mathcal{G}_{\text{disk}}(r)$, then $(p_{\text{goal}})_i$ belongs to $\overline{B}(\frac{p_i + p_j}{2}, \frac{r}{2})$;

  (ii) if p_i and p_j are neighbors in $\mathcal{G}_{\text{disk}}(r)$ and $p_i \leq p_j$, then $(p_{\text{goal}})_i \leq (p_{\text{goal}})_j$; and

Finally, discuss the implication of (i) and (ii) in the execution of the CRCMCNTR law on the 1-dimensional space $\mathbb{R}$.

Hint: *Express $(p_{\text{goal}})_i$ as a function of the position of the leftmost and rightmost points among the neighbors of p_i.*

E4.4   **(Enforcing range-limited links is unnecessary for the pll-crcmcntr law).** Let $\mathcal{P} = \{p_1, \ldots, p_n\} \in \mathbb{F}(\mathbb{R}^d)$ and $r \in \mathbb{R}_{>0}$. For $k \in \{1, \ldots, d\}$, denote by $\pi_k : \mathbb{R}^d \to \mathbb{R}$ the projection onto the kth component. Do the following tasks:

  (i) Show that p_i and p_j are neighbors in $\mathcal{G}_{\infty\text{-disk}}(r)$ if and only if, for all $k \in \{1, \ldots, d\}$, $\pi_k(p_i)$ and $\pi_k(p_j)$ are neighbors in $\mathcal{G}_{\text{disk}}(r)$.

  (ii) For $S \subset \mathbb{R}^d$, justify that the parallel circumcenter $\text{PCC}(S) \in \mathbb{R}^d$ of S can be described as

$$\pi_k(\text{PCC}(S)) = \text{CC}(\pi_k(S)), \quad \text{for } k \in \{1, \ldots, d\}.$$

  (iii) Use (i), (ii), and Exercise E4.3(i) to justify that no constraint is required to maintain connectivity of the ∞-disk graph in the PLL-CRCMCNTR law. In other words, show that if p_i and p_j are neighbors in the proximity graph $\mathcal{G}_{\infty\text{-disk}}(r)$, then also the points $\text{PCC}(\{p_i\} \cup \mathcal{N}_{\mathcal{G}_{\infty\text{-disk}}(r), p_i}(\mathcal{P}))$ and $\text{PCC}(\{p_j\} \cup \mathcal{N}_{\mathcal{G}_{\infty\text{-disk}}(r), p_j}(\mathcal{P}))$ are neighbors in the proximity graph $\mathcal{G}_{\infty\text{-disk}}(r)$.

E4.5   **(Finite-time convergence of the crcmcntr law on $\mathcal{S}_{\text{disk}}$).** For $u_{\max}, r \in \mathbb{R}_{>0}$, let $a = \min\{u_{\max}, \frac{r}{2}\}$. Let $\mathcal{P} = \{p_1, \ldots, p_n\} \in \mathbb{F}(\mathbb{R}^d)$, and assume that there exists $p \in \mathbb{R}^d$ such that

$$\{p_1, \ldots, p_n\} \subset \overline{B}(p, a).$$

Do the following tasks:

  (i) Show that $\mathcal{G}_{\text{disk}}(r)$ evaluated at $\{p_1, \ldots, p_n\}$ is the complete graph.

  (ii) Justify why $\|p_i - \text{CC}(\{p_1, \ldots, p_n\})\|_2 \leq a$, for all $i \in \{1, \ldots, n\}$.

  (iii) Show that $\text{CC}(\{p_1, \ldots, p_n\}) \in \mathcal{X}_{\text{disk}}(p_i, \mathcal{P}) \cap \overline{B}(p_i, u_{\max})$.

  (iv) What is the evolution of the pair $(\mathcal{S}_{\text{disk}}, \mathcal{CC}_{\text{CRCMCNTR}})$ (with control magnitude bounds) starting from $(p_1, \ldots, p_n)$?

E4.6   **(Variation of the crcmcntr law).** Let $\mathcal{P} = \{p_1, \ldots, p_n\} \in \mathbb{F}(\mathbb{R}^d)$. For $r \in \mathbb{R}_{>0}$, we work with the r-disk proximity graph $\mathcal{G}_{\text{disk}}(r)$ evaluated at $\mathcal{P}$. For each $i \in \{1, \ldots, n\}$, consider the circumcenter of the set comprised of p_i and of the mid-points with its neighbors:

$$(p_{\text{goal}})_i = \text{CC}\Big(\{p_i\} \cup \Big\{\frac{p_i + p_j}{2} \mid p_j \in \mathcal{N}_{\mathcal{G}_{\text{disk}}(r), p_i}(\mathcal{P})\Big\}\Big).$$

Do the following:

(i) Show that if p_i and p_j are neighbors in $\mathcal{G}_{\text{disk}}(r)$, then $(p_{\text{goal}})_i$ and $(p_{\text{goal}})_j$ are neighbors in $\mathcal{G}_{\text{disk}}(r)$.

(ii) Use (i) to design a control and communication law on the network $\mathcal{S}_{\text{disk}}$ in $\mathbb{R}^d$ that, while not enforcing any connectivity constraints, preserves all neighboring relationships in $\mathcal{G}_{\text{disk}}(r)$ and achieves the ε-rendezvous task $\mathcal{T}_{\varepsilon\text{-rndzvs}}$.

(iii) Justify why the law designed in (ii) does not achieve the exact rendezvous task $\mathcal{T}_{\text{rndzvs}}$.

E4.7 **(Asymptotics of the lower bound in Theorem 4.17(ii)).** Show that, as $n \to +\infty$,

$$\frac{\log(\varepsilon^{-1}n) - \log(5\sqrt{2})}{-\log\big(\cos(\frac{\pi}{n})\big)} = \frac{n^2}{\pi^2}\big(\log(\varepsilon^{-1}n) - \log(5\sqrt{2})\big) + O(1).$$

Use this fact to complete the proof of the lower bound in the proof of Theorem 4.17(ii).

Hint: *Use the Taylor series expansion of* $\log(\cos(x))$ *at* $x = 0$.

Chapter Five

Deployment

The aim of this chapter is to present various solutions to the deployment problem. The *deployment objective* is to optimally place a group of robots in an environment of interest. The approach taken here consists of identifying aggregate functions that measure the quality of deployment of a given network configuration and designing control and communication laws that optimize these measures.

The variety of algorithms presented in the chapter stems from two causes. First, different solutions arise from the interplay between the spatially distributed character of the coordination algorithms and the limited sensing and communication capabilities of the robotic network. As an example, different solutions are feasible when agents have range-limited communication capabilities or when agents have omnidirectional line-of-sight visibility sensors. Second, there is no universal notion of deployment. Different scenarios give rise to different ways of measuring what constitutes a good deployment. As an example, a robotic network might follow a different strategy depending on whether or not it has information about areas of importance in the environment: in the first case, by incorporating the knowledge on the environment; or in the second, by assuming a worst-case scenario, where important things can be happening precisely at the furthest-away location from the network configuration.

Our exposition here follows Cortés et al. (2004, 2005), and Cortés and Bullo (2005). Our approach makes extensive use of the multicenter functions from geometric optimization introduced in Chapter 2. It is not difficult to synthesize continuous-time gradient ascent algorithms using the smoothness results presented in Section 2.3, and characterize their asymptotic convergence properties (as we ask the reader to do in Exercises E2.14 and E2.15). However, following the robotic network model of Chapter 3, we are interested in discrete-time algorithms. In general, gradient ascent algorithms implemented in discrete time require the selection of appropriate step sizes that guarantee the monotonic evolution of the objective function. This is usually accomplished via line search procedures, (see e.g., Bertsekas and Tsitsiklis,

1997). In this chapter, we show that the special geometric properties of the multicenter functions and their gradients allow us to identify natural target locations for the robotic agents without the need to perform any line search.

The chapter is organized as follows. In the first section, we formally define the notions of deployment via task maps and multicenter functions. In the next section, we present motion coordination algorithms to achieve each deployment task. Specifically, we introduce control and communication laws based on various notions of geometric centers. We present convergence and complexity results for the proposed algorithms, along with simulations illustrating our analysis. The third section presents various simulations of the proposed motion coordination algorithms. We end the chapter with three sections on, respectively, bibliographic notes, proofs of the results presented in the chapter, and exercises. Throughout the exposition, we make extensive use of proximity graphs, multicenter functions, and geometric optimization. The convergence and complexity analyses are based on the LaSalle Invariance Principle and on linear dynamical systems defined by Toeplitz matrices.

5.1 PROBLEM STATEMENT

Here, we introduce various notions of deployment. We assume that $\mathcal{S} = (\{1, \ldots, n\}, \mathcal{R}, E_{\mathrm{cmm}})$ is a uniform robotic network, where the robots' physical state space is a (simple convex) polytope $Q \subset \mathbb{R}^d$ that describes an environment of interest. We define our notions of deployment relying upon the geometric optimization problems discussed in Section 2.3. Loosely speaking, we aim to deploy the robots in such a way as to optimize one of the multicenter functions, such as the expected-value multicenter function $\mathcal{H}_{\mathrm{exp}}$, the disk-covering multicenter function $\mathcal{H}_{\mathrm{dc}}$, or the sphere-packing multicenter function $\mathcal{H}_{\mathrm{sp}}$. Indeed, these functions can be interpreted as quality-of-service measures for different scenarios. In order to formally define the task maps encoding the deployment objective, we take the following approach: since the optimizers of these measures are critical points, and these critical points are network configurations that make the gradients vanish, we define the task map to take the `true` value at these configurations.

5.1.1 The distortion, area, and mixed distortion-area deployment tasks

In this section, we define various notions of deployment originating from the expected-value multicenter function $\mathcal{H}_{\mathrm{exp}}$. Recall the concepts of density and performance introduced in Section 2.3. Let $\phi : \mathbb{R}^d \to \mathbb{R}_{>0}$ be a density function on $\mathbb{R}^d$ with support Q. One can interpret ϕ as a function measuring the probability that some event takes place over the environment. Let

$f : \mathbb{R}_{\geq 0} \to \mathbb{R}$ be a performance, that is, a non-increasing and piecewise differentiable function possibly with finite jump discontinuities. Performance functions describe the utility of placing a robot at a certain distance from a location in the environment. Here, we will restrict our attention to the cases $f(x) = -x^2$ (distortion problem), $f(x) = 1_{[0,a]}(x)$, $a \in \mathbb{R}_{>0}$ (area problem), and $f(x) = -x^2 \, 1_{[0,a]}(x) - a^2 \cdot 1_{]a,+\infty[}(x)$, with $a \in \mathbb{R}_{>0}$ (mixed distortion-area problem).

For $\varepsilon \in \mathbb{R}_{>0}$, we define the ε-*distortion deployment task* $\mathcal{T}_{\varepsilon\text{-distor-dply}}$: $Q^n \to \{\texttt{true}, \texttt{false}\}$ by

$$\mathcal{T}_{\varepsilon\text{-distor-dply}}(P) = \begin{cases} \texttt{true}, & \text{if } \left\| p^{[i]} - \mathrm{CM}_\phi(V^{[i]}(P)) \right\|_2 \leq \varepsilon, \ i \in \{1, \dots, n\}, \\ \texttt{false}, & \text{otherwise}, \end{cases}$$

where $V^{[i]}(P)$ denotes the Voronoi cell of robot i, and $\mathrm{CM}_\phi(V^{[i]}(P))$ denotes its centroid computed according to ϕ (see Section 2.1). In other words, $\mathcal{T}_{\varepsilon\text{-distor-dply}}$ is $\texttt{true}$ for those network configurations where each robot is sufficiently close to the centroid of its Voronoi cell. According to Theorem 2.16, centroidal Voronoi configurations correspond to the critical points of the multicenter function $\mathcal{H}_{\mathrm{dist}}$.

For $r, \varepsilon \in \mathbb{R}_{>0}$, we define the ε-r-*area deployment task* $\mathcal{T}_{\varepsilon\text{-}r\text{-area-dply}} : Q^n \to \{\texttt{true}, \texttt{false}\}$ as follows: we define $\mathcal{T}_{\varepsilon\text{-}r\text{-area-dply}}(P) = \texttt{true}$ whenever

$$\left\| \int_{V^{[i]}(P) \cap \partial \overline{B}(p^{[i]}, \frac{r}{2})} \mathrm{n}_{\mathrm{out}}(q) \phi(q) dq \right\|_2 \leq \varepsilon, \quad i \in \{1, \dots, n\},$$

and we define $\mathcal{T}_{\varepsilon\text{-}r\text{-area-dply}}(P) = \texttt{true}$ otherwise. Here, the symbol $\mathrm{n}_{\mathrm{out}}$ denotes the outward normal vector to $\overline{B}(p^{[i]}, \frac{r}{2})$. In other words, $\mathcal{T}_{\varepsilon\text{-}r\text{-area-dply}}$ is $\texttt{true}$ for those network configurations where each agent is sufficiently close to a local maximum for the area of its $\frac{r}{2}$-limited Voronoi cell $V^{[i]}_{\frac{r}{2}}(P) = V^{[i]}(P) \cap \overline{B}(p^{[i]}, \frac{r}{2})$ at fixed $V^{[i]}(P)$. According to Theorem 2.16, the $\frac{r}{2}$-limited area-centered Voronoi configurations correspond to the critical points of the multicenter function $\mathcal{H}_{\mathrm{area}, \frac{r}{2}}$.

Finally, for $r, \varepsilon \in \mathbb{R}_{>0}$, we define the ε-r-*distortion-area deployment task* $\mathcal{T}_{\varepsilon\text{-}r\text{-distor-area-dply}} : Q^n \to \{\texttt{true}, \texttt{false}\}$ by

$$\mathcal{T}_{\varepsilon\text{-}r\text{-distor-area-dply}}(P)$$
$$= \begin{cases} \texttt{true}, & \text{if } \left\| p^{[i]} - \mathrm{CM}_\phi(V^{[i]}_{\frac{r}{2}}(P)) \right\|_2 \leq \varepsilon, \ i \in \{1, \dots, n\}, \\ \texttt{false}, & \text{otherwise}. \end{cases}$$

In other words, $\mathcal{T}_{\varepsilon\text{-}r\text{-distor-area-dply}}$ is $\texttt{true}$ for those network configurations where each robot is sufficiently close to the centroid of its $\frac{r}{2}$-limited Voronoi cell. According to Theorem 2.16, $\frac{r}{2}$-limited centroidal Voronoi configurations are the critical points of the multicenter function $\mathcal{H}_{\text{dist-area},\frac{r}{2}}$.

5.1.2 The disk-covering and sphere-packing deployment tasks

Here, we provide two additional notions of deployment based on the multi-center functions $\mathcal{H}_{\text{dc}}$ and $\mathcal{H}_{\text{sp}}$, respectively.

For $\varepsilon \in \mathbb{R}_{>0}$, the ε-*disk-covering deployment task* $\mathcal{T}_{\varepsilon\text{-dc-dply}} : Q^n \to \{\texttt{true}, \texttt{false}\}$ is defined as

$$\mathcal{T}_{\varepsilon\text{-dc-dply}}(P) = \begin{cases} \texttt{true}, & \text{if } \|p^{[i]} - \text{CC}(V^{[i]}(P))\|_2 \leq \varepsilon, \ i \in \{1, \dots, n\}, \\ \texttt{false}, & \text{otherwise}, \end{cases}$$

where $\text{CC}(V^{[i]}(P))$ denotes the circumcenter of the Voronoi cell of robot i. In other words, $\mathcal{T}_{\varepsilon\text{-dc-dply}}$ is $\texttt{true}$ for those network configurations where each robot is sufficiently close to the circumcenter of its Voronoi cell. According to Section 2.3.2, circumcenter Voronoi configurations are, under certain technical conditions, critical points of the multicenter function $\mathcal{H}_{\text{dc}}$.

For $\varepsilon \in \mathbb{R}_{>0}$, the ε-*sphere-packing deployment task* $\mathcal{T}_{\varepsilon\text{-sp-dply}} : Q^n \to \{\texttt{true}, \texttt{false}\}$ is defined as

$$\mathcal{T}_{\varepsilon\text{-sp-dply}}(P) = \begin{cases} \texttt{true}, & \text{if } \text{dist}_2(p^{[i]}, \text{IC}(V^{[i]}(P))) \leq \varepsilon, \ i \in \{1, \dots, n\}, \\ \texttt{false}, & \text{otherwise}, \end{cases}$$

where $\text{IC}(V^{[i]}(P))$ denotes the incenter set of the Voronoi cell of robot i. In other words, $\mathcal{T}_{\varepsilon\text{-sp-dply}}$ is $\texttt{true}$ for those network configurations where each robot is sufficiently close to the incenter set of its Voronoi cell. According to Section 2.3.3, incenter Voronoi configurations are, under certain technical conditions, critical points of the multicenter function $\mathcal{H}_{\text{sp}}$.

5.2 DEPLOYMENT ALGORITHMS

In this section, we present algorithms that can be used by a robotic network to achieve the various notions of deployment introduced in the previous section. Throughout the discussion, we use the uniform networks $\mathcal{S}_{\text{D}}$ and $\mathcal{S}_{\text{LD}}$ of locally connected first-order agents with the Delaunay and r-limited Delaunay communication, respectively, introduced in Example 3.4, and the uniform network $\mathcal{S}_{\text{vehicles}}$ of planar vehicle robots with Delaunay communication introduced in Example 3.5. The networks $\mathcal{S}_{\text{D}}$ and $\mathcal{S}_{\text{LD}}$ evolve in a polytope $Q \subset \mathbb{R}^d$, while the network $\mathcal{S}_{\text{vehicles}}$ evolves in a convex polygon $Q \subset \mathbb{R}^2$. For all the laws presented in this chapter, we assume that no two agents are initially at the same position, i.e., we assume that the initial network configuration always belongs to $Q^n \setminus \mathcal{S}_{\text{coinc}}$, where n denotes the number of robots.

All the laws presented in this chapter share a similar structure, which we loosely describe as follows:

> *[Informal description]* In each communication round, each agent performs the following tasks: (i) it transmits its position and receives its neighbors' positions; (ii) it computes a notion of the geometric center of its own cell, determined according to some notion of partition of the environment. Between communication rounds, each robot moves toward this center.

The notions of geometric center and of partition of the environment are different for each algorithm, and specifically tailored to the deployment task at hand. Let us examine them for each case.

5.2.1 Geometric-center laws

We present control and communication laws defined on the network $\mathcal{S}_D$. All the laws share in common the use of the notion of Voronoi partition of the environment Q. We first introduce the VRN-CNTRD law, which makes use of the notion of the centroid of a Voronoi cell. We then propose two sets of variations to this law. First, we present the VRN-CNTRD-DYNMCS law, which implements the same centroid strategy on a network of planar vehicles. Second, we introduce the VRN-CRCMCNTR and VRN-NCNTR laws, which instead make use of the notions of the circumcenter and incenter of a Voronoi cell, respectively.

5.2.1.1 Voronoi-centroid control and communication law

Here, we define the VRN-CNTRD control and communication law for the network $\mathcal{S}_D$, which we denote by $\mathcal{CC}_{\text{VRN-CNTRD}}$. This law was introduced by Cortés et al. (2004). We formulate the algorithm using the description model of Chapter 3. The law is uniform, static, and data-sampled, with standard message-generation function. (Recall from Definition 3.9 and Remark 3.11 that a control and coordination law (1) is uniform if processor state set, message-generation, state-transition and control functions are the same for each agent; (2) is static if the processor state set is a singleton, i.e., the law requires no memory; (3) is data-sampled if if the control functions are independent of the current position of the robot and depend only upon the robots position at the last sample time.)

Robotic Network: $\mathcal{S}_D$ with discrete-time motion model (4.1.1)
in Q, with absolute sensing of own position

Distributed Algorithm: VRN-CNTRD

Alphabet: $\mathbb{A} = \mathbb{R}^d \cup \{\texttt{null}\}$

function $\mathrm{msg}(p, i)$

  1: **return** p

function $\mathrm{ctl}(p, y)$

  1: $V := Q \cap \left(\bigcap \{H_{p, p_{\mathrm{rcvd}}} \mid \text{for all non-null } p_{\mathrm{rcvd}} \in y\} \right)$
  2: **return** $\mathrm{CM}_\phi(V) - p$

Recall that $H_{p,x}$ is the half-space of points q in $\mathbb{R}^d$ with the property that $\|q-p\|_2 \leq \|q-x\|_2$. Since the centroid of a Voronoi cell belongs to the interior of the cell itself, if the robots are at distinct locations at any one time, then they are at distinct locations after one step. Therefore, the set $Q^n \setminus \mathcal{S}_{\mathrm{coinc}}$ is positively invariant with respect to the control and communication law $\mathcal{CC}_{\text{VRN-CNTRD}}$. Moreover, note that the direction of motion specified by the control function ctl coincides with the gradient of the distortion multicenter function $\mathcal{H}_{\mathrm{dist}}$. Hence, this law prescribes a gradient ascent strategy for each robot that, as we will show later, monotonically optimizes $\mathcal{H}_{\mathrm{dist}}$.

5.2.1.2 Voronoi-centroid law on planar vehicles

Next, we provide an interesting variation of the VRN-CNTRD law defined on the network $\mathcal{S}_{\mathrm{vehicles}}$. Accordingly, we adopt the continuous-time motion model for the unicycle vehicle:

$$\dot{p}^{[i]}(t) = v^{[i]}(t) \left(\cos(\theta^{[i]}(t)), \sin(\theta^{[i]}(t)) \right),$$
$$\dot{\theta}^{[i]}(t) = \omega^{[i]}(t), \quad i \in \{1, \dots, n\}, \tag{5.2.1}$$

where we assume that forward and angular velocities are upper bounded. We refer to this control and communication law as the VRN-CNTRD-DYNMCS law, and we denote it by $\mathcal{CC}_{\text{VRN-CNTRD-DYNMCS}}$. The law was introduced by Cortés et al. (2004) and is uniform and static, but not data-sampled:

Robotic Network: $\mathcal{S}_{\mathrm{vehicles}}$ with motion model (5.2.1) in Q,
with absolute sensing of own position

Distributed Algorithm: VRN-CNTRD-DYNMCS

Alphabet: $\mathbb{A} = \mathbb{R}^2 \cup \{\text{null}\}$

function $\text{msg}((p, \theta), i)$

1: **return** p

function $\text{ctl}((p, \theta), (p_{\text{smpld}}, \theta_{\text{smpld}}), y)$

1: $V := Q \cap \left(\bigcap \{ H_{p_{\text{smpld}}, p_{\text{rcvd}}} \mid \text{for all non-null } p_{\text{rcvd}} \in y \} \right)$

2: $v := k_{\text{prop}} |(\cos\theta, \, \sin\theta) \cdot (p - \text{CM}_\phi(V))|$

3: $\omega := 2 k_{\text{prop}} \arctan \dfrac{(-\sin\theta, \, \cos\theta) \cdot (p - \text{CM}_\phi(V))}{(\cos\theta, \, \sin\theta) \cdot (p - \text{CM}_\phi(V))}$

4: **return** (v, ω)

This algorithm is illustrated in Figure 5.1.

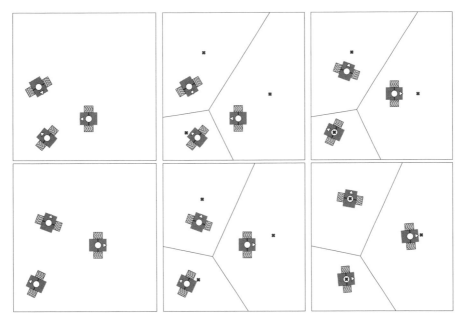

Figure 5.1 An illustration of the execution of VRN-CNTRD-DYNMCS. Each row of plots represents an iteration of the law. In each round, each agent first computes its Voronoi cell, then determines the centroid, and then moves towards it.

In the above description, we require the feedback gain k_{prop} to belong to the interval $]0, \frac{1}{\max\{\pi, \text{diam}(Q)\}}]$. This guarantees that the controls v, ω in the definition of ctl belong to the closed interval $[-1, 1]$, and are therefore, implementable in the unicycle and the differential drive robot models.

The definition of the control function ctl is based on the stabilizing feedback law of Astolfi (1999). When following this control law, the robot position p is guaranteed to monotonically approach the target position $\mathrm{CM}_\phi(V)$. Unfortunately, it is a only conjecture that this controller (or an appropriately modified controller) does not lead two agents to the same positions (indeed, it is possible that an agent move outside its Voronoi cell). Under this conjecture, the VRN-CNTRD-DYNMCS law enjoys the same convergence guarantees as the VRN-CNTRD law, that are described in Theorem 5.5.

Remark 5.1 (Vehicles with general dynamics). The general idea of moving towards the centroid of a robot's Voronoi region can be implemented over a network of vehicles with arbitrary dynamics, as long as these vehicles are capable of strictly decreasing the distance to any specified position in Q in the time intervals between communication rounds while remaining inside their Voronoi cells.   •

5.2.1.3 Voronoi-circumcenter control and communication law

Here, we define the VRN-CRCMCNTR control and communication law for the network $\mathcal{S}_\mathrm{D}$, which we denote by $\mathcal{CC}_\text{VRN-CRCMCNTR}$. This law was introduced by Cortés and Bullo (2005). The law is uniform, static, and data-sampled, with standard message-generation function:

Robotic Network: $\mathcal{S}_\mathrm{D}$ with discrete-time motion model (4.1.1)
in Q, with absolute sensing of own position

Distributed Algorithm: VRN-CRCMCNTR

Alphabet: $\mathbb{A} = \mathbb{R}^d \cup \{\texttt{null}\}$

function $\mathrm{msg}(p, i)$
   1: **return** p

function $\mathrm{ctl}(p, y)$
   1: $V := Q \cap \left(\bigcap \{ H_{p,p_\mathrm{rcvd}} \mid \text{for all non-null } p_\mathrm{rcvd} \in y \} \right)$
   2: **return** $\mathrm{CC}(V) - p$

Note that the circumcenter of a Voronoi cell belongs to the cell itself and therefore, robots evolving under the control and communication law $\mathcal{CC}_\text{VRN-CRCMCNTR}$ never leave the set Q. However, in general the set $Q^n \setminus \mathcal{S}_\mathrm{coinc}$ is not positively invariant, see Exercise E5.1. From a geometric perspective, this law makes sense as a strategy to optimize the disk-covering multicenter function $\mathcal{H}_\mathrm{dc}$. From Section 2.1.3, for fixed V, the circumcenter location minimizes the cost given by the maximum distance to all points in V. From Section 2.3.2, $\mathcal{H}_\mathrm{dc}$ can be expressed (2.3.12) as the maximum over the network of each robot's individual cost.

5.2.1.4 Voronoi-incenter control and communication law

Here, we define the VRN-NCNTR control and communication law for the network $\mathcal{S}_D$, which we denote by $\mathcal{CC}_{\text{VRN-NCNTR}}$. This law was introduced by Cortés and Bullo (2005). The law is uniform, static, and data-sampled, with standard message-generation function:

Robotic Network: $\mathcal{S}_D$ with discrete-time motion model (4.1.1) in Q, with absolute sensing of own position

Distributed Algorithm: VRN-NCNTR

Alphabet: $\mathbb{A} = \mathbb{R}^d \cup \{\texttt{null}\}$

function $\text{msg}(p, i)$
  1: **return** p

function $\text{ctl}(p, y)$
  1: $V := Q \cap \left(\bigcap \{ H_{p, p_{\text{rcvd}}} \mid \text{for all non-null } p_{\text{rcvd}} \in y \} \right)$
  2: **return** $x \in \text{IC}(V) - p$

Since the incenter set of a Voronoi cell belongs to the interior of the cell itself, if the robots are at distinct locations at any one time, then they are at distinct locations after one step. That is, the set $Q^n \setminus \mathcal{S}_{\text{coinc}}$ is positively invariant with respect to the control and communication law $\mathcal{CC}_{\text{VRN-NCNTR}}$. From a geometric perspective, this law makes sense as a strategy for optimizing the sphere-packing multicenter function $\mathcal{H}_{\text{sp}}$. From Section 2.1.3, for fixed V, the incenter locations maximize the cost given by the minimum distance to the boundary of V. From Section 2.3.3, $\mathcal{H}_{\text{sp}}$ can be expressed (2.3.15) as the minimum over the network of each robot's individual cost.

Remark 5.2 ("Move-toward-furthest-vertex" and "away-from-closest-neighbor" coordination algorithms). Consider the coordination algorithm where, at each time step, each robot moves towards the furthest-away vertex of its own Voronoi cell. Alternatively, consider the coordination algorithm where, at each time step, each robot moves away from its closest neighbor. Both coordination algorithms define maps which depend discontinuously on the robots' positions. Cortés and Bullo (2005) study the asymptotic behavior of these laws, and show that the "move-toward-furthest-vertex" algorithm monotonically optimizes the multicenter function $\mathcal{H}_{\text{dc}}$, while the "away-from-closest-neighbor" algorithm monotonically optimizes the multicenter function $\mathcal{H}_{\text{sp}}$.   •

5.2.2 Geometric-center laws with range-limited interactions

In the following, we present two control and communication laws on the network $\mathcal{S}_{\mathrm{LD}}$. Both laws prescribe a geometric centering strategy for each robot and accomplish specific forms of expected-value optimization. The LMTD-VRN-NRML law optimizes the area multicenter function $\mathcal{H}_{\mathrm{area},\frac{r}{2}}$, while the LMTD-VRN-CNTRD law optimizes the mixed distortion-area multicenter function $\mathcal{H}_{\mathrm{dist\text{-}area},\frac{r}{2}}$.

5.2.2.1 Limited-Voronoi-normal control and communication law

Here, we define the LMTD-VRN-NRML control and communication law for the network $\mathcal{S}_{\mathrm{LD}}$. This law was introduced by Cortés et al. (2005). The LMTD-VRN-NRML law, which we denote by $\mathcal{CC}_{\mathrm{LMTD\text{-}VRN\text{-}NRML}}$, uses the notion of $\frac{r}{2}$-limited Voronoi partition inside Q. The law is uniform, static, and data-sampled, with standard message-generation function:

Robotic Network: $\mathcal{S}_{\mathrm{LD}}$ with discrete-time motion model (4.1.1)
with absolute sensing of own position, and
with communication range r, in Q

Distributed Algorithm: LMTD-VRN-NRML

Alphabet: $\mathbb{A} = \mathbb{R}^d \cup \{\texttt{null}\}$

function $\mathrm{msg}(p, i)$
  1: **return** p

function $\mathrm{ctl}(p, y)$
  1: $V := Q \cap \left(\bigcap \{ H_{p,p_{\mathrm{rcvd}}} \mid \text{for all non-null } p_{\mathrm{rcvd}} \in y \} \right)$
  2: $v := \displaystyle\int_{V \cap \partial \overline{B}(p, \frac{r}{2})} \mathrm{n}_{\mathrm{out}}(q)\phi(q)dq$
  3: $\lambda_* := \max \left\{ \lambda \mid \delta \mapsto \displaystyle\int_{V \cap \overline{B}(p+\delta v, \frac{r}{2})} \phi(q)dq \text{ is strictly increasing on } [0, \lambda] \right\}$
  4: **return** $\lambda_* v$

In the above algorithm, $\mathrm{n}_{\mathrm{out}}$ denotes the outward normal vector to $\overline{B}(p, \frac{r}{2})$. Note that the direction of motion v specified by the control function ctl coincides with the gradient of the multicenter function $\mathcal{H}_{\mathrm{area},\frac{r}{2}}$. The parameter λ_* corresponds to performing a line search procedure along the direction of the vector v.

The control function has the property that the point $p + \text{ctl}(p, y)$ is guaranteed to be in the interior of V. This can be justified by noting that for fixed V, the gradient of the function $p \rightarrow \int_{V \cap \overline{B}(p, \frac{r}{2})} \phi(q) dq$ at points in the boundary of V is non-vanishing and points toward the interior of V (cf. Exercise E2.5). As a consequence, the line search procedure terminates before reaching the boundary of V. This discussion guarantees that the set $Q^n \setminus \mathcal{S}_{\text{coinc}}$ is positively invariant with respect to the control and communication law $\mathcal{CC}_{\text{LMTD-VRN-NRML}}$.

5.2.2.2 Limited-Voronoi-centroid control and communication law

Here, we define the LMTD-VRN-CNTRD control and communication law for the network $\mathcal{S}_{\text{LD}}$. This law was introduced by Cortés et al. (2005). The LMTD-VRN-CNTRD law, which we denote by $\mathcal{CC}_{\text{LMTD-VRN-CNTRD}}$, uses the notion of $\frac{r}{2}$-limited Voronoi partition inside Q and of centroid of the individual $\frac{r}{2}$-limited Voronoi cells. The law is uniform, static, and data-sampled, with standard message-generation function:

Robotic Network: $\mathcal{S}_{\text{LD}}$ with discrete-time motion model (4.1.1)
with absolute sensing of own position, and
with communication range r, in Q

Distributed Algorithm: LMTD-VRN-CNTRD

Alphabet: $\mathbb{A} = \mathbb{R}^d \cup \{\texttt{null}\}$

function $\text{msg}(p, i)$
  1: **return** p

function $\text{ctl}(p, y)$
  1: $V := Q \cap \overline{B}(p, \frac{r}{2}) \cap \left(\bigcap \{H_{p, p_{\text{rcvd}}} \mid \text{for all non-null } p_{\text{rcvd}} \in y\} \right)$
  2: **return** $\text{CM}_\phi(V) - p$

The centroid of a $\frac{r}{2}$-limited Voronoi cell belongs to the interior of the cell itself, and this fact guarantees that the set $Q^n \setminus \mathcal{S}_{\text{coinc}}$ is positively invariant with respect to the control and communication law $\mathcal{CC}_{\text{LMTD-VRN-CNTRD}}$. Moreover, note that the direction of motion specified by the control function ctl coincides with the gradient of the multicenter function $\mathcal{H}_{\text{dist-area}, \frac{r}{2}}$.

Remark 5.3 (Relative sensing version). It is possible to implement the limited-Voronoi-normal and limited-Voronoi-centroid laws as static relative-sensing control laws on the relative-sensing network $\mathcal{S}_{\text{disk}}^{\text{rs}}$. This is a consequence of the fact that the r-limited Delaunay graph is spatially distributed over the r-disk graph (cf., Theorem 2.7(iii)). Let us present one of these examples for completeness:

Relative Sensing Network: $\mathcal{S}_{\mathrm{disk}}^{\mathrm{rs}}$ with motion model (4.1.2) in Q, no communication, relative sensing for robot i given by: robot measurements y contains $p_i^{[j]} \in \overline{B}(\mathbf{0}_2, r)$ for all $j \neq i$ environment measurement is $y_{\mathrm{env}} = (Q_\varepsilon)_i \cap \overline{B}(\mathbf{0}_d, r)$

Distributed Algorithm: RELATIVE-SENSING LMTD-VRN-CNTRD

function ctl(y, y_{env})
  1: $V := y_{\mathrm{env}} \cap \overline{B}(\mathbf{0}_d, \frac{r}{2}) \cap \left(\bigcap \{ H_{\mathbf{0}_d, p_{\mathrm{snsd}}} \mid \text{for all non-null } p_{\mathrm{snsd}} \in y \} \right)$
  2: **return** $\mathrm{CM}_\phi(V)$

Note that only the positions of neighboring robots in the r-limited Delaunay graph have an effect on the computation of the set V. •

Remark 5.4 (Range-limited version of Vrn-cntrd). The LMTD-VRN-NRML and LMTD-VRN-CNTRD laws can be combined into a single control and communication law to synthesize an algorithm that monotonically optimizes the function $\mathcal{H}_{\mathrm{dist\text{-}area}, \frac{r}{2}, b}$, with $b = -\operatorname{diam}(Q)^2$. This law, which we term RNG-VRN-CNTRD, is uniform, static, and data-sampled, with standard message-generation function:

Robotic Network: $\mathcal{S}_{\mathrm{LD}}$ with discrete-time motion model (4.1.1) in Q, with absolute sensing of own position, and with communication range r

Distributed Algorithm: RNG-VRN-CNTRD

Alphabet: $\mathbb{A} = \mathbb{R}^d \cup \{\texttt{null}\}$

function msg(p, i)
  1: **return** p

function ctl(p, y)
  1: $V := Q \cap \left(\bigcap \{ H_{p, p_{\mathrm{rcvd}}} \mid \text{for all non-null } p_{\mathrm{rcvd}} \in y \} \right)$
  2: $v_1 := 2 \, \mathrm{A}_\phi(V \cap \overline{B}(p, \frac{r}{2}))(\mathrm{CM}_\phi(V \cap \overline{B}(p, \frac{r}{2})) - p)$
  3: $v_2 := (\operatorname{diam}(Q)^2 - \frac{r^2}{4}) \int_{V \cap \partial \overline{B}(p, \frac{r}{2})} \mathrm{n}_{\mathrm{out}}(q) \phi(q) dq$
  4: $\lambda_* := \max \left\{ \lambda \mid \delta \mapsto \mathcal{H}_V(p + \delta(v_1 + v_2), \overline{B}(p + \delta(v_1 + v_2), \frac{r}{2})) \right.$
    $\left. \text{is strictly increasing on } (0, \lambda) \right\}$
  5: **return** $\lambda_*(v_1 + v_2)$

In the above algorithm, n_{out} denotes the outward normal vector to $\overline{B}(p, \frac{r}{2})$ and, for a point $p \in V$ and a closed ball $\overline{B}$ centered at a point in V with radius $\frac{r}{2}$, $\mathcal{H}_V$ is defined as

$$\mathcal{H}_V(p, \overline{B}) = -\int_{V \cap \overline{B}} \|q - p\|_2^2 \phi(q)dq - \operatorname{diam}(Q)^2 \int_{V \cap (Q \setminus \overline{B})} \phi(q)dq.$$

The RNG-VRN-CNTRD law is relevant because of the following discussion. Recall from Proposition 2.17 that the general mixed distortion-area multicenter function can be used to provide constant-factor approximations of the distortion function $\mathcal{H}_{dist}$. As we discussed in Section 2.3.1, robots with range-limited interactions cannot implement VRN-CNTRD because, for a given $r \in \mathbb{R}_{>0}$, $\mathcal{G}_D$ is not in general spatially distributed over $\mathcal{G}_{disk}(r)$ (cf., Remark 2.10). However, robotic agents with range-limited interactions can implement the computations involved in LMTD-VRN-NRML and LMTD-VRN-CNTRD, and hence can optimize $\mathcal{H}_{dist-area, \frac{r}{2}, b}$, with $b = -\operatorname{diam} Q^2$. Assuming $r \leq 2 \operatorname{diam}(Q)$, it is fair to say that the above algorithm can be understood as a range-limited version of the VRN-CNTRD law.   •

5.2.3 Correctness and complexity of geometric-center laws

In this section, we characterize the convergence and complexity properties of the geometric-center laws. The asynchronous execution of the Voronoi-centroid control and communication law can be studied as an asynchronous gradient dynamical system (see Cortés et al., 2004).

The following theorem summarizes the results known in the literature about the asymptotic properties of these laws.

Theorem 5.5 (Correctness of the geometric-center algorithms). *For $d \in \mathbb{N}$, $r \in \mathbb{R}_{>0}$, and $\varepsilon \in \mathbb{R}_{>0}$, the following statements hold for any execution that starts from a configuration in $Q^n \setminus \mathcal{S}_{coinc}$:*

(i) *On the network $\mathcal{S}_D$, the law $\mathcal{CC}_{VRN-CNTRD}$ achieves the ε-distortion deployment task $\mathcal{T}_{\varepsilon-distor-dply}$. Moreover, any execution of the law $\mathcal{CC}_{VRN-CNTRD}$ monotonically optimizes the multicenter function $\mathcal{H}_{dist}$.*

(ii) *On the network $\mathcal{S}_D$, any execution of the law $\mathcal{CC}_{VRN-CRCMCNTR}$ monotonically optimizes the multicenter function $\mathcal{H}_{dc}$.*

(iii) *On the network $\mathcal{S}_D$, any execution of the law $\mathcal{CC}_{VRN-NCNTR}$ monotonically optimizes the multicenter function $\mathcal{H}_{sp}$.*

(iv) *On the network* $\mathcal{S}_{\mathrm{LD}}$, *the law* $\mathcal{CC}_{\mathrm{LMTD\text{-}VRN\text{-}NRML}}$ *achieves the* ε-r-*area deployment task* $\mathcal{T}_{\varepsilon\text{-}r\text{-area-dply}}$. *Moreover, any execution of the law* $\mathcal{CC}_{\mathrm{LMTD\text{-}VRN\text{-}NRML}}$ *monotonically optimizes the multicenter function* $\mathcal{H}_{\mathrm{area},\frac{r}{2}}$.

(v) *On the network* $\mathcal{S}_{\mathrm{LD}}$, *the law* $\mathcal{CC}_{\mathrm{LMTD\text{-}VRN\text{-}CNTRD}}$ *achieves the* ε-r-*distortion-area deployment task* $\mathcal{T}_{\varepsilon\text{-}r\text{-distor-area-dply}}$. *Moreover, any execution of* $\mathcal{CC}_{\mathrm{LMTD\text{-}VRN\text{-}CNTRD}}$ *monotonically optimizes the multicenter function* $\mathcal{H}_{\mathrm{dist\text{-}area},\frac{r}{2}}$.

The proof of this theorem is given in Section 5.5.1. The results on $\mathcal{CC}_{\mathrm{VRN\text{-}CNTRD}}$ appeared originally in Cortés et al. (2004). Note that an execution of $\mathcal{CC}_{\mathrm{VRN\text{-}CNTRD}}$ can be viewed as an alternating sequence of configuration of points and partitions of the space, with the properties that (i) each configuration of points corresponds to the set of centroid locations of the immediately preceding partition in the sequence, and (ii) each partition corresponds to the Voronoi partition determined by the immediately preceding configuration of points in the sequence. The monotonic behavior of $\mathcal{H}_{\mathrm{dist}}$ now follows from Propositions 2.13 and 2.14. Similar interpretations can be given to all other laws. In particular, the monotonic behavior of $\mathcal{H}_{\mathrm{dc}}$ along executions of $\mathcal{CC}_{\mathrm{VRN\text{-}CRCMCNTR}}$ can be established via Proposition 2.19, and the monotonic behavior of $\mathcal{H}_{\mathrm{sp}}$ along executions of $\mathcal{CC}_{\mathrm{VRN\text{-}NCNTR}}$ can be established via Proposition 2.21. Continuous-time versions of these laws are studied by Cortés and Bullo (2005) via nonsmooth stability analysis, where the following convergence properties are established (recall the notion of active and passive nodes introduced in Sections 2.3.2 and 2.3.3): all active agents are guaranteed to asymptotically reach the circumcenter (resp., incenter) of their Voronoi region, whereas it is not known if the same conclusion holds for the passive agents. Depending on the polytope Q, there exist circumcenter and incenter Voronoi configurations where not all agents are active, and simulations show that in some cases the continuous-time versions of $\mathcal{CC}_{\mathrm{VRN\text{-}CRCMCNTR}}$ and $\mathcal{CC}_{\mathrm{VRN\text{-}NCNTR}}$ converge to them. It is an open research question to show that $\mathcal{CC}_{\mathrm{VRN\text{-}CRCMCNTR}}$ and $\mathcal{CC}_{\mathrm{VRN\text{-}NCNTR}}$ achieve the ε-disk-covering deployment task $\mathcal{T}_{\varepsilon\text{-dc-dply}}$ and the ε-sphere-packing deployment task $\mathcal{T}_{\varepsilon\text{-sp-dply}}$, respectively. Finally, the results on $\mathcal{CC}_{\mathrm{LMTD\text{-}VRN\text{-}NRML}}$ and $\mathcal{CC}_{\mathrm{LMTD\text{-}VRN\text{-}CNTRD}}$ appeared in Cortés et al. (2005).

Next, we analyze the time complexity of $\mathcal{CC}_{\mathrm{LMTD\text{-}VRN\text{-}CNTRD}}$. We provide complete results only for the case $d = 1$ and uniform density. We assume that diam(Q) is independent of n, r, and ε.

Theorem 5.6 (Time complexity of Lmtd-Vrn-cntrd law). *Assume that the robots evolve in a closed interval* $Q \subset \mathbb{R}$, *that is,* $d = 1$, *and assume that the density is uniform, that is,* $\phi \equiv 1$. *For* $r \in \mathbb{R}_{>0}$ *and* $\varepsilon \in \mathbb{R}_{>0}$, *on the network* $\mathcal{S}_{\mathrm{LD}}$, $\mathrm{TC}(\mathcal{T}_{\varepsilon\text{-}r\text{-distor-area-dply}}, \mathcal{CC}_{\mathrm{LMTD\text{-}VRN\text{-}CNTRD}}) \in O(n^3 \log(n\varepsilon^{-1}))$.

The proof of this result is given in Section 5.5.2 following the treatment in Martínez et al. (2007b).

Remark 5.7 (Congestion effects). Interestingly, Theorem 5.6 also holds if, motivated by wireless congestion considerations, we take the communication range r to be a monotone non-increasing function $r : \mathbb{N} \to]0, 2\pi[$ of the number of robotic agents n.   •

5.3 SIMULATION RESULTS

In this section, we illustrate the execution of the various control and communication laws introduced in this chapter.

Geometric-center algorithms for expected-value optimization

The VRN-CNTRD, LMTD-VRN-NRML, and LMTD-VRN-CNTRD control and communication laws are implemented in Mathematica® as a library of routines and a main program running the simulation. The objective of a first routine is to compute the $\frac{r}{2}$-limited Voronoi partition and parameterize each cell $V_{i,\frac{r}{2}}$, $i \in \{1, \dots, n\}$ in polar coordinates. The objective of a second routine is to compute the surface integrals on these sets and the line integrals on their boundaries via the numerical integration routine `NIntegrate`. We pay careful attention to numerical accuracy issues in the computation of the Voronoi diagram and in the integration.

Measuring displacements in meters, we consider the polygon Q determined by the vertices

$$\{(0,0), (2.125, 0), (2.9325, 1.5), (2.975, 1.6),$$
$$(2.9325, 1.7), (2.295, 2.1), (0.85, 2.3), (0.17, 1.2)\}.$$

The diameter of Q is $\text{diam}(Q) \approx 3.378$. In all figures, the density function ϕ is the sum of four Gaussian functions of the form $11 \exp(6(-(x - x_{\text{center}})^2 - (y - y_{\text{center}})^2))$ and is represented by means of its contour plot. Darker-colored areas correspond to higher values of the density function. The four centers $(x_{\text{center}}, y_{\text{center}})$ of the Gaussian functions are the points $(2.15, 0.75)$, $(1.0, 0.25)$, $(0.725, 1.75)$ and $(0.25, 0.7)$, respectively. The area of the polygon is $A_\phi(Q) = 17.6352$.

We show evolutions of $(\mathcal{S}_D, \text{VRN-CNTRD})$ and $(\mathcal{S}_D, \text{VRN-CNTRD-DYNMCS})$ in Figures 5.2 and 5.3, respectively. One can verify that the final network configurations is a centroidal Voronoi configuration. In other words, the task $\mathcal{T}_{\varepsilon\text{-distor-dply}}$ is achieved, as guaranteed by Theorem 5.5(i) for the VRN-

CNTRD ALGORITHM. For each evolution we depict the initial positions, the trajectories, and the final positions of all robots.

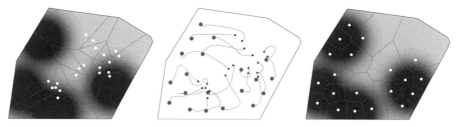

Figure 5.2 The evolution of $(\mathcal{S}_\mathrm{D}, \mathrm{VRN\text{-}CNTRD})$ with $n = 20$ robots. The left-hand (resp., right-hand) figure illustrates the initial (resp., final) locations and Voronoi partition. The central figure illustrates the evolution of the robots. After 13 seconds, the value of $\mathcal{H}_\mathrm{dist}$ has monotonically increased to approximately -0.515.

Figure 5.3 The evolution of $(\mathcal{S}_\mathrm{D}, \mathrm{VRN\text{-}CNTRD\text{-}DYNMCS})$ with $n = 20$ robots and with feedback gain $k_\mathrm{prop} = 3.5$. The left-hand (resp., right-hand) figure illustrates the initial (resp., final) locations and Voronoi partition. The central figure illustrates the evolution of the robots. After 20 seconds, the value of $\mathcal{H}_\mathrm{dist}$ has monotonically increased to approximately -0.555.

We show an evolution of $(\mathcal{S}_\mathrm{LD}, \mathrm{LMTD\text{-}VRN\text{-}NRML})$ in Figure 5.4. One can verify that the final network configuration is an $\frac{r}{2}$-limited area-centered Voronoi configuration. In other words, the task $\mathcal{T}_{\varepsilon\text{-}r\text{-area\text{-}dply}}$ is achieved, as guaranteed by Theorem 5.5(ii).

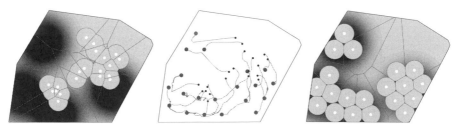

Figure 5.4 The evolution of $(\mathcal{S}_\mathrm{LD}, \mathrm{LMTD\text{-}VRN\text{-}NRML})$ with $n = 20$ robots and $r = 0.4$. The left-hand (resp., right-hand) figure illustrates the initial (respectively, final) locations and Voronoi partition. The central figure illustrates the evolution of the robots. The $\frac{r}{2}$-limited Voronoi cell of each robot is plotted in light gray. After 36 seconds, the value of $\mathcal{H}_{\mathrm{area},\frac{r}{2}}$ has monotonically increased to approximately 14.141.

We show an evolution of $(\mathcal{S}_{\mathrm{LD}}, \textsc{Lmtd-Vrn-cntrd})$ in Figure 5.5. One can verify that the final network configuration is a $\frac{r}{2}$-limited centroidal Voronoi configuration. In other words, the task $\mathcal{T}_{\varepsilon\text{-}r\text{-distor-area-dply}}$ is achieved, as guaranteed by Theorem 5.5(iii).

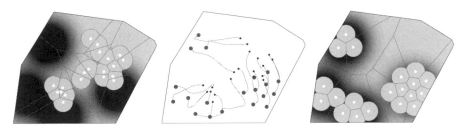

Figure 5.5 The evolution of $(\mathcal{S}_{\mathrm{LD}}, \textsc{Lmtd-Vrn-cntrd})$ with $n = 20$ robots and $r = 0.4$. The left-hand (resp., right-hand) figure illustrates the initial (resp., final) locations and Voronoi partition. The central figure illustrates the evolution of the robots. The $\frac{r}{2}$-limited Voronoi cell of each robot is plotted in light gray. After 90 seconds, the value of $\mathcal{H}_{\mathrm{dist\text{-}area},\frac{r}{2}}$ reaches approximately -0.386.

We show an evolution of $(\mathcal{S}_{\mathrm{LD}}, \textsc{Rng-Vrn-cntrd})$ in Figure 5.6. One can verify that the final network configuration corresponds to a critical point of the mixed distortion-area multicenter function $\mathcal{H}_{\mathrm{dist\text{-}area},\frac{r}{2},b}$, with $b = -\operatorname{diam}(Q)^2$ (see Exercise E5.4).

Figure 5.6 The evolution of $(\mathcal{S}_{\mathrm{LD}}, \textsc{Rng-Vrn-cntrd})$ with $n = 20$ robots and $r = 0.47$. The left-hand (resp., right-hand) figure illustrates the initial (respectively, final) locations and Voronoi partition. The central figure illustrates the evolution of the robots. The $\frac{r}{2}$-limited Voronoi cell of each robot is plotted in light gray. After 13 seconds, the value of $\mathcal{H}_{\mathrm{dist\text{-}area},\frac{r}{2},b}$, with $b = -\operatorname{diam}(Q)^2$, is approximately -4.794.

As discussed in Remark 5.4, $\textsc{Rng-Vrn-cntrd}$ can be understood as a range-limited implementation of $\textsc{Vrn-cntrd}$ in a network of robots with range-limited interactions. Let us briefly compare the evolutions depicted in Figures 5.2 and 5.6. According to Proposition 2.17, we compute

$$\beta = \frac{\frac{r}{2}}{\operatorname{diam} Q} \approx 0.06957.$$

From the constant-factor approximation (2.3.7), the absolute error is guaranteed to be less than or equal to $(\beta^2 - 1)\mathcal{H}_{\text{dist-area},\frac{r}{2},b}(P_{\text{final}}) \approx 4.77$, where P_{final} denotes the final configuration in Figure 5.6. The percentage error in the value of the multicenter function $\mathcal{H}_{\text{dist}}$ between the final configuration of the evolution in Figure 5.2 and the final configuration of the evolution in Figure 5.6 is approximately equal to 3.277%. As expected, one can verify in simulations that the percentage error of the performance of the range-limited implementation improves with higher values of the ratio $\frac{r}{\text{diam}\, Q}$.

Geometric-center algorithms for disk-covering and sphere-packing

The VRN-CRCMCNTR and VRN-NCNTR control and communication laws are implemented in Mathematica® as a single centralized program running the simulation. We compute the bounded Voronoi diagram of a collection of points using the package `ComputationalGeometry`. We compute the circumcenter of a polygon via the algorithm in Skyum (1991) and the incenter set via the `LinearProgramming` solver in Mathematica®.

Measuring displacements in meters, we consider the polygon determined by the vertices

$$\{(0,0), (2.5, 0), (3.45, 1.5), (3.5, 1.6),$$
$$(3.45, 1.7), (2.7, 2.1), (1.0, 2.4), (0.2, 1.2)\}.$$

We show an evolution of $(\mathcal{S}_D, \text{VRN-CRCMCNTR})$ in Figure 5.7. One can verify that in the final configuration all robots are at the circumcenter of their own Voronoi cell. In other words, the task $\mathcal{T}_{\varepsilon\text{-dc-dply}}$ is achieved by this evolution. As stated in Section 5.2.3, it is an open research question to show that this fact holds in general for $\mathcal{CC}_{\text{VRN-CRCMCNTR}}$. Cortés and Bullo (2005) prove a similar result for a continuous-time implementation of this law.

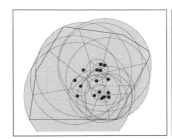  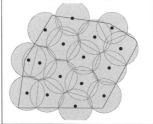

Figure 5.7 The evolution of $(\mathcal{S}_D, \text{VRN-CRCMCNTR})$ with $n = 16$ robots. The left-hand (resp., right-hand) figure illustrates the initial (resp., final) locations and Voronoi partition. The central figure illustrates the evolution of the robots. After 20 seconds, the value of $\mathcal{H}_{\text{dc}}$ has monotonically decreased to approximately 0.43273 meters.

We show an evolution of $(\mathcal{S}_D, \text{VRN-NCNTR})$ in Figure 5.8. One can verify that in the final configuration all robots are at the incenter of their own Voronoi cell. In other words, the task $\mathcal{T}_{\varepsilon\text{-sp-dply}}$ is achieved by this evolution. As stated in Section 5.2.3, it is an open research question to show that this fact holds in general for $\mathcal{CC}_{\text{VRN-NCNTR}}$. Cortés and Bullo (2005) prove a similar result for a continuous-time implementation of this law.

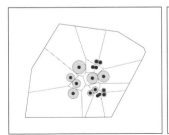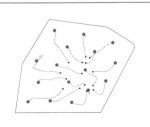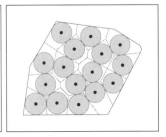

Figure 5.8 The evolution of $(\mathcal{S}_D, \text{VRN-NCNTR})$ with $n = 16$ robots. The left-hand (resp., right-hand) figure illustrates the initial (resp., final) locations and Voronoi partition. The central figure illustrates the evolution of the robots. After 20 seconds, the value of $\mathcal{H}_{\text{sp}}$ has monotonically increased to approximately 0.2498 meters.

5.4 NOTES

The deployment problem studied in this chapter is related to the literature on facility location (Drezner, 1995; Okabe et al., 2000; Du et al., 1999) and geometric optimization (Agarwal and Sharir, 1998; Boltyanski et al., 1999) (see also Section 2.4). These disciplines study spatial resource allocation problems and play an important role in quantization theory, mesh and grid optimization methods, clustering analysis, data compression, and statistical pattern recognition. Indeed, our algorithms are closely related to some early work by Lloyd (1982) on "centering and partitioning" algorithms for optimal quantizer design.

Dispersion laws have been traditionally studied in behavior control (see, e.g., (Arkin, 1998; Schultz and Parker, 2002; Balch and Parker, 2002)). Deployment algorithms that make use of potential field methods are proposed by Payton et al. (2001) and Howard et al. (2002). Other works include (Bulusu et al., 2001) on adaptive beacon placement for localization, Poduri and Sukhatme (2004) on network deployments that satisfy a pre-specified constraint in the number of neighbors of each robot, Arsie et al. (2009) on sensor-based deployment strategies that minimize the expected service time for newly appearing target points, and Hussein and Stipanovic̀ (2007) on dynamically surveying a known environment.

Deployment algorithms for coverage control are a subject of active research. Among the most recent works, Martínez (2009a) and Schwager et al. (2009) consider coverage problems where the density function is unknown, Lekien and Leonard (2009) propose centralized laws for non-uniform coverage using cartograms, de Silva and Ghrist (2007) study static coverage problems with minimal assumptions on the capabilities of individual sensors using algebraic topology, Kwok and Martínez (2009) propose distributed deployment strategies for energy-constrained networks, Laventall and Cortés (2009) design distributed algorithms for networks of robots whose sensors have range-limited wedge-shaped footprints, Gao et al. (2008) consider discrete coverage problems, Schwager et al. (2008) consider joint exploration and deployment problems, and Zhong and Cassandras (2008), Pimenta et al. (2008), and Caicedo-Nùñez and Žefran (2008) deal with centroidal Voronoi tessellations in nonconvex environments. Graham and Cortés (2009) study the optimality of circumcenter and incenter Voronoi configurations for the estimation of stochastic spatial fields. Susca et al. (2009) consider some planar interpolation problems. Finally, Cortés (2008a); Pavone et al. (2008) consider equitable partitioning policies in which the workspace is divided into subregions of equal area and their application to vehicle routing problems.

Deployment problems play a relevant role in other coordination tasks, such as surveillance, search and rescue, and exploration and map building of unknown environments. Choset (2001) considers sweep coverage problems, where one or more robots equipped with limited footprint sensors have to visit all points in the environment. In Simmons et al. (2000), deployment locations for a network of heterogeneous robots are user-specified after an initial map of the unknown environment has been built. Gupta et al. (2006a) consider a combined sensor coverage and selection problem.

Deployment of robotic agents with visibility sensors has been studied under a variety of assumptions. When the environment is known *a priori*, the problem can be cast as the classical Art Gallery Problem (Chvátal, 1975) from computational geometry, where one is interested in achieving complete visibility with the minimum number of agents possible. The Art Gallery Problem is computationally hard (Lee and Lin, 1986; Eidenbenz et al., 2001) and the best-known approximation algorithms yield solutions within a logarithmic factor of the optimum number of agents (Ghosh, 1987; Efrat and Har-Peled, 2006). Pinciu (2003) and Hernández-Peñalver (1994) study the problem of achieving full visibility while guaranteeing that the final network configuration will have a connected visibility graph. Recent works on multi-robot exploration of unknown environments include (Batalin and Sukhatme, 2004), Burgard et al. (2005), and Howard et al. (2006). Topological exploration of graph-like environments by single and multiple robots is studied in Rekleitis et al. (2001), Fraigniaud et al. (2004), and Dynia et al. (2006).

A simple one-step strategy for visibility deployment, without the need for synchronization, achieving the worst-case optimal bounds in terms of the number of robots required, and under limited communication, is presented in Ganguli et al. (2007).

5.5 PROOFS

This section gathers the proofs of the main results presented in the chapter.

5.5.1 Proof of Theorem 5.5

Proof. Let $P_0 = (p^{[1]}(0), \ldots, p^{[n]}(0)) \in Q^n \setminus \mathcal{S}_{\text{coinc}}$ denote the initial condition. The proof strategy for all five facts is similar and is based on the application of the LaSalle Invariance Principle with Lyapunov function given by an appropriate multicenter function. In other words, we need to establish the monotonic behavior of the certain multicenter functions along the executions of the control and communication laws and we need to characterize certain invariant sets using geometric properties of the multicenter functions. Additionally, in order to apply the LaSalle Invariance Principle, we need to work with the set $Q^n \setminus \mathcal{S}_{\text{coinc}}$ which is not closed and, therefore, we rely upon the extension of Theorem 1.19 given in Exercise E1.8(ii).

In what follows, we discuss in detail the proof of fact *(i)* regarding the control and communication law $\mathcal{CC}_{\text{VRN-CNTRD}}$ for the network $\mathcal{S}_{\text{D}}$. We leave it to the reader to fill out some of the proof details for the other laws.

Fact (i). First, note that, starting from a configuration in $Q^n \setminus \mathcal{S}_{\text{coinc}}$, one step of the law $\mathcal{CC}_{\text{VRN-CNTRD}}$ leads the network to another configuration in $Q^n \setminus \mathcal{S}_{\text{coinc}}$. Therefore, it is convenient to let $f_{\text{VRN-CNTRD}} : Q^n \setminus \mathcal{S}_{\text{coinc}} \to Q^n \setminus \mathcal{S}_{\text{coinc}}$ denote the map induced by the execution of one step of the control and communication law $\mathcal{CC}_{\text{VRN-CNTRD}}$.

To apply Theorem 1.19, we work with the set $W = Q^n \setminus \mathcal{S}_{\text{coinc}}$. Clearly, this set is positively invariant for $f_{\text{VRN-CNTRD}}$ and it is bounded. Therefore, assumptions (i) and (iii) of Theorem 1.19 are satisfied, except for the closedness of W. Next, we show that executions of the law $\mathcal{CC}_{\text{VRN-CNTRD}}$ monotonically optimize the function $\mathcal{H}_{\text{dist}}$. Using the extension of the multicenter function defined over the set of points and partitions of Q, we deduce from Proposition 2.13 that, for $P \in Q^n \setminus \mathcal{S}_{\text{coinc}}$,

$$
\begin{aligned}
\mathcal{H}_{\text{dist}}(f_{\text{VRN-CNTRD}}(P)) &= \mathcal{H}_{\text{dist}}(f_{\text{VRN-CNTRD}}(P), \mathcal{V}(f_{\text{VRN-CNTRD}}(P))) \\
&\geq \mathcal{H}_{\text{dist}}(f_{\text{VRN-CNTRD}}(P), \mathcal{V}(P)).
\end{aligned}
$$

The application of Proposition 2.14 yields

$$\mathcal{H}_{\mathrm{dist}}(f_{\mathrm{VRN\text{-}CNTRD}}(P), \mathcal{V}(P)) \geq \mathcal{H}_{\mathrm{dist}}(P, \mathcal{V}(P)),$$

and, therefore, $\mathcal{H}_{\mathrm{dist}}(f_{\mathrm{VRN\text{-}CNTRD}}(P)) \geq \mathcal{H}_{\mathrm{dist}}(P)$. Additionally, recall from Proposition 2.14, that, for $P \in Q^n \setminus \mathcal{S}_{\mathrm{coinc}}$, the inequality is strict unless $f_{\mathrm{VRN\text{-}CNTRD}}(P) = P$. This discussion establishes assumption (ii) of Theorem 1.19. The continuity of the map $f_{\mathrm{VRN\text{-}CNTRD}} : Q^n \setminus \mathcal{S}_{\mathrm{coinc}} \to Q^n \setminus \mathcal{S}_{\mathrm{coinc}}$, is a consequence of the following two facts. First, one can verify that each Voronoi cell is a convex set whose boundary is a piecewise continuously differentiable function of the positions of the robots. Second, given a convex set whose boundary depends upon a parameter in a piecewise continuously differentiable fashion, Proposition 2.23 guarantees that the centroid of that set is a continuously differentiable function of the parameter. This discussion and the continuity of $\mathcal{H}_{\mathrm{dist}}$ establishes assumption (iv) of Theorem 1.19.

In the following, we consider an evolution $\gamma : \mathbb{Z}_{\geq 0} \to Q^n \setminus \mathcal{S}_{\mathrm{coinc}}$ of $f_{\mathrm{VRN\text{-}CNTRD}}$ and we prove that no point in $\mathcal{S}_{\mathrm{coinc}}$ may be an accumulation point of γ. By contradiction, we assume that $P = (p_1, \ldots, p_n) \in \mathcal{S}_{\mathrm{coinc}}$ is an accumulation point for γ. Our first claim is that there exists a sequence of increasing times $\{\ell_k \mid k \in \mathbb{N}\}$ and unit-length vectors $u_{ij} \in \mathbb{R}^d$, for $i, j \in \{1, \ldots, n\}$, such that $\gamma(\ell_k) \to P$ and simultaneously $\mathrm{vers}(\gamma_i(\ell_k) - \gamma_j(\ell_k)) \to u_{ij}$ as $k \to \infty$. Here, the versor operator $\mathrm{vers} : \mathbb{R}^d \to \mathbb{R}^d$ is defined by $\mathrm{vers}(\mathbf{0}_d) = \mathbf{0}_d$ and $\mathrm{vers}(v) = v/\|v\|_2$ for $v \neq \mathbf{0}_d$. This first claim is true because P is an accumulation point and because the sequences $\ell \mapsto \mathrm{vers}(\gamma_i(\ell) - \gamma_j(\ell))$ take value in a compact set. Our second claim is that, as $k \to \infty$, the sequence of partitions $\mathcal{V}(\gamma(\ell_k))$ has a limiting partition, say $\{V_1^\infty, \ldots, V_n^\infty\}$. This second claim is true because, for each pair of robots i and j converging to the same position $p_i = p_j$, the bisector of the segment connecting them admits a limit that is equal to the line through the point $p_i = p_j$ and perpendicular to the unit-length vector u_{ij}. Therefore, each of the edges of each of the polygons $\mathcal{V}(\gamma(\ell_k))$ has a limit for $k \to \infty$. Finally, note that each polygon V_i^∞ has a positive measure.

We know that $\ell \mapsto \mathcal{H}_{\mathrm{dist}}(\gamma(\ell))$ is monotonically non-increasing and lower-bounded. Therefore, we must have that

$$\lim_{\ell \to \infty} \Big(\mathcal{H}_{\mathrm{dist}}(\gamma(\ell)) - \mathcal{H}_{\mathrm{dist}}(\gamma(\ell+1)) \Big) = 0. \qquad (5.5.1)$$

Define the short-hand $W_{k,1} = V_1(\gamma(\ell_k))$, and compute

$$\lim_{k \to \infty} \Big(\mathcal{H}_{\text{dist}}(\gamma(\ell_k)) - \mathcal{H}_{\text{dist}}(\gamma(\ell_k + 1)) \Big)$$

$$\geq \lim_{k \to \infty} \Big(\int_{W_{k,1}} \|q - \gamma_1(\ell_k)\|_2^2 \phi(q) dq - \int_{W_{k,1}} \|q - \text{CM}_\phi(W_{k,1})\|_2^2 \phi(q) dq \Big)$$
$$(5.5.2)$$

$$= \int_{V_1^\infty} \|q - p_1\|_2^2 \phi(q) dq - \int_{V_1^\infty} \|q - \text{CM}_\phi(V_1^\infty)\|_2^2 \phi(q) dq \qquad (5.5.3)$$

$$= \text{A}_\phi(V_1^\infty) \|p_1 - \text{CM}_\phi(V_1^\infty)\|_2, \qquad (5.5.4)$$

where inequality (5.5.2) follows from Proposition 2.14, equality (5.5.3) follows from the definition of the limiting partition $\{V_1^\infty, \ldots, V_n^\infty\}$, and equation (5.5.4) follows from the Parallel Axis Theorem. Now, the quantity $\text{A}_\phi(V_1^\infty)$ is strictly positive, as mentioned above, and the quantity $\|p_1 - \text{CM}_\phi(V_1^\infty)\|_2$ is strictly positive because $p_1 \in \partial V_1^\infty$ and $\text{CM}_\phi(V_1^\infty)$ belongs to the interior of V_1^∞ by Exercise (E2.2). The fact that the last limit is lower bounded by a positive is a contradiction with equation (5.5.1). Therefore, we now know that no point in $\mathcal{S}_{\text{coinc}}$ may be an accumulation point of γ.

Finally, we are now ready to apply the LaSalle Invariance Principle as stated in Exercise E1.8(ii) and deduce that the execution of $\mathcal{CC}_{\text{VRN-CNTRD}}$ starting from $P_0 \in Q^n \setminus \mathcal{S}_{\text{coinc}}$ tends to the largest positively invariant set S contained in

$$\{P \in Q^n \mid \mathcal{H}_{\text{dist}}(f_{\text{VRN-CNTRD}}(P)) = \mathcal{H}_{\text{dist}}(P)\}.$$

The set S is precisely the set of centroidal Voronoi configurations. This result is a consequence of the fact that $\mathcal{H}_{\text{dist}}(f_{\text{VRN-CNTRD}}(P)) = \mathcal{H}_{\text{dist}}(P)$ implies that $f_{\text{VRN-CNTRD}}(P) = P$, that is, P is a centroidal Voronoi configuration.

Facts (ii) and (iii). The proofs of these facts run parallel to the proof of fact *(i)*. Propositions 2.19 and 2.21 are key in establishing the monotonic evolution of $\mathcal{H}_{\text{dc}}$ and $\mathcal{H}_{\text{sp}}$, respectively.

Fact (iv). Let $f_{\text{LMTD-VRN-NRML}} : Q^n \setminus \mathcal{S}_{\text{coinc}} \to Q^n \setminus \mathcal{S}_{\text{coinc}}$ denote the map induced by the execution of one step of the law $\mathcal{CC}_{\text{LMTD-VRN-NRML}}$. Let us show that executions of $\mathcal{CC}_{\text{LMTD-VRN-NRML}}$ monotonically optimize the function $\mathcal{H}_{\text{area},\frac{r}{2}}$. Using the extension of the multicenter function defined over the set of points and partitions of Q, we deduce from Proposition 2.13 that, for $P \in Q^n \setminus \mathcal{S}_{\text{coinc}}$,

$$\mathcal{H}_{\text{area},\frac{r}{2}}(f_{\text{LMTD-VRN-NRML}}(P))$$
$$= \mathcal{H}_{\text{area},\frac{r}{2}}(f_{\text{LMTD-VRN-NRML}}(P), \mathcal{V}(f_{\text{LMTD-VRN-NRML}}(P)))$$
$$\geq \mathcal{H}_{\text{area},\frac{r}{2}}(f_{\text{LMTD-VRN-NRML}}(P), \mathcal{V}(P)).$$

The line search procedure for each robot embedded in the definition of the control function of $\mathcal{CC}_{\text{LMTD-VRN-NRML}}$ ensures that

$$\mathcal{H}_{\text{area},\frac{r}{2}}(f_{\text{LMTD-VRN-NRML}}(P), \mathcal{V}(P)) \geq \mathcal{H}_{\text{area},\frac{r}{2}}(P, \mathcal{V}(P)),$$

and hence, $\mathcal{H}_{\text{area},\frac{r}{2}}(f_{\text{LMTD-VRN-NRML}}(P)) \geq \mathcal{H}_{\text{area},\frac{r}{2}}(P)$. Note that the inequality is strict unless $f_{\text{LMTD-VRN-NRML}}(P) = P$. We leave it to the interested reader to prove, similarly to what we did for fact (i), that the map $f_{\text{LMTD-VRN-NRML}}$ is continuous and that no point in $\mathcal{S}_{\text{coinc}}$ may be an accumulation point of any trajectory of $f_{\text{LMTD-VRN-NRML}}$. Finally, the application of the LaSalle Invariance Principle as in the proof of fact (i) leads us to the result stated in fact (iv).

Fact (v). The proof of this fact runs parallel to the proofs of facts (i) and (iv). Propositions 2.13 and 2.15 are key in establishing the monotonic evolution of the cost function $\mathcal{H}_{\text{dist-area},\frac{r}{2}}$. ∎

5.5.2 Proof of Theorem 5.6

Proof. For $d = 1$, Q is a compact interval on $\mathbb{R}$—say $Q = [q_-, q_+]$. We start with a brief discussion about connectivity. In the r-limited Delaunay graph, two agents that are at most at a distance r from each other are neighbors if and only if there are no other agents between them. Additionally, we claim that if agents i and j are neighbors, then $|\,\text{CM}_\phi(V^{[i]}) - \text{CM}_\phi(V^{[j]})| \leq r$, where $V^{[i]}$ denotes the set defined by the control function ctl when evaluated by agent i. To show this fact, let us assume without loss of generality that $p^{[i]} \leq p^{[j]}$. Let us consider the case where the agents have neighbors on both sides (the other cases can be treated analogously). Let $p^{[i]}_-$ (resp., $p^{[j]}_+$) denote the position of the neighbor of agent i to the left (resp., of agent j to the right). Now,

$$\text{CM}_\phi(V^{[i]}) = \frac{1}{4}(p^{[i]}_- + 2p^{[i]} + p^{[j]}),$$

$$\text{CM}_\phi(V^{[j]}) = \frac{1}{4}(p^{[i]} + 2p^{[j]} + p^{[j]}_+),$$

where we have used the fact that $\phi \equiv 1$. Therefore,

$$|\,\text{CM}_\phi(V^{[i]}) - \text{CM}_\phi(V^{[j]})| \leq \frac{1}{4}(|p^{[i]}_- - p^{[i]}| + 2|p^{[i]} - p^{[j]}| + |p^{[j]} - p^{[j]}_+|) \leq r.$$

This implies that agents i and j belong to the same connected component of the r-limited Delaunay graph at the next time step.

Next, let us consider the case when $\mathcal{G}_{\text{LD}}(r)$ is connected at the initial network configuration $P_0 = (p^{[1]}(0), \dots, p^{[n]}(0))$. Without loss of generality, assume that the agents are ordered from left to right according to their unique identifier, that is, $p^{[1]}(0) \leq \cdots \leq p^{[n]}(0)$. We distinguish three cases depending on the proximity of the leftmost and rightmost agents 1 and n, respectively, to the boundary of the environment: in case (**a**) both agents are within a distance $\frac{r}{2}$ of ∂Q; in case (**b**), neither of the two is within a distance $\frac{r}{2}$ of ∂Q; and in case (**c**) only one of the agents is within a distance $\frac{r}{2}$ of ∂Q. Here is an important observation: from one time instant to the next, the network configuration can fall into any of the cases described above. However, because of the discussion on connectivity, transitions can only occur from case (**b**) to either case (**a**) or case (**c**); and from case (**c**) to case (**a**). As we show below, for each of these cases, the network evolution under $\mathcal{CC}_{\text{VRN-CNTRD}}$ can be described as a discrete-time linear dynamical system which respects agents' ordering.

Let us consider case (**a**). In this case, we have

$$p^{[1]}(\ell + 1) = \frac{1}{4}(p^{[1]}(\ell) + p^{[2]}(\ell)) + \frac{1}{2}q_-,$$

$$p^{[2]}(\ell + 1) = \frac{1}{4}(p^{[1]}(\ell) + 2p^{[2]}(\ell) + p^{[3]}(\ell)),$$

$$\vdots$$

$$p^{[n-1]}(\ell + 1) = \frac{1}{4}(p^{[n-2]}(\ell) + 2p^{[n-1]}(\ell) + p^{[n]}(\ell)),$$

$$p^{[n]}(\ell + 1) = \frac{1}{4}(p^{[n-1]}(\ell) + p^{[n]}(\ell)) + \frac{1}{2}q_+.$$

Equivalently, we can write $P(\ell + 1) = A_{(\mathbf{a})} \cdot P(\ell) + b_{(\mathbf{a})}$, where the matrix $A_{(\mathbf{a})} \in \mathbb{R}^{n \times n}$ and the vector $b_{(\mathbf{a})} \in \mathbb{R}^n$ are given by

$$A_{(\mathbf{a})} = \begin{bmatrix} \frac{1}{4} & \frac{1}{4} & 0 & \cdots & \cdots & 0 \\ \frac{1}{4} & \frac{1}{2} & \frac{1}{4} & \cdots & \cdots & 0 \\ 0 & \frac{1}{4} & \frac{1}{2} & \frac{1}{4} & \cdots & 0 \\ \vdots & & \ddots & \ddots & \ddots & \vdots \\ 0 & \cdots & \cdots & \frac{1}{4} & \frac{1}{2} & \frac{1}{4} \\ 0 & \cdots & \cdots & 0 & \frac{1}{4} & \frac{1}{4} \end{bmatrix}, \quad b_{(\mathbf{a})} = \begin{bmatrix} \frac{1}{2}q_- \\ 0 \\ \vdots \\ 0 \\ \frac{1}{2}q_+ \end{bmatrix}.$$

Note that the only equilibrium network configuration P_* respecting the ordering of the agents is given by

$$p_*^{[i]} = q_- + \frac{1}{2n}(1 + 2(i - 1))(q_+ - q_-), \quad i \in \{1, \ldots, n\},$$

and note that this is a $\frac{r}{2}$-centroidal Voronoi configuration (under the assumption of case (**a**)). We can therefore write $(P(\ell+1) - P_*) = A_{(\mathbf{a})}(P(\ell) - P_*)$. Now, note that $A_{(\mathbf{a})} = \text{ATrid}_n^-\left(\frac{1}{4}, \frac{1}{2}\right)$. Theorem 1.80(ii) implies that $\lim_{\ell \to +\infty}\left(P(\ell) - P_*\right) = \mathbf{0}_n$, and that the maximum time required for $\|P(\ell) -$

$P_*\big\|_2 \leq \varepsilon\|P_0 - P_*\|_2$ (over all initial conditions in $\mathbb{R}^n$) is $\Theta(n^2 \log \varepsilon^{-1})$. It is not obvious, but it can be verified, that the initial condition providing the lower bound in the time complexity estimate does indeed have the property of respecting the agents' ordering; this fact holds for all three possible cases (**a**), (**b**), and (**c**).

Case (**b**) can be treated in the same way. The network evolution now takes the form $P(\ell + 1) = A_{(\mathbf{b})} \cdot P(\ell) + b_{(\mathbf{b})}$, where the matrix $A_{(\mathbf{b})} \in \mathbb{R}^{n \times n}$ and the vector $b_{(\mathbf{b})} \in \mathbb{R}^n$ are given by

$$
A_{(\mathbf{b})} = \begin{bmatrix}
\frac{3}{4} & \frac{1}{4} & 0 & \cdots & \cdots & 0 \\
\frac{1}{4} & \frac{1}{2} & \frac{1}{4} & \cdots & \cdots & 0 \\
0 & \frac{1}{4} & \frac{1}{2} & \frac{1}{4} & \cdots & 0 \\
\vdots & & \ddots & \ddots & \ddots & \vdots \\
0 & \cdots & \cdots & \frac{1}{4} & \frac{1}{2} & \frac{1}{4} \\
0 & \cdots & \cdots & 0 & \frac{1}{4} & \frac{3}{4}
\end{bmatrix}, \quad
b_{(\mathbf{b})} = \begin{bmatrix}
-\frac{1}{4}r \\
0 \\
\vdots \\
0 \\
\frac{1}{4}r
\end{bmatrix}.
$$

In this case, a (non-unique) equilibrium network configuration respecting the ordering of the agents is of the form

$$
p_*^{[i]} = ir - \frac{1+n}{2}r, \quad i \in \{1, \ldots, n\}.
$$

Note that this is a $\frac{r}{2}$-centroidal Voronoi configuration (under the assumption of case (**b**)). We can therefore write $(P(\ell + 1) - P_*) = A_{(\mathbf{b})}(P(\ell) - P_*)$. Now, observe that $A_{(\mathbf{b})} = \mathrm{ATrid}_n^+\left(\frac{1}{4}, \frac{1}{2}\right)$. We compute that $P_{\mathrm{ave}} = \frac{1}{n}\mathbf{1}_n^T(P_0 - P_*) = \frac{1}{n}\mathbf{1}_n^T P_0$. With this calculation, Theorem 1.80(i) implies that $\lim_{\ell \to +\infty} (P(\ell) - P_* - P_{\mathrm{ave}}\mathbf{1}_n) = \mathbf{0}_n$, and that the maximum time required for $\|P(\ell) - P_* - P_{\mathrm{ave}}\mathbf{1}_n\|_2 \leq \varepsilon\|P_0 - P_* - P_{\mathrm{ave}}\mathbf{1}_n\|_2$ (over all initial conditions in $\mathbb{R}^n$) is $\Theta(n^2 \log \varepsilon^{-1})$.

Case (**c**) needs to be handled differently. Without loss of generality, assume that agent 1 is within distance $\frac{r}{2}$ of ∂Q and agent n is not (the other case is treated analogously). Then, the network evolution now takes the form $P(\ell + 1) = A_{(\mathbf{c})} \cdot P(\ell) + b_{(\mathbf{c})}$, where the matrix $A_{(\mathbf{c})} \in \mathbb{R}^{n \times n}$ and the vector $b_{(\mathbf{c})} \in \mathbb{R}^n$ are given by

$$
A_{(\mathbf{c})} = \begin{bmatrix}
\frac{1}{4} & \frac{1}{4} & 0 & \cdots & \cdots & 0 \\
\frac{1}{4} & \frac{1}{2} & \frac{1}{4} & \cdots & \cdots & 0 \\
0 & \frac{1}{4} & \frac{1}{2} & \frac{1}{4} & \cdots & 0 \\
\vdots & & \ddots & \ddots & \ddots & \vdots \\
0 & \cdots & \cdots & \frac{1}{4} & \frac{1}{2} & \frac{1}{4} \\
0 & \cdots & \cdots & 0 & \frac{1}{4} & \frac{3}{4}
\end{bmatrix}, \quad
b_{(\mathbf{c})} = \begin{bmatrix}
\frac{1}{2}q- \\
0 \\
\vdots \\
0 \\
\frac{1}{4}r
\end{bmatrix}.
$$

Note that the only equilibrium network configuration P_* respecting the ordering of the agents is given by

$$p_*^{[i]} = q_- + \frac{1}{2}(2i-1)r, \quad i \in \{1, \dots, n\},$$

and note that this is a $\frac{r}{2}$-centroidal Voronoi configuration (under the assumption of case (**c**)). In order to analyze $A_{(\mathbf{c})}$, we recast the n-dimensional discrete-time dynamical system as a $2n$-dimensional one. To do this, we define a $2n$-dimensional vector y by

$$y^{[i]} = p^{[i]}, \ i \in \{1, \dots, n\}, \quad \text{and} \quad y^{[n+i]} = p^{[n-i+1]}, \ i \in \{1, \dots, n\}. \quad (5.5.5)$$

Now, one can see that the network evolution can be alternatively described in the variables $(y^{[1]}, \dots, y^{[2n]})$ as a linear dynamical system determined by the $2n \times 2n$ matrix $\mathrm{ATrid}_{2n}^-(\frac{1}{4}, \frac{1}{2})$. Using Theorem 1.80(ii), and exploiting the chain of equalities (5.5.5), it is possible to infer that, in case (**c**), the maximum time required for $\|P(\ell) - P_*\|_2 \leq \varepsilon \|P_0 - P_*\|_2$ (over all initial conditions in $\mathbb{R}^n$) is $\Theta(n^2 \log \varepsilon^{-1})$.

In summary, for all three cases (**a**), (**b**), and (**c**), our calculations show that, in time $O(n^2 \log \varepsilon^{-1})$, the error 2-norm satisfies the contraction inequality $\|P(\ell) - P_*\|_2 \leq \varepsilon \|P_0 - P_*\|_2$. We convert this inequality on 2-norms into an appropriate inequality on ∞-norms as follows. Note that $\|P_0 - P_*\|_\infty = \max_{i \in \{1, \dots, n\}} |p^{[i]}(0) - p_*^{[i]}| \leq (q_+ - q_-)$. For ℓ of order $n^2 \log \eta^{-1}$, we have that

$$\begin{aligned} \|P(\ell) - P_*\|_\infty &\leq \|P(\ell) - P_*\|_2 \leq \eta \|P_0 - P_*\|_2 \\ &\leq \eta\sqrt{n}\|P_0 - P_*\|_\infty \leq \eta\sqrt{n}(q_+ - q_-). \end{aligned}$$

This means that ε-r-deployment is achieved for $\eta\sqrt{n}(q_+ - q_-) = \varepsilon$, that is, in time $O(n^2 \log \eta^{-1}) = O(n^2 \log(n\varepsilon^{-1}))$.

Up to here, we have proved that if the graph $\mathcal{G}_{\mathrm{LD}}(r)$ is connected at P_0, then $\mathrm{TC}(\mathcal{T}_{\varepsilon\text{-}r\text{-dply}}, \mathcal{CC}_{\mathrm{VRN\text{-}CNTRD}}, P_0) \in O(n^2 \log(n\varepsilon^{-1}))$. If $\mathcal{G}_{\mathrm{LD}}(r)$ is not connected at P_0, note that along the network evolution there can only be a finite number of time instants, at most $n-1$ where a merging of two connected components occurs. Therefore, the time complexity is at most $O(n^3 \log(n\varepsilon^{-1}))$, as claimed. ■

5.6 EXERCISES

E5.1   (**The Vrn-crcmcntr law is not positively invariant on $Q^n \backslash \mathcal{S}_{\mathbf{coinc}}$**). Consider the network $\mathcal{S}_{\mathrm{D}}$ composed by 2 robots evolving in the convex polygon depicted in Figure E5.1. Describe the evolution of the network starting from the configuration depicted in Figure E5.1 and discuss its implication on the positive invariance of the set $Q^2 \setminus \mathcal{S}_{\mathrm{coinc}}$ with respect to $\mathcal{CC}_{\mathrm{VRN\text{-}CRCMCNTR}}$.

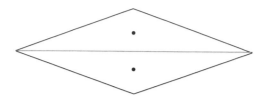

Figure E5.1 Convex polygon for Exercise E5.1. The height of the polygon is strictly less
        that its width.

E5.2   **(Monotonic evolution of $\mathcal{H}_{\mathbf{dc}}$ and $\mathcal{H}_{\mathbf{sp}}$).** Prove the facts relative to statements
        (ii) and (iii) in Theorem 5.5.
        *Hint: Make use of the optimality of the Voronoi partition and of center locations
        stated in Propositions 2.19 and 2.21.*

E5.3   **(Correctness of Lmtd-Vrn-cntrd).** Prove Theorem 5.5(v).
        *Hint: To establish the monotonic evolution of the multicenter function, make use
        of the optimality of the Voronoi partition stated in Proposition 2.13 and of the
        centroid locations stated in Proposition 2.15. To establish the convergence result,
        make use of the LaSalle Invariance Principle stated in Theorem 1.19.*

E5.4   **(Correctness of Rng-Vrn-cntrd).** Mimic the proof of Theorem 5.5(iv) to
        show that the evolutions of RNG-VRN-CNTRD monotonically optimize the mixed
        distortion-area multicenter function

$$\mathcal{H}_{\text{dist-area}, \frac{r}{2}, b}, \quad \text{with } b = -\operatorname{diam}(Q)^2,$$

        and asymptotically approach its set of critical points.

E5.5   **(The "n-bugs problem" and cyclic interactions: cont'd).** Consider n
        robots at counterclockwise-ordered positions $\theta_1, \ldots, \theta_n$. First, consider the cyclic
        balancing system described in Exercise E1.30 with parameter $k = 1/4$, and given
        by

$$\theta_i(\ell + 1) = \frac{1}{4}\theta_{i+1}(\ell) + \frac{1}{2}\theta_i(\ell) + \frac{1}{4}\theta_{i-1}(\ell), \quad \ell \in \mathbb{Z}_{\geq 0}.$$

        Second, consider the Voronoi-centroid law on the circle (with uniform density) in
        which each robot computes its Voronoi partition of the circle (see Figure 2.5) and
        then moves to the midpoint of its arc. Show that the two behaviors are identical.

Chapter Six

Boundary estimation and tracking

The aim of this chapter is to provide an example of a motion coordination algorithm that can be used in a specific sensing task. This is the task of detection and estimation of an evolving boundary in two dimensions by a robotic sensor network. This type of operation can be of interest in the validation of oceanographic and atmospheric models, as well as for the demarcation of hazardous environments. In the ocean, a boundary can delimit areas where there are abrupt changes in temperature, which can influence the marine biodiversity in those areas. In the atmosphere, a boundary can establish the front of a highly polluting expanding substance. The containment of a spreading fire is another situation that can translate into the specific task of boundary estimation and tracking.

Under full knowledge and centralized computation, various methods exist in the literature to solve the boundary estimation task. A first challenge that we face when designing coordination algorithms for robotic networks is the determination of the extent to which these tasks can be performed in a distributed way and under limited information. In this regard, the algorithm presented in this chapter to track environmental boundaries is distributed, in the sense that it does not require the use of a central station or "fusion center." Our algorithm builds on basic notions from interpolation theory and employs distributed linear iterations and consensus algorithms. The algorithm can be seen as part of a general effort to investigate distributed filters for estimation tasks.

A second challenge is posed by sudden events that may occur when performing sensing tasks, such as the detection of an intruder or an abrupt change in the concentration of some chemical. These events require a specific action on the part of the network. Since the timing of such events is not known *a priori*, this requires coordination algorithms that specify event-driven, asynchronous responses of the robotic network. We deal with this issue by building on the robotic network model proposed in Chapter 3. Our exposition here on boundary estimation relies on Susca (2007) and Susca et al. (2008).

The chapter is organized as follows. The first section extends the synchronous model proposed in Chapter 3 to include the event-driven, asynchronous operation of a robotic network. The second section reviews some basic facts on interpolation theory for boundaries. In the third section, we introduce the ESTIMATE UPDATE AND BALANCING LAW to solve the boundary estimation task and analyzes its correctness. We end the chapter with three sections on, respectively, bibliographic notes, proofs of the results presented in the chapter, and exercises. Throughout the exposition, we make extensive use of polygonal approximations, geometric decompositions, and consensus algorithms. The convergence analysis is based on the LaSalle Invariance Principle and on distributed linear iterations.

6.1 EVENT-DRIVEN ASYNCHRONOUS ROBOTIC NETWORKS

In what follows, we model "event-driven asynchronous" robotic networks. This model describes groups of agents that reset their processor states upon certain asynchronous events, that is, events that do not necessarily happen simultaneously for all agents. For example, a relevant event might be a robot reaching a location or leaving a region. The following event-driven model is convenient to describe our algorithm for boundary estimation, but its applicability extends beyond this particular scenario. Following our discussion of synchronous robotic networks in Section 3.1, the event-driven model consists of the following ingredients: a robotic network, as in Definition 3.2, and an event-driven control and communication law, as defined next.

Definition 6.1 (Event-driven control and communication law). An *event-driven control and communication law* $\mathcal{ECC}$ for a robotic network $\mathcal{S}$ consists of the following sets:

(i) $\mathbb{A}$, a set containing the `null` element, called the *communication alphabet*—elements of $\mathbb{A}$ are called *messages*;

(ii) $W^{[i]}$, $i \in I$, called the *processor state sets*; and

(iii) $W_0^{[i]} \subseteq W^{[i]}$, $i \in I$, sets of *allowable initial values*;

and the following maps:

(i) $(\text{msg-trig}^{[i]}, \text{msg-gen}^{[i]}, \text{msg-rec}^{[i]})$, $i \in I$, called the *message-trigger function*, *message-generation function*, and *message-reception function*, respectively, such that

(a) $\text{msg-trig}^{[i]} : X^{[i]} \times W^{[i]} \to \{\texttt{true}, \texttt{false}\}$,

(b) $\text{msg-gen}^{[i]} : X^{[i]} \times W^{[i]} \times I \to \mathbb{A}$,

(c) $\text{msg-rec}^{[i]} : X^{[i]} \times W^{[i]} \times \mathbb{A} \times I \to W^{[i]}$;

(ii) $(\text{stf-trig}_k^{[i]}, \text{stf}_k^{[i]})$, $i \in I$, $k \in \{1, \ldots, K_{\text{stf}}^{[i]}\}$, called the kth *state-transition trigger function* and the kth *(processor) state-transition function*, respectively, such that

(a) $\text{stf-trig}_k^{[i]} : X^{[i]} \times W^{[i]} \to \{\texttt{true}, \texttt{false}\}$,

(b) $\text{stf}_k^{[i]} : X^{[i]} \times W^{[i]} \to W^{[i]}$; and

(iii) $\text{ctl}^{[i]} : X^{[i]} \times W^{[i]} \to U^{[i]}$, $i \in I$, called *(motion) control functions*.

If the network $\mathcal{S}$ is uniform and all sets and maps of the law $\mathcal{ECC}$ are independent of the identifier, that is, for all $i \in I$ and $k \in \{1, \ldots, K_{\text{stf}} = K_{\text{stf}}^{[i]}\}$,

$$W^{[i]} = W, \quad (\text{stf-trig}_k^{[i]}, \text{stf}_k^{[i]}) = (\text{stf-trig}_k, \text{stf}_k), \text{ctl}^{[i]} = \text{ctl},$$

$$(\text{msg-trig}^{[i]}, \text{msg-gen}^{[i]}, \text{msg-rec}^{[i]}) = (\text{msg-trig}, \text{msg-gen}, \text{msg-rec}),$$

then $\mathcal{ECC}$ is said to be *uniform* and is described by the tuple

$$(\mathbb{A}, W, \{W_0^{[i]}\}_{i \in I}, (\text{msg-trig}, \text{msg-gen}, \text{msg-rec}), \{\text{stf-trig}_k, \text{stf}_k\}_{k=1}^{K_{\text{stf}}}, \text{ctl}). \quad \bullet$$

Observe that a key difference between Definitions 3.9 and 6.1 is that the message-generation and state-transition functions are substituted by sets of maps $(\text{msg-trig}^{[i]}, \text{msg-gen}^{[i]}, \text{msg-rec}^{[i]})$ and $(\text{stf-trig}_k^{[i]}, \text{stf}_k^{[i]})$. A second difference is that the control function depends only upon the current robot position, and not the position at last sample time.

The event-driven control and communication law models situations in which the agent physical and processor states need to satisfy certain constraints before a message should be sent. For example, in each triplet $(\text{msg-trig}^{[i]}, \text{msg-gen}^{[i]}, \text{msg-rec}^{[i]})$, the map $\text{msg-trig}^{[i]}$ acts as a trigger for agent i to send a message to its neighbors, the map $\text{msg-gen}^{[i]}$ computes the message to be sent, and $\text{msg-rec}^{[i]}$ specifies how agent i updates its processor state when receiving a message. In hybrid systems terminology (van der Schaft and Schumacher, 2000), the map $\text{msg-trig}^{[i]}$ can be seen as a guard map. Similarly, in the pair $(\text{stf-trig}_k^{[i]}, \text{stf}_k^{[i]})$, the map $\text{stf-trig}_k^{[i]}$ acts as a trigger for agent i to update its processor state. If several $\text{stf-trig}_k^{[i]}$ are satisfied at the same time, then the agent can freely choose among the corresponding state transition functions to update the processor state. This freedom means that our dynamical system is described by a set-valued map and leads to non-deterministic evolutions.

The evolution of a robotic network dictated by an event-driven control and communication law is asynchronous: there is no common time schedule for all robots to send messages, receive messages, and update their processor states. Only when an event happens, an agent sends a message or updates its state. The asynchronous event-driven evolution can be loosely described in the following way:

(i) Starting from the initial conditions, the physical state of each agent evolves in continuous time according to the control function $\mathrm{ctl}^{[i]}$.

(ii) At every instant of time $t_1 \in \mathbb{R}_{\geq 0}$ such that the message-trigger function for agent i satisfies $\mathrm{msg\text{-}trig}^{[i]}(x^{[i]}(t_1), w^{[i]}(t_1)) = \texttt{true}$, agent i generates a non-null message according to $\mathrm{msg\text{-}gen}^{[i]}$ and sends it to all its out-neighbors. At time t_1, each out-neighbor j of agent i receives a messages and processes it according to $\mathrm{msg\text{-}rec}^{[j]}$. If multiple messages are received at the same time, then we allow all possible orders of execution of the message-reception function.

(iii) Additionally, at every instant of time $t_2 \in \mathbb{R}_{\geq 0}$ such that one of the state-transition triggers satisfies $\mathrm{stf\text{-}trig}_k^{[i]}(x^{[i]}(t_2), w^{[i]}(t_2), y^{[i]}(t_2)) = \texttt{true}$, agent i updates its processor state $w^{[i]}$ according to $\mathrm{stf}_k^{[i]}$. If multiple state transitions are triggered at the same time, then we allow all possible orders of execution of the state-transition functions.

(iv) If one or multiple state-transition and message triggers are equal to \texttt{true} at the same time, then we assume that all state transitions take place first, and immediately after the messages are generated and transmitted.

(v) If one or multiple state-transition triggers are equal to \texttt{true} at the same time at which messages are received, then we allow all possible orders of execution of the state-transition and message-reception functions.

(vi) In order to avoid the possibility of an infinite number of message transmissions or state transitions in finite time, we introduce a "dwell logic." Let $\delta > 0$ be a *dwell time* common to all agents. For each agent i, if a message was generated at time t_1 by agent i, then no additional message is to be generated before time $t_1 + \varepsilon$ by agent i independently of the value of its message-trigger function. Similarly, for each agent i, if a state-transition function was executed at time t_2 by agent i, then no additional state-transition function is to be executed before time $t_2 + \varepsilon$ by agent i independently of the value of its state-transition-trigger function.

Remark 6.2 (Dwell time prevents Zeno behavior). Note that: (1) a dwell time is introduced only for the purpose of properly defining an execution for a general event-drive control and communication law; (2) infinite numbers of message transmissions or state transitions in finite time will not take place during the execution of the algorithm that we present later in the chapter; and (3) we refer the reader interested in comprehensive treatments of the so-called Zeno behavior to van der Schaft and Schumacher (2000); Johansson et al. (1999).   •

Finally, as we did in Chapter 3, we now give a formal definition of the asynchronous evolution of an event-driven control and communication law on a robotic network.

Definition 6.3 (Asynchronous event-driven evolution with dwell time). Let $\mathcal{ECC}$ be an event-driven control and communication law for the robotic network $\mathcal{S}$. For $\delta \in \mathbb{R}_{>0}$, the *evolution of $(\mathcal{S}, \mathcal{ECC})$ with dwell time* δ from initial conditions $x_0^{[i]} \in X_0^{[i]}$ and $w_0^{[i]} \in W_0^{[i]}$, $i \in I$, is the collection of absolutely continuous curves $x^{[i]} : \mathbb{R}_{\geq 0} \to X^{[i]}$, $i \in I$, and piecewise-constant curves $w^{[i]} : \mathbb{R}_{\geq 0} \to W^{[i]}$, $i \in I$, such that at almost all times,

$$\dot{x}^{[i]}(t) = f\big(x^{[i]}(t), \mathrm{ctl}^{[i]}\big(x^{[i]}(t), w^{[i]}(t)\big)\big), \tag{6.1.1}$$
$$\dot{w}^{[i]}(t) = 0, \tag{6.1.2}$$

with $x^{[i]}(0) = x_0^{[i]}$ and $w^{[i]}(0) = w_0^{[i]}$, $i \in I$, and such that:

(i) For every $i \in I$ and $t_1 \in \mathbb{R}_{>0}$, a message is generated by agent i and received by all its out-neighbors j, that is,

$$y_i^{[j]}(t_1) = \mathrm{msg\text{-}gen}^{[i]}\big(x^{[i]}(t_1), w^{[i]}(t_1), j\big),$$
$$w^{[j]}(t_1) = \mathrm{msg\text{-}rec}^{[j]}\big(x^{[j]}(t_1), \lim_{t \to t_1^-} w^{[j]}(t), y_i^{[j]}(t_1), i\big),$$

if $\mathrm{msg\text{-}trig}^{[i]}(x^{[i]}(t_1), w^{[i]}(t_1)) = \texttt{true}$ and agent i has not transmitted any message to its out-neighbors during the time interval $]t_1 - \delta, t_1[\cap \mathbb{R}_{>0}$. Here, agent j is an out-neighbor of agent i at time t_1 if $(i,j) \in E_{\mathrm{cmm}}\big(x^{[1]}(t_1), \dots, x^{[n]}(t_1)\big)$.

(ii) For every $i \in I$, $k \in \{1, \dots, K_{\mathrm{stf}}^{[i]}\}$, and $t_2 \in \mathbb{R}_{>0}$, the state-transition function $\mathrm{stf}_k^{[i]}$ is executed, that is,

$$w^{[i]}(t_2) = \mathrm{stf}_k^{[i]}\big(x^{[i]}(t_2), \lim_{t \to t_2^-} w^{[i]}(t)\big),$$

if $\mathrm{stf\text{-}trig}_k^{[i]}(x^{[i]}(t_2), w^{[i]}(t_2)) = \texttt{true}$ and there has been no execution of $\mathrm{stf}_k^{[i]}$ during the time interval $]t_2 - \delta, t_2[\cap \mathbb{R}_{>0}$.   •

This model of event-driven control and communication law and of asynchronous evolution is adopted in the rest of this chapter.

6.2 PROBLEM STATEMENT

In this section, we formalize the network objective. We begin by reviewing some important facts on interpolations of planar boundaries by means of inscribed polygons. After introducing the robotic network model, we make use of notions from the theory of linear interpolations to formalize the boundary estimation task.

6.2.1 Linear interpolations for boundary estimation

Consider a simply connected set Q in $\mathbb{R}^2$, that we term the *body*. We are interested in obtaining a description of the boundary ∂Q of a convex body. In particular, we will consider the *symmetric difference* error metric (Gruber, 1983) to measure the goodness of an approximation to ∂Q. The symmetric difference δ^S between two compact bodies C, $B \subseteq \mathbb{R}^2$ is defined by

$$\delta^S(C, B) = A(C \cup B) - A(C \cap B),$$

where, given a set $S \subset \mathbb{R}^2$, $A(S)$ is its Lebesgue measure. This definition is illustrated in Figure 6.1. We note that the symmetric difference admits alternative definitions (see Exercise E6.1). In what follows, we search for

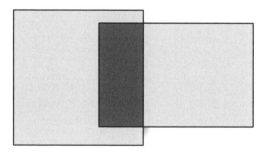

Figure 6.1 The symmetric difference between the two quadrilaterals is the area corresponding to the region colored in light gray.

approximations to a convex body Q by means of *inscribed polygons*. A convex polygon is inscribed in Q if all its vertices belong to the boundary of Q. We denote an inscribed polygon with m vertices by Q_m. The symmetric difference between the body Q and the polygon Q_m takes the simpler form

$$\delta^S(Q, Q_m) = A(Q) - A(Q_m).$$

The inscribed polygons that are critical points of δ^S can be characterized as follows.

Lemma 6.4 (Characterization of critical inscribed polygons for the symmetric difference). *Let Q be a convex planar body with a continuously differentiable boundary. Let Q_m be an inscribed polygon with vertices $\{q_1, \ldots, q_m\}$ in counterclockwise order. For $i \in \{1, \ldots, m\}$, let $t(q_i)$ be the tangent vector to ∂Q at q_i. Then, Q_m is a critical point of δ^S if and only if*

$$t(q_i) \text{ is parallel to } (q_{i+1} - q_{i-1}), \tag{6.2.1}$$

for all $i \in \{1, \ldots, m\}$, where $q_0 = q_m$ and $q_{m+1} = q_1$.

Note that the characterization of critical inscribed polygons in Lemma 6.4 can be satisfied not only by polygons that are maximizers, but also by saddle points (see Exercise E6.2). On the other hand, we would like to make use of a characterization that can be extended to nonconvex bodies and that relies as much as possible on local information that agents can collect.

In what follows, we describe the *method of empirical distributions*, based on the asymptotic formula provided in the following lemma. We start with some useful notation. As in Section 1.1, let ∂Q be twice continuously differentiable and let $\gamma_{\text{arc}} : [0, L] \to \partial Q$ be a counterclockwise arc-length parametrization of ∂Q. Additionally, let $\kappa_{\text{signed}} : [0, L] \to \mathbb{R}$, $\kappa_{\text{abs}} : [0, L] \to \mathbb{R}_{\geq 0}$, and $\rho : [0, L] \to \mathbb{R}_{\geq 0}$ be, respectively, the signed curvature, the absolute curvature, and radius of curvature of the boundary. For convex bodies, the following result is proved in McLure and Vitale (1975), and Gruber (1983).

Lemma 6.5 (Optimal polygonal approximation of a convex body). *Let Q be a convex planar body whose boundary is twice continuously differentiable and has strictly positive signed curvature κ_{signed}. If Q_m^* is an optimal approximating polygon of Q, then*

$$\lim_{m \to +\infty} m^2 \delta^S(Q, Q_m^*) = \frac{1}{12} \int_0^L \rho(s)^{2/3} ds.$$

To compute an optimal approximating polygon for a strictly convex body, McLure and Vitale (1975) suggest the following method of empirical distributions. Let $q_1, \ldots, q_m$ be consecutive points on ∂Q ordered counterclockwise and, for $i \in \{1, \ldots, m\}$, define $s_i \in [0, L]$ by requiring $q_i = \gamma_{\text{arc}}(s_i)$. The positions q_i, $i \in \{1, \ldots, m\}$, along ∂Q are said to obey the *method of empirical distributions* if

$$\int_{s_{i-1}}^{s_i} \rho(s)^{2/3} ds = \int_{s_i}^{s_{i+1}} \rho(s)^{2/3} ds \tag{6.2.2}$$

for all $i \in \{1, \ldots, m\}$, where we set $s_0 = s_m$ and $s_{m+1} = s_1$. Interpolating polygons computed according to the method of empirical distributions converge to an optimal polygon approximation Q_m^* as $m \to \infty$. Roughly speaking, this property translates into the placement of more interpolation points on those parts of the boundary that have higher curvature. Figure 6.2.1 illustrates an approximating polygon with empirically distributed vertices. As final comment about convex bodies, it is useful to know from Gruber

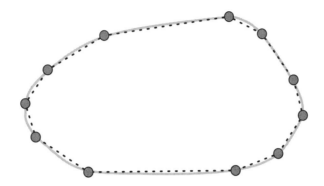

Figure 6.2 Equidistant interpolation points according to the integral in equation (6.2.2). The solid line represents the boundary and the dashed line represents the optimal approximating polygon.

(1983) that, for $\alpha > 0$,

$$\int_0^L \rho(s)^\alpha ds = \int_0^L \kappa_{\text{abs}}(s)^{1-\alpha} ds. \tag{6.2.3}$$

Next, we discuss the case of nonconvex bodies whose boundary can be parameterized by a twice continuously differentiable curve. We begin with a definition: given a twice continuously differentiable curve $\gamma : [0, L] \to \mathbb{R}^2$, an *inflection point* of γ is a point $q \in \gamma([0, L])$ with the property that, for $q = \gamma(s_q)$, $\text{sign}(\kappa_{\text{signed}}(s_q - \varepsilon)) \neq \text{sign}(\kappa_{\text{signed}}(s_q + \varepsilon))$ for every $\varepsilon \in \mathbb{R}_{>0}$ sufficiently small. Nonconvex bodies have an arbitrary number of inflection points; we restrict our attention to nonconvex bodies with a finite number of them. Because the radius of curvature of a nonconvex body is unbounded at inflection points, equality (6.2.2) is ill posed in general. Therefore, in order to extend the method of empirical distributions to nonconvex bodies, we introduce the following notions of distance along a boundary. Given two points $q_i = \gamma_{\text{arc}}(s_i)$ and $q_j = \gamma_{\text{arc}}(s_j)$, with $s_i < s_j$, we define

$$\mathcal{D}_{\text{curvature}}(q_i, q_j) = \int_{s_i}^{s_j} \kappa_{\text{abs}}(s)^{1/3} ds,$$

$$\mathcal{D}_{\text{arc}}(q_i, q_j) = s_j - s_i.$$

Note that the quantity $\mathcal{D}_{\text{arc}}(q_i, q_j)$ is always strictly positive for $q_i \neq q_j$, whereas the quantity $\mathcal{D}_{\text{curvature}}(q_i, q_j)$ vanishes if the points q_i and q_j are connected by a straight line. Additionally, for $\lambda \in [0, 1]$, we define the pseudo-distance $\mathcal{D}_\lambda$ between the vertices q_i and q_j as

$$\mathcal{D}_\lambda(q_i, q_{i+1}) = \lambda \mathcal{D}_{\text{curvature}}(q_i, q_j) + (1 - \lambda)\mathcal{D}_{\text{arc}}(q_i, q_{i+1}).$$

The empirical distribution criterion (6.2.2) is substituted by the following one when the boundary is nonconvex. We look for approximations of ∂Q, $\{q_1, \ldots, q_m\}$, such that $\mathcal{D}_\lambda(q_{i-1}, q_i) = \mathcal{D}_\lambda(q_i, q_{i+1})$ for all $i \in \{1, \ldots, m\}$. This choice has the following interpretation. Taking $\lambda \approx 1$ leads to an interpolation that satisfies a modified method of empirical distributions. The method is modified in the sense that we adopt the distance $\mathcal{D}_{\text{curvature}}$ instead of the integral of the curvature radius; our informal justification for this step is equality (6.2.3). Instead, taking $\lambda \approx 0$ leads to an interpolation that divides the boundary into segments of equal arc-length. A choice of $\lambda \in (0, 1)$ leads to a polygon approximation that is midway between these two options. For sufficiently large $\lambda < 1$, the resulting polygon has a higher number of vertices in the portions of the boundary with higher curvature and the distance between any two consecutive interpolation points is guaranteed to be positive.

6.2.2 Network model and boundary estimation task

Next, we formulate a robotic network model and the boundary estimation objective. Assume that Q is a simply connected subset of $\mathbb{R}^2$ with differentiable boundary ∂Q. Consider the network $\mathcal{S}_{\text{bndry}} = (I, \mathcal{R}, E_{\text{cmm}})$, with $I = \{1, \ldots, n\}$. In this network, each robot is described by a tuple

$$(\partial Q, [-v_{\min}, v_{\max}], \partial Q, (0, \mathbf{e})), \tag{6.2.4}$$

where $\mathbf{e}$ is the vector field tangent to ∂Q describing counterclockwise motion at unit speed; we assume that unit speed is an admissible speed, that is, we assume that $1 \in [-v_{\min}, v_{\max}]$. We assume that each robot can sense its own location $p^{[i]} \in \partial Q$, $i \in I$, and can communicate with its clockwise and counterclockwise neighbors along ∂Q. In other words, the communication graph E_{cmm} is the ring graph or the Delaunay graph on ∂Q. Later, we shall assume that Q varies in a continuously differentiable way with time, and that, therefore, agents move along its time-varying boundary.

Next, assume that the processor state of each agent contains a set of n_{ip} interpolation points used to approximate ∂Q, that is, the processor state is given by $q^{[i]} \in (\mathbb{R}^2)^{n_{\text{ip}}}$, for $i \in I$. We illustrate the combination of agents and interpolation points along the boundary in Figure 6.3.

For $\varepsilon \in \mathbb{R}_{>0}$ and $\lambda \in [0, 1]$, the *boundary estimation task* $\mathcal{T}_{\varepsilon\text{-bndry}} : (\partial Q)^n \times$

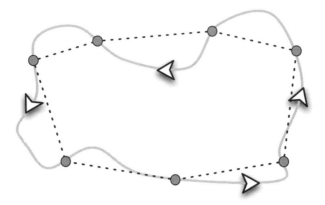

Figure 6.3 Agents and interpolation points on the boundary ∂Q.

$((\mathbb{R}^2)^{n_{\mathrm{ip}}})^n \to \{\mathtt{true}, \mathtt{false}\}$ for $\mathcal{S}_{\mathrm{bndry}}$ is the coordination task

$$\mathcal{T}_{\varepsilon\text{-bndry}}(p^{[1]}, \ldots, p^{[n]}, q^{[1]}, \ldots, q^{[n]}) = \mathtt{true} \quad \text{if and only if}$$

$$\left| \mathcal{D}_\lambda(q^{[i]}_{\alpha-1}, q^{[i]}_\alpha) - \mathcal{D}_\lambda(q^{[i]}_\alpha, q^{[i]}_{\alpha+1})) \right| < \varepsilon, \quad \alpha \in \{1, \ldots, n_{\mathrm{ip}}\} \text{ and } i \in I.$$

Roughly speaking, this task is achieved when the n_{ip} interpolation points are approximately uniformly placed along the boundary according to the counterclockwise pseudo-distance $\mathcal{D}_\lambda$.

A second objective is our desire to space the agents equally far apart along the boundary. As in Example 3.22, for $\varepsilon > 0$, we define the *agent equidistance task* $\mathcal{T}_{\varepsilon\text{-eqdstnc}} : (\partial Q)^n \to \{\mathtt{true}, \mathtt{false}\}$ to be $\mathtt{true}$ if and only if

$$\left| \mathcal{D}_{\mathrm{arc}}(p^{[i-1]}, p^{[i]}) - \mathcal{D}_{\mathrm{arc}}(p^{[i]}, p^{[i+1]}) \right| < \varepsilon, \quad \text{for all } i \in I,$$

where $\mathcal{D}_{\mathrm{arc}}$ is the counterclockwise arc-length distance along ∂Q. In other words, $\mathcal{T}_{\varepsilon\text{-eqdstnc}}$ is true when, for every agent, the (unsigned) distances to the closest clockwise neighbor and to the closest counterclockwise neighbor are approximately equal.

6.3 ESTIMATE UPDATE AND CYCLIC BALANCING LAW

Here, we propose a coordination algorithm for a robotic network to achieve the boundary estimation task. The algorithm requires individual agents to maintain and continuously update an approximation of the boundary that asymptotically meets the criterion of the method of empirical distributions. The algorithm is an event-driven control and communication law, as defined in Section 6.1. To facilitate the understanding, the algorithm is presented in an incremental way. First, we specify an estimate update law for a single robot. Second, we consider multiple robots cooperatively performing the estimate update law to achieve the boundary estimation task. Third and finally, we introduce a cyclic balancing algorithm to achieve a robot equidistance task.

6.3.1 Single-robot estimate update law

Let Q be a simply connected subset of $\mathbb{R}^2$ with a differentiable moving boundary ∂Q. Consider a single robot described by (6.2.4) that moves along ∂Q. Assume that the processor state contains a set of interpolation points $\{q_1, \ldots, q_{n_{\mathrm{ip}}}\}$ used to approximate ∂Q. We begin with an informal description of the SINGLE-ROBOT ESTIMATE UPDATE LAW and we illustrate in Figure 6.4 the two actions characterizing this law:

> *[Informal description]* The agent moves counterclockwise along the moving boundary ∂Q, collecting estimates of its tangent and curvature. Using these estimates, the agent executes the following two actions. First, it updates the positions of the interpolation points so that they take value on the estimate of ∂Q. In other words, as sufficient information is available, each interpolation point q_α, $\alpha \in \{1, \ldots, n_{\mathrm{ip}}\}$, is projected onto the estimated boundary. Second, after an interpolation point q_α has been projected, the agent collects sufficient information so that it can locally optimize the location of q_α along the estimate of ∂Q. Here, by an estimate of the time-varying ∂Q, we mean the trajectory of the agent along the moving boundary.

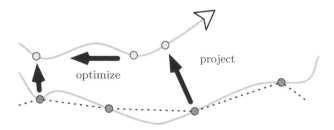

Figure 6.4 The two actions characterizing the SINGLE-ROBOT ESTIMATE UPDATE LAW.

Next, we begin our detailed description of the algorithm by specifying what variables the agent maintains in its memory. The processor state of the agent consists of the following variables:

(i) A counter **nxt** taking values in $\{1, \ldots, n_{\mathrm{ip}}\}$ that specifies which interpolation point the agent is going to project next.

(ii) A boundary representation comprised of the pairs

$$\{(q_\alpha, v_\alpha) \in (\mathbb{R}^2)^2 \mid \alpha \in \{1, \dots, n_{\mathrm{ip}}\}\},$$

where q_α is the position of the interpolation point α and v_α represents the tangent vector of ∂Q at q_α.

(iii) A curve of the form $\texttt{path} : [0, t] \to \mathbb{R}^2$. This curve is the trajectory followed by the agent from initial time until present time t. We assume that the agent updates the variable $\texttt{path}$ continuously. We let $\mathcal{C}(\mathbb{R}^2)$ be the set of planar curves, that is, twice differentiable functions from an interval to $\mathbb{R}^2$. With this notation, we may write $\texttt{path} \in \mathcal{C}(\mathbb{R}^2)$.

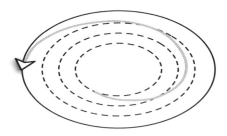

Figure 6.5 The agent moves along the time-varying boundary ∂Q, here depicted as a sequence of growing ellipses, and its trajectory is an approximation to ∂Q.

Remark 6.6 (Boundary approximation). For simplicity, we assume that, at every instant of time, the agent is located exactly on top of the boundary. This assumption implies that if the boundary is time-invariant, then the agent trajectory $\texttt{path}$ is locally equal to ∂Q. Furthermore, if the boundary is slowly time-varying, then the agent's trajectory $\texttt{path}$ is an estimate of the moving boundary ∂Q, as illustrated in Figure 6.5.   •

The agent updates its processor state according to the following two rules:

Rule #1: When and how to project onto ∂Q the interpolation point q_{nxt}. Let q_{nxt} denote the interpolation point about to be projected and let v_{nxt} denote the corresponding tangent vector. The projection takes place when the agent crosses the line, denoted by $\texttt{line}_{\mathrm{nxt}}$, that passes through q_{nxt} and is perpendicular to v_{nxt}. At this crossing time, we define the updated values for the interpolation point $\texttt{nxt}$, denoted by q_{nxt}^+, to be the point on $\texttt{path}$ where the agent's trajectory crosses the line $\texttt{line}_{\mathrm{nxt}}$. This projection operation is illustrated in Figure 6.6. We refer to this operation by the map $\texttt{perp-proj} : (\mathbb{R}^2)^2 \times \mathcal{C}(\mathbb{R}^2) \to \mathbb{R}^2$; in other words, we write

$$q_{\mathrm{nxt}}^+ := \texttt{perp-proj}(q_{\mathrm{nxt}}, v_{\mathrm{nxt}}, \texttt{path}).$$

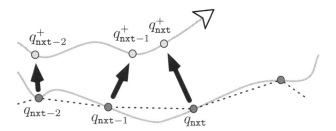

Figure 6.6 The projection of interpolation point q_{nxt} onto the curve **path**.

Rule #2: When and how to optimize the interpolation point $q_{\text{nxt}-1}$. The local optimization of the interpolation point $(\text{nxt} - 1)$ takes place immediately after the projection of the interpolation point **nxt** onto the estimated boundary. The interpolation point $(\text{nxt} - 1)$ is moved along the curve **path** in order to balance its two pseudodistances to its clockwise and counterclockwise neighboring interpolation points (recall that **path** is an estimate of the boundary ∂Q as discussed in Remark 6.6). Specifically, we define the map $\texttt{cyclic-balance} : (\mathbb{R}^2)^3 \times \mathcal{C}(\mathbb{R}^2) \to \mathbb{R}^2$ by

$\texttt{cyclic-balance}(q_{\text{nxt}-2}, q_{\text{nxt}-1}, q_{\text{nxt}}, \texttt{path})$ is point q^* in the curve **path**

such that $\mathcal{D}_\lambda(q_{\text{nxt}-2}, q^*) = \dfrac{3}{4}\mathcal{D}_\lambda(q_{\text{nxt}-2}, q_{\text{nxt}-1}) + \dfrac{1}{4}\mathcal{D}_\lambda(q_{\text{nxt}-1}, q_{\text{nxt}})$.

This map is illustrated in Figure 6.7. With this definition, we update the interpolation $(\text{nxt} - 1)$ to be

$$q_{\text{nxt}-1}^+ := \texttt{cyclic-balance}(q_{\text{nxt}-2}, q_{\text{nxt}-1}, q_{\text{nxt}}, \texttt{path}).$$

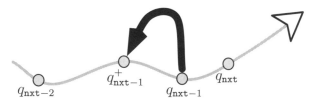

Figure 6.7 Optimal placement of the interpolation point $q_{\text{nxt}-1}$ along the curve **path**.

Remark 6.7 (Balancing property of the optimal placement). The optimal placement $q_{\text{nxt}-1}^+$ can be equivalently defined by

$$\mathcal{D}_\lambda(q_{\text{nxt}-1}^+, q_{\text{nxt}}) = \frac{1}{4}\mathcal{D}_\lambda(q_{\text{nxt}-2}, q_{\text{nxt}-1}) + \frac{3}{4}\mathcal{D}_\lambda(q_{\text{nxt}-1}, q_{\text{nxt}}),$$

so that it achieves the balancing property that

$$\begin{bmatrix} \mathcal{D}_\lambda(q_{\mathrm{nxt}-2}, q^+_{\mathrm{nxt}-1}) \\ \mathcal{D}_\lambda(q^+_{\mathrm{nxt}-1}, q_{\mathrm{nxt}}) \end{bmatrix} = \frac{1}{4} \begin{bmatrix} 3 & 1 \\ 1 & 3 \end{bmatrix} \begin{bmatrix} \mathcal{D}_\lambda(q_{\mathrm{nxt}-2}, q_{\mathrm{nxt}-1}) \\ \mathcal{D}_\lambda(q_{\mathrm{nxt}-1}, q_{\mathrm{nxt}}). \end{bmatrix}$$

This iteration is the same as the cyclic balancing system with parameter $k = 1/4$ studied in Exercises E1.30, E5.5, and E6.3.   •

Finally, we define one last useful operation. Given a point on the curve path, it is useful to be able to compute the tangent of the curve path at the point q. Specifically, given a point q on the curve path, we shall write

$$v := \mathtt{tangentat}(\mathtt{path}, q).$$

In summary, the SINGLE-ROBOT ESTIMATE UPDATE LAW is formally described as follows:

Robot: single robot moving at constant speed along ∂Q, continuously recording its trajectory

Event-driven Algorithm: SINGLE-ROBOT ESTIMATE UPDATE LAW

Processor State: $w = (\mathrm{nxt}, \{(q_\alpha, v_\alpha)\}_{\alpha=1}^{n_{\mathrm{ip}}}, \mathtt{path})$, where

nxt	$\in \{1, \ldots, n_{\mathrm{ip}}\}$, initially equal to index of interpolation point closest to robot moving counterclockwise
$\{(q_\alpha, v_\alpha)\}_{\alpha=1}^{n_{\mathrm{ip}}} \subset \mathbb{R}^2 \times \mathbb{R}^2,$	initially counterclockwise along boundary
path $\in \mathcal{C}(\mathbb{R}^2),$	continuously recording agent's trajectory

% A state transition is triggered when the agent crosses a certain line
function stf-trig(p, w)

1: $\mathtt{line}_{\mathrm{nxt}} :=$ line through point q_{nxt} perpendicular to direction v_{nxt}
2: **if** $p \in \mathtt{line}_{\mathrm{nxt}}$ **then**
3:    **return true**
4: **else**
5:    **return false**

% The current interpolation point and tangent vector are projected and the previous interpolation point is optimized along the new boundary
function stf(p, w)

1: $\{(q^+_\alpha, v^+_\alpha)\}_{\alpha=1}^{n_{\mathrm{ip}}} := \{(q_\alpha, v_\alpha)\}_{\alpha=1}^{n_{\mathrm{ip}}}$
2: $q^+_{\mathrm{nxt}} := \mathtt{perp\text{-}proj}(q_{\mathrm{nxt}}, v_{\mathrm{nxt}}, \mathtt{path})$
3: $q^+_{\mathrm{nxt}-1} := \mathtt{cyclic\text{-}balance}(q_{\mathrm{nxt}-2}, q_{\mathrm{nxt}-1}, q^+_{\mathrm{nxt}}, \mathtt{path})$
4: $v^+_{\mathrm{nxt}} := \mathtt{tangentat}(\mathtt{path}, q^+_{\mathrm{nxt}})$
5: $v^+_{\mathrm{nxt}-1} := \mathtt{tangentat}(\mathtt{path}, q^+_{\mathrm{nxt}-1})$
6: **return** $(\mathrm{nxt} + 1, \{(q^+_\alpha, v^+_\alpha)\}_{\alpha=1}^{n_{\mathrm{ip}}}, \mathtt{path})$

The SINGLE-ROBOT ESTIMATE UPDATE LAW may be improved in a number of ways; here, we present some important algorithmic clarifications.

Remarks 6.8 (Content and representation of the variable path).

(i) The above discussion assumes that, with the information provided by the variable path, the agent can compute the tangent vector and the curvature along its trajectory in order to perform the calculation of the pseudodistance $\mathcal{D}_\lambda$.

(ii) It is not necessary for the agent to keep in the variable path its entire trajectory since initial time. In fact, when the agent updates the location of the interpolation point $(\mathtt{nxt} - 1)$ in instruction 4: of the state-transition function, it is sufficient that path contains the trajectory of the agent starting from interpolation point $(\mathtt{nxt} - 2)$ until the current agent position. This "limited-length" requirement may be implemented as follows: the variable path is a curve of the form $\mathtt{path} : [\mathtt{t_{path}}, t] \to \mathbb{R}^2$, the variable t denotes the present time, the variable $\mathtt{t_{path}}$ is initially set equal to 0, and the instruction

$$\mathtt{t_{path}} := t^* \text{ such that } q_{\mathtt{nxt}-1} = \mathtt{path}(t^*),$$

is executed before instruction 6: in the state-transition function.

(iii) In a realistic implementation, the path variable and its first two derivatives are to be represented with finite resolution over a discrete time domain. The interested reader is referred to Susca et al. (2008) for a discussion of the SINGLE-ROBOT ESTIMATE UPDATE LAW algorithm with a realistic implementation of the path variable in a finite-length finite-resolution manner.   ●

Remark 6.9 (Timeout for the projection of interpolation points).
In the definition of the SINGLE-ROBOT ESTIMATE UPDATE LAW, we have implicitly assumed that the agent crosses the line through $q_{\mathtt{nxt}}$ perpendicular to $v_{\mathtt{nxt}}$. This is certainly the case if ∂Q is static or slowly time-varying. However, if ∂Q changes drastically, it is conceivable that the agent never crosses the line through $q_{\mathtt{nxt}}$ perpendicular to $v_{\mathtt{nxt}}$. In other words, it can happen that the state-transition trigger function is always false. This situation can be prevented by prescribing a timeout such that, if the agent has not crossed the line after a certain time has elapsed, then the interpolation point nxt is projected onto its trajectory anyway. Formally, let t^* be implicitly defined by

$$\mathcal{D}_\lambda(q_{\mathtt{nxt}-1}^+, p(t)) = 2\mathcal{D}_\lambda(q_{\mathtt{nxt}-1}, q_{\mathtt{nxt}}).$$

If no crossing has happened at time t, then $q_{\mathtt{nxt}}^+$ is set equal to the point on path that is closest to $q_{\mathtt{nxt}}$. The tangent vector $v_{\mathtt{nxt}}^+$ is set equal to `tangentat(path, `q_{nxt}^+`)`. This projection is well-defined and has the property that if ∂Q is time-invariant, then $q_{\text{nxt}}^+ = q_{\text{nxt}}$. The algorithm described by Susca et al. (2008) explicitly incorporates this timeout.   •

6.3.2 Cooperative estimate update law

In the previous section, we presented an event-driven algorithm for a single robot to monitor a boundary. We consider the robotic network $\mathcal{S}_{\text{bndry}}$ with ring communication topology, described in Section 6.2.2, and we develop a parallel version of the SINGLE-ROBOT ESTIMATE UPDATE LAW that allows the network to monitor the boundary efficiently (i.e., with more accuracy than a single robot could). We begin with an informal description of the COOPERATIVE ESTIMATE UPDATE LAW:

> *[Informal description]* Each agent moves counterclockwise along the moving boundary ∂Q, has its individual copy of the boundary representation, including the interpolation points, and executes the SINGLE-ROBOT ESTIMATE UPDATE LAW. Because the agents are spatially distributed, each agent updates its individual boundary representation separately. On top of the SINGLE-ROBOT ESTIMATE UPDATE LAW, the agents run a communication protocol that transmits the updated interpolation points along the ring topology. Specifically, every time an agent updates two interpolation points (using the state-transition function of the SINGLE-ROBOT ESTIMATE UPDATE LAW and, thus, the functions `perp-proj` and `cyclic-balance`), this agent transmits these updated interpolation points to its clockwise and counterclockwise neighbors. In turn, the neighbors record the updates in their individual boundary representation.

Next, we give a more detailed description of the algorithm. We assume that each robot i has a processor state with its local copy of $w^{[i]}$ containing a counter $\text{nxt}^{[i]}$, a boundary representation $\{(q_\alpha^{[i]}, v_\alpha^{[i]})\}_{\alpha=1}^{n_{\text{ip}}}$, where n_{ip} is equal for all robots, and its $\text{path}^{[i]}$. The COOPERATIVE ESTIMATE UPDATE LAW is formally described as follows:

Robotic Network: $\mathcal{S}_{\text{bndry}}$, first-order agents moving at unit speed along ∂Q with absolute sensing of own position, communicating with clockwise and counterclockwise neighbors

Event-driven Algorithm: COOPERATIVE ESTIMATE UPDATE LAW

Alphabet: $\mathbb{A} = \{1, \ldots, n_{\text{ip}}\} \times (\mathbb{R}^2)^2 \times (\mathbb{R}^2)^2 \cup \{\text{null}\}$

`Processor State`, `function` stf-trig, and `function` stf
same as in SINGLE-ROBOT ESTIMATE UPDATE LAW

% A transmission is triggered right after the interpolation points are updated
`function` msg-trig(p, w)

   1: **return** stf-trig(p, w)

% The updated interpolation points (and reference label) are transmitted
`function` msg-gen(p, w, i)

   1: **return** $\big(\mathtt{nxt}, (q_{\mathtt{nxt}-1}, v_{\mathtt{nxt}-1}), (q_{\mathtt{nxt}-2}, v_{\mathtt{nxt}-2})\big)$

% The received updated interpolation points are stored
`function` msg-rec(p, w, y, i)

   1: $\{(q_\alpha^+, v_\alpha^+)\}_{\alpha=1}^{n_{\mathrm{ip}}} := \{(q_\alpha, v_\alpha)\}_{\alpha=1}^{n_{\mathrm{ip}}}$
   2: $(\mathtt{nxtrec}, y_1, y_2) := y$
   3: $(q_{\mathtt{nxtrec}-1}^+, v_{\mathtt{nxtrec}-1}^+) := y_1$
   4: $(q_{\mathtt{nxtrec}-2}^+, v_{\mathtt{nxtrec}-2}^+) := y_2$
   5: **return** $(\mathtt{nxt}, \{(q_\alpha^+, v_\alpha^+)\}_{\alpha=1}^{n_{\mathrm{ip}}}, \mathtt{path})$

We conclude this section with an important clarification. We begin with a useful definition and then give two related remarks.

Definition 6.10 (Two-hop separation). A group of $n \geq 2$ agents implementing the COOPERATIVE ESTIMATE UPDATE LAW is *two-hop separated along the interpolation points* if $\mathtt{nxt}^{[i-1]} \leq \mathtt{nxt}^{[i]} - 2$ for all $i \in I$ at all times during the execution of the algorithm.   •

Remarks 6.11 (Well-posedness of the Cooperative Estimate Update Law).

  (i) The inequality $\mathtt{nxt}^{[i-1]} \leq \mathtt{nxt}^{[i]} - 2$ guarantees that each robot can correctly implement the algorithm. Indeed, if this inequality is violated, then the `cyclic-balance` function performed by robot i during the state-transition function is invoked with an interpolation point $q_{\mathtt{nxt}-2}^{[i]}$ which does not take value in the curve $\mathtt{path}^{[i]}$ (because the boundary might be time-varying). This inequality therefore guarantees that the algorithm is well-posed. Assuming two-hop separation guarantees that the projection and optimization events happen in the following order: each interpolation point $\mathtt{nxt}^{[i]}$ is projected and later optimized by robot i, strictly before it is projected by robot $(i - 1)$.

  (ii) The two-hop separation property is easily seen to hold when (1) the number of interpolation points n_{ip} is much larger than the number of robots n, (2) the robots are approximately equidistant along ∂Q, and (3) the distances between pairs of consecutive interpolation points are much less than the length of ∂Q divided by n.   •

6.3.3 Cyclic balancing algorithm for agent equidistance task

Here, we propose a motion coordination controller to achieve the agent equidistance task $\mathcal{T}_{\varepsilon\text{-eqdstnc}}$ among the agents moving along the boundary. This task also leads to the orderly interactions mentioned in the last remark. Specifically, we extend the COOPERATIVE ESTIMATE UPDATE LAW to include a motion coordination component that makes the agents achieve the agent equidistance task while moving at approximately unit speed along the boundary. The control design is straightforward: for robot i at position $p^{[i]}$ moving in continuous time with speed $v^{[i]}$ along ∂Q, we define

$$v^{[i]} = 1 + k_{\mathrm{prop}}\big(\mathcal{D}_{\mathrm{arc}}(p^{[i]}, p^{[i+1]}) - \mathcal{D}_{\mathrm{arc}}(p^{[i-1]}, p^{[i]})\big), \qquad (6.3.1)$$

where $k_{\mathrm{prop}} \in \mathbb{R}_{>0}$ is a fixed control gain. In other words, the agent speeds up or slows down depending upon whether it is closer to the following or to the preceding agent, respectively. This simple motion control law is the continuous-time analog of the cyclic balancing system described in Exercise E1.30; recall that this system was adopted also in the SINGLE-ROBOT ESTIMATE UPDATE LAW for the purpose of balancing pseudodistances among interpolation points.

To handle the lower and upper bounds constraints on the velocity, that is, the constraint $v \in [-v_{\min}, v_{\max}]$, we introduce a saturation function in the design (6.3.1). Specifically, we implement

$$v^{[i]} = \mathrm{sat}_{[v_{\min}, v_{\max}]}\left(1 + k_{\mathrm{prop}}\big(\mathcal{D}_{\mathrm{arc}}(p^{[i]}, p^{[i+1]}) - \mathcal{D}_{\mathrm{arc}}(p^{[i-1]}, p^{[i]})\big)\right), \quad (6.3.2)$$

where the saturation function $\mathrm{sat}_{[a,b]} : \mathbb{R} \to [a, b]$, for $a < b$, is defined by

$$\mathrm{sat}_{[a,b]}(x) = \begin{cases} a, & \text{if } x < a, \\ x, & \text{if } x \in [a, b], \\ b, & \text{if } x > b. \end{cases}$$

The difficulty in implementing controller (6.3.2) in the COOPERATIVE ESTIMATE UPDATE LAW is how to measure the counterclockwise arc-length distance between robots. To tackle this difficulty, let us begin with a useful observation. Given the interpolation points $\{q_1, \ldots, q_{n_{\mathrm{ip}}}\}$ and two points on the boundary r_1, r_2, assume that sufficient information is available to compute the indices $[r_1]$ and $[r_2]$ of the counterclockwise-closest interpolation points from r_1, r_2, respectively. With this notation and the assumption, the counterclockwise pseudodistance from r_1 to r_2 and one of its possible approximations are as follows:

$$\mathcal{D}_{\mathrm{arc}}(r_1, r_2) = \mathcal{D}_{\mathrm{arc}}(r_1, q_{[r_1]}) + \sum_{\alpha=[r_1]}^{[r_2]-2} \mathcal{D}_{\mathrm{arc}}(q_\alpha, q_{\alpha+1}) + \mathcal{D}_{\mathrm{arc}}(q_{[r_2]-1}, r_2)$$

(6.3.3)

$$\approx \mathrm{dist}_2(r_1, q_{[r_1]}) + \sum_{\alpha=[r_1]}^{[r_2]-2} \mathrm{dist}_2(q_\alpha, q_{\alpha+1}) + \mathrm{dist}_2(q_{[r_2]-1}, r_2).$$

(6.3.4)

Based on this equality and on this approximation, we propose two methods to implement the controller (6.3.2). One may implement either of the following:

(i) The approximation proposed in (6.3.4); this approximated pairwise counterclockwise arc-length distance may be computed with the information available to the agents in the COOPERATIVE ESTIMATE UPDATE LAW.

(ii) The exact computation proposed in (6.3.3); in order to perform this computation, however, the robots require more information. The processor state is required to store a collection of arc-length distances $\mathcal{D}_{\mathrm{arc}}(q_\alpha, q_{\alpha+1})$, $\alpha \in \{1, \ldots, n_{\mathrm{ip}}\}$ that are measured by the agents as they move, and that are maintained accurate via communication. In the interest of brevity, we omit a detailed discussion of this point here.

Finally, independently of the computation or approximation of the arc-length distances, the implementation of controller (6.3.2) requires each agent to have a continuous-time estimate of the location of its clockwise and counterclockwise neighbors: this information may be acquired by either a dedicated message-exchanging protocol or, possibly, by proximity sensors mounted on the robots. In the interest of brevity, we omit a detailed discussion of this point here.

6.3.4 Correctness of the estimate update and cyclic balancing law

We call the ESTIMATE UPDATE AND BALANCING LAW the combination of the COOPERATIVE ESTIMATE UPDATE LAW with the cyclic balancing control law (6.3.2), with exact arc-length distance computation between robots.

We call the APPROXIMATE ESTIMATE AND BALANCING LAW the combination of the COOPERATIVE ESTIMATE UPDATE LAW with the cyclic balancing control law (6.3.2), with (1) finite-resolution finite-length representation of the `path` variable in the robots, and (2) approximate arc-length distance computation between robots.

We state the properties of these laws in the following theorem, whose proof is postponed to Section 6.6.

Theorem 6.12 (Correctness of the exact and approximate laws).
On the network $\mathcal{S}_{\mathrm{bndry}}$, along evolutions with the two-hop separation property:

(i) *the* ESTIMATE UPDATE AND BALANCING LAW *achieves the boundary estimation task* $\mathcal{T}_{\varepsilon\text{-bndry}}$ *and the agent equidistance task* $\mathcal{T}_{\varepsilon\text{-eqdstnc}}$ *for any* $\varepsilon \in \mathbb{R}_{>0}$ *if the boundary is time-independent; and*

(ii) *the* APPROXIMATE ESTIMATE AND BALANCING LAW *achieves the boundary estimation task* $\mathcal{T}_{\varepsilon\text{-bndry}}$ *and the agent equidistance task* $\mathcal{T}_{\varepsilon\text{-eqdstnc}}$ *for some* $\varepsilon \in \mathbb{R}_{>0}$ *if the boundary varies in a continuously differentiable way and sufficiently slowly with time, and its length is upper bounded.*

Remark 6.13 (Error induced by the evolution of the boundary and its discretization). In the second statement in the theorem, the constant ε depends upon the rate of change of the boundary and upon the accuracy of the various approximations made in the algorithm. •

6.4 SIMULATION RESULTS

In order to illustrate the performance of the algorithms, we include here different simulation results of the APPROXIMATE ESTIMATE AND BALANCING LAW. In the first simulation, the boundary to be estimated is time invariant, while in the second it is time-varying.

Time-invariant boundary

As a first simulation, we assume that $n = 3$ agents aim to approximate the time-invariant boundary ∂Q described by the closed curve

$$\gamma(\theta) = \big(2 + \cos(5\theta) + 0.5\sin(2\theta)\big) \begin{bmatrix} \cos(\theta) \\ \sin(\theta) \end{bmatrix}, \quad \theta \in [0, 2\pi].$$

The control gain is $k_{\mathrm{prop}} = 0.05$. The minimum and maximum velocities are $v_{\min} = 0.5$ and $v_{\max} = 2$. The number of interpolation points is $n_{\mathrm{ip}} = 30$.

Pseudodistances are computed with $\lambda = \frac{10}{11}$. The simulation time is 50 seconds. At initial time, the interpolation points are selected to be the positions of the agents and other randomly distributed points on the boundary. Finally, each robot maintains a discretized representation of its trajectory `path` with a resolution of 0.01 seconds.

The behavior of the APPROXIMATE ESTIMATE AND BALANCING LAW is illustrated in Figure 6.8. The left- and right-hand figures correspond to the positions of the interpolation points and the agents at the initial and final configurations, respectively. In the right-hand figure one can see the approximating polygon and how close it is to the actual boundary.

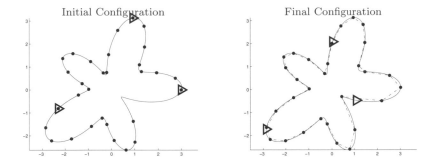

Figure 6.8 The APPROXIMATE ESTIMATE AND BALANCING LAW for a time-invariant boundary: initial and final configuration of agents (drawn as triangles) and interpolation points. The right-hand figure shows also the approximating polygon.

Figure 6.9 illustrates the convergence of the algorithm. Although the APPROXIMATE ESTIMATE AND BALANCING LAW uses an approximated version of the pseudodistances between interpolation points and of the arc-length distance between agents, we illustrate the performance of the algorithm by plotting the exact versions of the pseudodistances and arc-length distances. Regarding the boundary estimation task, the left-hand figure illustrates how the quantity $\max_{\alpha \in \{1,\dots,n_{\mathrm{ip}}\}} \mathcal{D}_\lambda(q_\alpha, q_{\alpha+1}) - \min_{\alpha \in \{1,\dots,n_{\mathrm{ip}}\}} \mathcal{D}_\lambda(q_\alpha, q_{\alpha+1})$ does indeed decrease towards zero, even though it does not vanish because of the adopted approximations. Regarding the equidistance task, the right figure illustrates how the agents become uniformly spaced along the boundary. Again, the arc-length distances converge toward a common steady-state value, even though the convergence is not exact because of the adopted approximations.

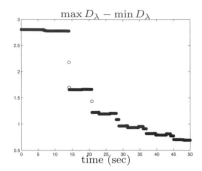
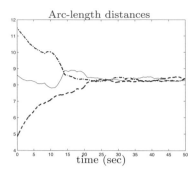

Figure 6.9 The APPROXIMATE ESTIMATE AND BALANCING LAW for a time-invariant
boundary: the left-hand figure shows the largest minus the smallest pseudodistance between neighboring interpolation points. The right-hand figure
shows the three arc-length distances between the three agents.

Slowly time-varying boundary

As a second simulation, we assume that $n = 4$ agents aim to approximate
the time-varying boundary ∂Q described by the time-varying closed curve

$$\gamma(\theta, t) = \left(2\frac{t_{\text{final}} - t}{t_{\text{final}}} + \left(2 + \cos(5\theta) + 0.5\sin(2\theta) \right)\frac{t}{t_{\text{final}}} \right) \begin{bmatrix} \cos(\theta) \\ \sin(\theta) \end{bmatrix},$$

with $\theta \in [0, 1]$, $t \in [0, t_{\text{final}}]$, and $t_{\text{final}} = 200$ seconds, as shown in Figure 6.10.
The parameters and initial conditions of the APPROXIMATE ESTIMATE AND
BALANCING LAW are the same as in the time-invariant case. The four plots
in Figure 6.10 show the positions of the interpolation points and of the
agents at the four time instants 0, 50, 100, and 200 seconds, respectively.
The last plot also illustrates how close the approximating polygon is to the
actual boundary. From the frames in Figure 6.10, it is clear that the agents
can adapt as ∂Q changes.

6.5 NOTES

For a discussion of hybrid systems we refer to van der Schaft and Schumacher
(2000). Other relevant references include Lygeros et al. (2003), Goebel et al.
(2004), and Sanfelice et al. (2007).

Many methods are currently available (Mehaute et al., 1993) for the approximation of planar curves; this fact is largely motivated by computational
and signal-processing applications. Among them, the use of interpolated
curves is a standard and important approach. In their most simple version,
interpolations provide polygonal approximations of curves, with generalizations that make use of splines, or combinations of functions in a certain basis.
In particular, the problem of characterizing the polygons that optimally ap-

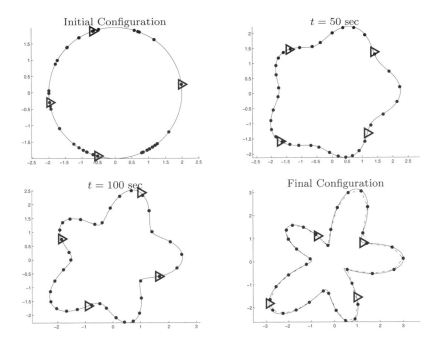

Figure 6.10 The APPROXIMATE ESTIMATE AND BALANCING LAW for a time-varying boundary: the configuration of the agents (drawn as triangles) and interpolation points at time instants 0, 50, 100, and 200 seconds. The last figure also shows the approximating polygon.

proximate a closed, convex body is a classical one; see the survey by Gruber (1983). In particular, the asymptotic formula in Lemma 6.5 was extended in (Gruber, 1983) to higher dimensions in terms of the Gauss curvature.

Boundary estimation and tracking is useful is numerous applications such as the detection of harmful algal blooms (Marthaler and Bertozzi, 2003; Bertozzi et al., 2004), oil spills (Clark and Fierro, 2007), and fire containment (Casbeer et al., 2005, 2006). Marthaler and Bertozzi (2003) adopt the so-called "snake algorithm" (from the computer vision literature) to detect and track the boundary of harmful algal bloom. Each agent is equipped with a chemical sensor that is able to measure the concentration gradient and with a communication system that is able to exchange information with a data fusion center. Bertozzi et al. (2004) suggest an algorithm that requires only a concentration sensor: the agents repeatedly cross the region boundary using a bang–bang angular velocity controller. Clark and Fierro (2007) use a random coverage controller, a collision avoidance controller, and a bang–bang angular velocity controller to detect and surround an oil spill. Casbeer et al. (2006) describe an algorithm that allows Low Altitude Short Endurance Unmanned Vehicles (LASEUVs) to closely monitor the boundary of a fire. Each of the LASEUVs has an infrared camera and a short-range communication device to exchange information with other agents, and to download the information collected to the base station. In Zhang and Leonard (2005), a formation of four robots tracks at unitary speed, the level sets of a field. Their relative positions change, so that they can optimally measure the gradient and estimate the curvature of the field in the center of the formation. In Zhang and Leonard (2007), a controller is proposed to steer a group of constant-speed robots onto an equally spaced configuration along a close curve.

6.6 PROOFS

This section presents the main result of the chapter on the correctness of Theorem 6.12. For the sake of completeness, we review first some notations and main concepts for Input-to-State-Stability (ISS) of discrete-time systems, as introduced in Angeli (1999), Jiang and Wang (2001), and Angeli (1999); Sontag (2008). We then make use of these to prove the result of Theorem 6.12.

6.6.1 Review of ISS concepts

A function $\gamma : \mathbb{R}_{\geq 0} \to \mathbb{R}_{\geq 0}$ is a $\mathcal{K}$-*function* if it is continuous, strictly increasing, and $\gamma(0) = 0$. A function $\beta : \mathbb{R}_{\geq 0} \times \mathbb{R}_{\geq 0} \to \mathbb{R}_{\geq 0}$ is a $\mathcal{KL}$-*function* if, for each $t \in \mathbb{R}_{\geq 0}$, the function $s \mapsto \beta(s, t)$ is a $\mathcal{K}$-function, and for each $s \in \mathbb{R}_{\geq 0}$, $t \mapsto \beta(s, t)$ is decreasing and $\beta(s, t) \to 0$ as $t \to +\infty$.

Consider the discrete-time nonlinear system

$$x(\ell + 1) = f(x(\ell), u(\ell)), \tag{6.6.1}$$

where ℓ takes values in $\mathbb{Z}_{\geq 0}$, x takes values in $\mathbb{R}^n$, and u takes values in $\mathbb{R}^m$. We assume that $f : \mathbb{Z}_{\geq 0} \times \mathbb{R}^n \times \mathbb{R}^m \to \mathbb{R}^n$ is continuous. In what follows, we let $\|u\|_{2,\infty} = \sup\{\|u(\ell)\|_2 \mid \ell \in \mathbb{Z}_{\geq 0}\} \leq +\infty$.

Definition 6.14 (Input-to-state stability). The system (6.6.1) is *input-to-state stable (ISS)* if there exist a $\mathcal{KL}$-function β and a $\mathcal{K}$-function γ such that, for each initial condition $x_0 \in \mathbb{R}^n$ at time $\ell_0 \in \mathbb{Z}_{\geq 0}$ and for each bounded input $u : \mathbb{Z}_{\geq 0} \to \mathbb{R}^m$, the system evolution x satisfies, for each $\ell \geq \ell_0$,

$$\|x(\ell)\|_2 \leq \beta(\|x_0\|_2, \ell - \ell_0) + \gamma(\|u\|_{2,\infty}). \qquad \bullet$$

Definition 6.15 (ISS-Lyapunov function). A function $V : \mathbb{R}^n \to \mathbb{R}_{\geq 0}$ is an *ISS-Lyapunov function* for system (6.6.1) if:

(i) it is continuously differentiable;

(ii) there exist $\mathcal{K}_\infty$-functions, α_1, α_2, such that $\alpha_1(\|x\|_2) \leq V(x) \leq \alpha_2(\|x\|_2)$; and

(iii) there exist a $\mathcal{K}_\infty$-function α_3 and a $\mathcal{K}$-function σ such that

$$V(f(x,u)) - V(x) \leq -\alpha_3(\|x\|_2) + \sigma(\|u\|_2). \qquad \bullet$$

We refer to Jiang and Wang (2001) for a proof of the following result.

Theorem 6.16 (ISS and Lyapunov functions). *System* (6.6.1) *is ISS if and only if it admits an ISS-Lyapunov function.*

6.6.2 Proof of Theorem 6.12

Proof. In the interest of brevity, we prove only the statements that pertain to the boundary estimation task $\mathcal{T}_{\varepsilon\text{-bndry}}$, that is, to the task

$$\left| \mathcal{D}_\lambda(q_{\alpha-1}^{[i]}, q_\alpha^{[i]}) - \mathcal{D}_\lambda(q_\alpha^{[i]}, q_{\alpha+1}^{[i]})) \right| < \varepsilon,$$

for all $\alpha \in \{1, \ldots, n_{\mathrm{ip}}\}$ and $i \in I$. We refer to Susca et al. (2008) for the proof of the statements regarding the agent equidistance task.

We begin our analysis with the case of a single robot, that is, we consider the ESTIMATE UPDATE AND BALANCING LAW algorithm, and we first consider the case of a time-invariant boundary with exact `path` representation and with no approximations in any computation. Define the shorthand $\mathcal{D}_\alpha = \mathcal{D}_\lambda(q_\alpha, q_{\alpha+1})$, for $\alpha \in \{1, \ldots, n_{\mathrm{ip}}\}$, and the positive vector $\mathcal{D} = (\mathcal{D}_1, \ldots, \mathcal{D}_{n_{\mathrm{ip}}}) \in \mathbb{R}_{>0}^{n_{\mathrm{ip}}}$. We now characterize how the vector $\mathcal{D}$ changes after one application of the state-transition function with counter `nxt` in the SINGLE-ROBOT ESTIMATE UPDATE LAW. We refer to an application of the state-transition function as a projection-and-placement event. Because the boundary is time-invariant, the projection operation (performed by the function `perp-proj`) leaves the interpolation point q_{nxt} unchanged. Furthermore, as discussed in Remark 6.7, the placement operation (performed by the function `cyclic-balance`) modifies the interpolation point $q_{\mathrm{nxt}-1}$ so that

$$\begin{bmatrix} \mathcal{D}_{\mathrm{nxt}-2} \\ \mathcal{D}_{\mathrm{nxt}-1} \end{bmatrix}^+ = \frac{1}{4} \begin{bmatrix} 3 & 1 \\ 1 & 3 \end{bmatrix} \begin{bmatrix} \mathcal{D}_{\mathrm{nxt}-2} \\ \mathcal{D}_{\mathrm{nxt}-1} \end{bmatrix},$$

where we adopt the shorthand $\mathcal{D}_\alpha^+ = \mathcal{D}_\lambda(q_\alpha^+, q_{\alpha+1}^+)$, for $\alpha \in \{1, \ldots, n_{\mathrm{ip}}\}$. For

$\mathtt{nxt} \in \{1, \dots, n_{\mathrm{ip}}\}$, define $A_{\mathtt{nxt}} \in \mathbb{R}^{n_{\mathrm{ip}} \times n_{\mathrm{ip}}}$ by

$$(A_{\mathtt{nxt}})_{jk} = \begin{cases} 3/4, & \text{if } (j,k) \text{ equals } (\mathtt{nxt}-1, \mathtt{nxt}-1) \text{ or } (\mathtt{nxt}-2, \mathtt{nxt}-2), \\ 1/4, & \text{if } (j,k) \text{ or } (j,k) \text{ equals } (\mathtt{nxt}-1, \mathtt{nxt}-2), \\ \delta_{jk}, & \text{otherwise} \end{cases}$$

and define the undirected graph $G_{\mathtt{nxt}}$ with vertices $\{1, \dots, n_{\mathrm{ip}}\}$ and with the single edge $(\mathtt{nxt}-1, \mathtt{nxt}-2)$. In summary, the matrix $A_{\mathtt{nxt}}$ determines the change of state $\mathcal{D}^+ = A_{\mathtt{nxt}} \mathcal{D}$ when a projection-and-placement event takes place with counter $\mathtt{nxt}$ and its associated graph is $G_{\mathtt{nxt}}$.

Because the boundary has finite length and the agent moves at lower-bounded speed, an infinite number of projection-and-placement events take place for each interpolation point. After a re-parametrization of time, let $\ell \in \mathbb{N}$ denote the times at which projection-and-placements events take place and let $\mathtt{nxt}(\ell) \in \{1, \dots, n_{\mathrm{ip}}\}$ denote the index corresponding to the event taking place at time ℓ. At each time $\ell \in \mathbb{N}$, we write

$$\mathcal{D}(\ell) = A_{\mathtt{nxt}(\ell)} \mathcal{D}(\ell - 1). \tag{6.6.2}$$

Next, note that $A_{\mathtt{nxt}(\ell)}$, for $\ell \in \mathbb{N}$, is a non-degenerate sequence of symmetric and doubly stochastic matrices. Additionally, note that the undirected graph $\cup_{\tau \geq \ell} G_{\mathtt{nxt}(\tau)}$ is connected. Therefore, by Theorem 1.65 and Corollary 1.70, we know that, for all $\alpha \in \{1, \dots, n_{\mathrm{ip}}\}$,

$$\lim_{\ell \to +\infty} \mathcal{D}_\alpha(\ell) = \frac{1}{n_{\mathrm{ip}}} \sum_{\alpha=1}^{n_{\mathrm{ip}}} \mathcal{D}_\alpha(0) = \frac{1}{n_{\mathrm{ip}}} \big(\text{total pseudodistance length of } \partial Q\big).$$

This proves that the interpolation points become equally spaced along ∂Q with respect to pseudodistance and that the boundary estimation task is achieved in the time-invariant case. In other words, this concludes the proof of the boundary estimation part of statement (i) for a single robot in Theorem 6.12.

Next, we consider the COOPERATIVE ESTIMATE UPDATE LAW for networks of $n \geq 2$ agents. Each agent has maintains a local copy of the pseudodistance vector and of the interpolation points (which always take value in ∂Q because the boundary is time-invariant). Specifically, for $i \in \{1, \dots, n\}$, agent i maintains vector $\mathcal{D}^{[i]}$ and interpolation points $q_\alpha^{[i]}$, for $\alpha \in \{1, \dots, n_{\mathrm{ip}}\}$. We define the *aggregate pseudodistance vector* $\mathcal{D}$ as follows: we let $\mathcal{D}_\alpha$ equal the most recently updated element of the vector $\{\mathcal{D}_\alpha^{[1]}, \dots, \mathcal{D}_\alpha^{[n]}\}$, that is, the pseudodistance between the most recently updated interpolation points $q_\alpha^{[i]}$ and $q_{\alpha+1}^{[i]}$, for $i \in \{1, \dots, n\}$. As before, after a re-parametrization of time, let $\ell \in \mathbb{N}$ denote the times at which projection-and-placements events take place (independently of which agent i executes the event) and let $\mathtt{nxt}(\ell) \in \{1, \dots, n_{\mathrm{ip}}\}$ denote the index corresponding to the event taking place at time ℓ (independently of which agent i executes the event). Note that when agent i updates the interpolation points $q_{\mathtt{nxt}(\ell)-1}$ and $q_{\mathtt{nxt}(\ell)-2}$, agent i then transmits the updated points to its immediately following agent $i-1$, so that the processor state of agent $i-1$ contains the correct updated information. Also note that since the boundary is time-invariant, the updated interpolation points belong to the trajectory $\mathtt{path}^{[i-1]}$ that agent $i-1$ maintains in its memory: this fact guarantees that agent $i-i$ can properly perform the $\mathtt{cyclic\text{-}balance}$ operation. In summary, equation (6.6.2) is the correct model not only for the SINGLE-ROBOT ESTIMATE UPDATE LAW but also for the COOPERATIVE ESTIMATE UPDATE LAW. This concludes the proof of the boundary estimation part of statement (i) for $n \geq 2$ robots in Theorem 6.12.

Let us now relax the assumption on the boundary and consider a time-varying $t \mapsto \partial Q(t)$; as before, we first consider the case of a single robot. We assume that pseudodistances between interpolation points along the agent $\mathtt{path}$ curve are computed exactly. By assumption, the boundary ∂Q varies in a continuously differentiable way and slowly in time and, therefore, the projection of the interpolation points is well defined and unique. For the case of a time-varying boundary, the state trajectory in continuous time is a curve of the form $\mathcal{D} : \mathbb{R}_{\geq 0} \to \mathbb{R}^{n_{\mathrm{ip}}}$ defined as follows. Note that, in general, the interpolation points lie outside ∂Q at almost all times, and therefore it makes no sense to define $\mathcal{D}_\alpha$ as the pseudodistance from point q_α to $q_{\alpha+1}$ along ∂Q. Rather, we give the following definition: $\mathcal{D}$ is the vector of pseudodistances computed by the robot along the curve $\mathtt{path}$. As a consequence, the state trajectory $\mathcal{D}$ is constant for almost all times and it changes only at projection-and-placement events. Specifically, let ℓ denote the time at which a projection-and-placements event takes place with corresponding index $\mathtt{nxt}(\ell) \in \{1, \ldots, n_{\mathrm{ip}}\}$. At time ℓ, the SINGLE-ROBOT ESTIMATE UPDATE LAW computes new values for the pseudodistances $\mathcal{D}_{\mathtt{nxt}-2}$ and $\mathcal{D}_{\mathtt{nxt}-1}$ based on the processor state of the agent (i.e., based on the interpolation points and the $\mathtt{path}$ variable); these values are the new values of the state $\mathcal{D}$. Because the boundary has upper-bounded length uniformly in time and because the agent moves at constant speed, an infinite number of projection-and-placement events takes place for each interpolation point and the duration of time between two consecutive events is uniformly upper bounded. Given this fact, we may let $\ell \in \mathbb{N}$ serve as index for all projection-and-placement times. Clearly, if the boundary does not vary with time, then the transition $\mathcal{D}(\ell) = A_{\mathtt{nxt}(\ell)}\mathcal{D}(\ell-1)$ describes the projection-and-placement event at instant ℓ. Because, instead, the boundary is time-varying, we model the change in $\mathcal{D}$ due to the boundary motion by

$$\mathcal{D}(\ell) = A_{\mathtt{nxt}(\ell)}\big(\mathcal{D}(\ell-1) + \mathcal{U}(\ell)\big), \qquad (6.6.3)$$

where $\mathcal{U}(\ell) \in \mathbb{R}^{n_{\mathrm{ip}}}$ is a disturbance. By design, $\mathcal{U}(\ell)$ is nonzero only on components $(\mathtt{nxt}(\ell) - 1)$ and $(\mathtt{nxt}(\ell) - 2)$ of $\mathcal{D}$. By the assumptions that the boundary varies in a continuously differentiable way and slowly with time, and that its length is upper bounded, we know that $\mathcal{U}$ is vanishing in the rate of change of the boundary. Finally, define the disagreement vector $\ell \to \mathbf{d}(\ell) \in \operatorname{span}\{\mathbf{1}_{n_{\mathrm{ip}}}\}^{\perp}$ by

$$\mathbf{d}(\ell) = \mathcal{D}(\ell) - \frac{\mathbf{1}_{n_{\mathrm{ip}}}^T \mathcal{D}(\ell)}{n_{\mathrm{ip}}} \mathbf{1}_{n_{\mathrm{ip}}}. \tag{6.6.4}$$

From equation (6.6.3) and from the fact that $A_{\mathtt{nxt}(\ell)}$ is doubly stochastic, the update law for $\mathbf{d}$ is

$$\mathbf{d}(\ell) = A_{\mathtt{nxt}(\ell)}\mathbf{d}(\ell - 1) + \mathbf{u}(\ell), \quad \ell \in \mathbb{N}, \tag{6.6.5}$$

where $\mathbf{u}(\ell) = \mathcal{U}(\ell) - \frac{1}{n_{\mathrm{ip}}}\mathbf{1}_{n_{\mathrm{ip}}}^T \mathcal{U}(\ell)\mathbf{1}_{n_{\mathrm{ip}}}$.

Equation (6.6.5) is the correct update equation even in the case of $n \geq 2$ robots moving along a time-varying boundary. This fact is a consequence of the two-hop separation assumption (see Definition 6.10). Indeed, as explained in Remarks 6.11, the inequality $\mathtt{nxt}^{[i-1]} \leq \mathtt{nxt}^{[i]} - 2$ guarantees that each robot can correctly perform each projection-and-placement event. Given the sequence $\ell \in \mathbb{N}$, define a new sequence $\ell_k \in \mathbb{N}$, for $k \in \mathbb{N}$, as follows: set $\ell_1 = 1$, assume without loss of generality that agent 1 is the agent executing the first projection-and-placement event with index $\mathtt{nxt}(1)$, and let $\ell_k \geq 2$ be the k-th time when agent 1 performs the projection-and-placement event with same index $\mathtt{nxt}(1)$. Reasoning about the possible positions of all agents at time ℓ_{k-1} and ℓ_k, one can see that $\ell_k - \ell_{k-1} \leq 2n \cdot n_{\mathrm{ip}}$. Define sequence $\mathcal{A}_{\ell_k} \in \mathbb{R}^{n_{\mathrm{ip}} \times n_{\mathrm{ip}}}$, for $k \in \mathbb{N}$, by $\mathcal{A}(1) = A_{\mathtt{nxt}(1)}$ and

$$\mathcal{A}(\ell_k) = A_{\mathtt{nxt}(\ell_k)} \cdots A_{\mathtt{nxt}(\ell_{k-1}+2)} A_{\mathtt{nxt}(\ell_{k-1}+1)}, \quad \text{for } k \geq 2.$$

By Exercise E1.17, each matrix $\mathcal{A}(\ell_k)$ is doubly stochastic and irreducible, because it is the product of doubly stochastic matrices and because the union of the undirected graphs associated with the matrices defining $\mathcal{A}(\ell_k)$ is connected. By definition, equation (6.6.5) becomes, for $k \in \mathbb{N}$,

$$\mathbf{d}(\ell_k) = \mathcal{A}(\ell_k)\mathbf{d}(\ell_{k-1}) + \sum_{\ell=\ell_{k-1}+1}^{\ell_k} A_{\mathtt{nxt}(\ell_k)} \cdots A_{\mathtt{nxt}(\ell+1)}\mathbf{u}(\ell)$$

$$= \mathcal{A}(\ell_k)\mathbf{d}(\ell_{k-1}) + \mathcal{B}(\ell_k)\mathbf{u}_{\mathrm{stacked}}(\ell_k), \tag{6.6.6}$$

where the vector $\mathbf{u}_{\mathrm{stacked}}(\ell_k)$ contains all vectors $\mathbf{u}(\ell_{k-1}+1), \ldots, \mathbf{u}(\ell_k)$, and the matrix $\mathcal{B}(\ell_k)$ is defined in the trivial corresponding way.

Define $V : \mathbb{R}^{n_{\mathrm{ip}}} \to \mathbb{R}_{\geq 0}$ by $V(x) = x^T x$ and adopt this function as a candidate ISS-Lyapunov function for system (6.6.6). We compute

$$V(\mathbf{d}(\ell_{k+1})) - V(\mathbf{d}(\ell_k)) = -\mathbf{d}(\ell_k)^T R(\ell_k)\mathbf{d}(\ell_k)$$
$$+ \mathbf{u}_{\text{stacked}}^T(\ell_k)\mathbf{u}_{\text{stacked}}(\ell_k) + 2\mathbf{u}_{\text{stacked}}^T(\ell_k)\mathcal{A}(\ell_k)\mathbf{d}(\ell_k),$$

where $R(\ell_k) = I_{n_{\text{ip}}} - \mathcal{A}(\ell_k)^T \mathcal{A}(\ell_k)$. Because $\mathcal{A}(\ell_k)$ is doubly stochastic and irreducible, we know (from Exercise E1.5) that $R(\ell_k)$ is positive semidefinite and that its simple eigenvalue 0 is associated with the eigenvector $\mathbf{1}_{n_{\text{ip}}}$. This fact implies that the quantity $-x^T R(\ell_k)x$ is strictly negative for all $x \notin \text{span}\{\mathbf{1}_{n_{\text{ip}}}\}^\perp$. To upper bound this quantity by a negative number, we let $\mathcal{A}_s$ be a generic element of the set of all the possible matrices $\mathcal{A}(\ell_k)$; such matrices are the iterated products of at most $2n \cdot n_{\text{ip}}$ matrices of the form A_{nxt}, where each matrix A_{nxt}, $\text{nxt} \in \{1, \ldots, n_{\text{ip}}\}$, appears at least once. Define the set of nonzero eigenvalues of $\mathcal{A}_s$ by

$$S_s = \{\lambda \in \mathbb{R} \mid \det\left(\lambda I_{n_{\text{ip}}} - (\mathcal{A}_s^T \mathcal{A}_s - I_{n_{\text{ip}}})\right) = 0\} \setminus \{0\}$$

and define the eigenvalue with smallest magnitude among all matrices by $\bar{r} = \min_s \min\{|\lambda| \mid \lambda \in S_s\}$. Note that $\bar{r} > 0$, because we are considering a finite set of matrices. We can then write

$$V(\mathbf{d}(\ell_k + 1)) - V(\mathbf{d}(\ell_k)) \le -\alpha_3(\|\mathbf{d}(\ell_k)\|) + \sigma(\|\mathbf{u}_{\text{stacked}}(\ell_k)\|),$$

where $\alpha_3(\|\mathbf{d}\|) = \frac{1}{2}\bar{r}\|\mathbf{d}\|^2$ and $\sigma(\|\mathbf{u}_{\text{stacked}}\|) = (\frac{2}{\bar{r}} + 1)\|\mathbf{u}_{\text{stacked}}\|^2$. By Definition 6.15, the system described by (6.6.6) is input-to-state stable. The input-to-state stability implies the existence of a positive ε, as in the boundary estimation part of statement (ii) for $n \ge 2$ robots in Theorem 6.12. ∎

6.7 EXERCISES

E6.1 **(Alternative expression of the symmetric difference).** Show that the symmetric difference δ^S between two compact bodies $C, B \subseteq \mathbb{R}^3$ can be alternatively expressed as

$$\delta^S(C, B) = \mu(C \setminus B) + \mu(B \setminus C).$$

   Hint: Use the expressions $C = (C \setminus B) \cup (C \cap B)$ and $B = (B \setminus C) \cup (C \cap B)$.

E6.2 **(Characterization of critical inscribed polygons for the symmetric difference).** Prove Lemma 6.4. Also, show that not all critical configurations are optimal. Specifically, consider the convex body and the gray inscribed triangle depicted in Figure E6.1(a). Show that the gray triangle is a saddle configuration for δ^S by establishing that modifications of the triangle as in Figure E6.1(b) decrease its area (and hence increase δ^S), whereas modifications of the gray triangle as in Figure E6.1(c) increase its area (and hence decrease δ^S).

E6.3 **(The "n-bugs problem" and cyclic interactions: cont'd).** Consider n robots at counterclockwise-ordered positions $\theta_1, \ldots, \theta_n$ following the cyclic bal-

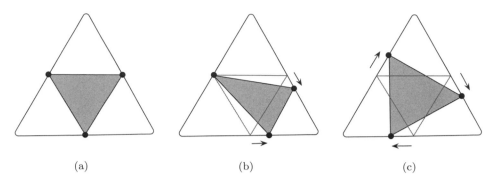

$$\text{(a)} \qquad\qquad\qquad \text{(b)} \qquad\qquad\qquad \text{(c)}$$

Figure E6.1 An illustration of the fact that polygons satisfying (6.2.1) may not be optimal for δ^S.

ancing system described in Exercise E1.30, with parameter $k = 1/4$:

$$\theta_i(\ell + 1) = \frac{1}{4}\theta_{i+1}(\ell) + \frac{1}{2}\theta_i(\ell) + \frac{1}{4}\theta_{i-1}(\ell), \quad \ell \in \mathbb{Z}_{\geq 0}.$$

Show that

$$\operatorname{dist}_{\mathrm{cc}}\big(\theta_{i-1}(\ell), \theta_i(\ell+1)\big) = \frac{3}{4}\operatorname{dist}_{\mathrm{cc}}\big(\theta_{i-1}(\ell), \theta_i(\ell)\big) + \frac{1}{4}\operatorname{dist}_{\mathrm{cc}}\big(\theta_i(\ell), \theta_{i+1}(\ell)\big).$$

E6.4   **(ISS properties of averaging algorithms with inputs and outputs).** This is a guided exercise to prove some of the ISS properties of averaging algorithms with inputs and outputs. An *averaging algorithm with inputs* associated to a sequence of stochastic matrices $\{F(\ell) \mid \ell \in \mathbb{Z}_{\geq 0}\} \subseteq \mathbb{R}^{n \times n}$, a sequence of input gains $\{D(\ell) \mid \ell \in \mathbb{Z}_{\geq 0}\} \subseteq \mathbb{R}^{n \times k}$, and a sequence of disturbances $u : \mathbb{Z}_{\geq 0} \to \mathbb{R}^k$ is the discrete-time dynamical system

$$x(\ell + 1) = F(\ell)x(\ell) + D(\ell)u(\ell), \quad \ell \in \mathbb{Z}_{\geq 0}. \qquad (E6.1)$$

A natural question to ask is how the evolution of the trajectory x is affected by the noise u. Let us address this in the following. Define the matrix

$$P = \begin{bmatrix} 1 & -1 & 0 & \dots & 0 \\ 0 & 1 & -1 & \dots & 0 \\ \vdots & & \ddots & \ddots & \vdots \\ 0 & \dots & 0 & 1 & -1 \end{bmatrix} \in \mathbb{R}^{(n-1) \times n}.$$

Note that, with the notation of Exercise E1.7, one can write

$$T = \begin{bmatrix} P \\ \frac{1}{n}\mathbf{1}_n^T \end{bmatrix}.$$

Define the following output for the dynamical system (E6.1):

$$y_{\mathrm{err}} = Px = \begin{pmatrix} x_1 - x_2 \\ \vdots \\ x_{n-1} - x_n \end{pmatrix} \in \mathbb{R}^{n-1}.$$

This output can be thought of as an error signal that quantifies the disagreement among the components of the state. Now, consider the change of variables

$$z = Tx = \begin{pmatrix} y_{\text{err}} \\ x_{\text{ave}} \end{pmatrix},$$

where $x_{\text{ave}} \in \mathbb{R}$ is the average of the components of x.

Verify that system (E6.1) reads in the new variable z as

$$z(\ell + 1) = TF(\ell)T^{-1}\, z(\ell) + TD(\ell)u(\ell).$$

The previous result is a formal statement of the following intuition. Because of the definition of z and of the special structure of $\{F(\ell) \mid \ell \in \mathbb{Z}_{\geq 0}\}$, the variable x_{ave} plays no role in the evolution of y_{err}. Accordingly, we define the *error system* by

$$y_{\text{err}}(\ell + 1) = F_{\text{err}}(\ell)y_{\text{err}}(\ell) + D_{\text{err}}(\ell)u(\ell), \tag{E6.2}$$

and the *average system* by

$$x_{\text{ave}}(\ell + 1) = x_{\text{ave}}(\ell) + c_{\text{err}}(\ell)y_{\text{err}}(\ell) + D_{\text{ave}}(\ell)u(\ell), \tag{E6.3}$$

for $D_{\text{err}}(\ell) = PD(\ell)$ and $D_{\text{ave}}(\ell) = \frac{1}{n}1_n^T D(\ell)$.

Assume now that:

(a) The sequence $\{F(\ell) \mid \ell \in \mathbb{Z}_{\geq 0}\}$ is a non-degenerate sequence of stochastic matrices.

(b) For $\ell \in \mathbb{Z}_{\geq 0}$, let $G(\ell)$ be the unweighted digraph associated with $F(\ell)$. There exists a duration $\delta \in \mathbb{N}$ such that, for all $\ell \in \mathbb{Z}_{\geq 0}$ the digraph $G(\ell + 1) \cup \ldots \cup G(\ell + \delta)$ contains a globally reachable node.

(c) The induced norm of $\{D(\ell) \mid \ell \in \mathbb{Z}_{\geq 0}\}$, for $\ell \in \mathbb{Z}_{\geq 0}$, is uniformly bounded.

Prove that, under assumptions (a), (b) and (c) on the averaging system with inputs, the following equivalent statements hold:

(i) the system (E6.1) with output y_{err} is IOS; and

(ii) the error system (E6.2) is ISS.

Bibliography

Abdallah, C. T. and Tanner, H. G. (2007) Complex networked control systems: introduction to the special section, *IEEE Control Systems Magazine*, **27**(4), 30–32.

Agarwal, P. K. and Sharir, M. (1998) Efficient algorithms for geometric optimization, *ACM Computing Surveys*, **30**(4), 412–458.

Agmon, N. and Peleg, D. (2006) Fault-tolerant gathering algorithms for autonomous mobile robots, *SIAM Journal on Computing*, **36**(1), 56–82.

Alighanbari, M. and How, J. P. (2006) Robust decentralized task assignment for cooperative UAVs, in *AIAA Conference on Guidance, Navigation and Control*, Keystone, CO.

Ando, H., Oasa, Y., Suzuki, I., and Yamashita, M. (1999) Distributed memoryless point convergence algorithm for mobile robots with limited visibility, *IEEE Transactions on Robotics and Automation*, **15**(5), 818–828.

Angeli, D. (1999) Intrinsic robustness of global asymptotic stability, *Systems & Control Letters*, **38**(4-5), 297–307.

Angeli, D. and Bliman, P.-A. (2006) Stability of leaderless discrete-time multi-agent systems, *Mathematics of Control, Signals and Systems*, **18**(4), 293–322.

Arai, T., Pagello, E., and Parker, L. E. (2002) Guest editorial: Advances in multirobot systems, *IEEE Transactions on Robotics and Automation*, **18**(5), 655–661.

Arkin, R. C. (1998) *Behavior-Based Robotics*, MIT Press, ISBN 0262011654.

Arsie, A., Savla, K., and Frazzoli, E. (2009) Efficient routing algorithms for multiple vehicles with no explicit communications, *IEEE Transactions on Automatic Control*, to appear.

Arslan, G., Marden, J. R., and Shamma, J. S. (2007) Autonomous vehicle-target assignment: A game theoretic formulation, *ASME Journal on Dynamic Systems, Measurement, and Control*, **129**(5), 584–596.

Asama, H. (1992) Distributed autonomous robotic system configured with multiple agents and its cooperative behaviors, *Journal of Robotics and Mechatronics*, **4**(3), 199–204.

Aspnes, J., Eren, T., Goldenberg, D. K., Morse, A. S., Whiteley, W., Yang, Y. R., Anderson, B. D. O., and Belhumeur, P. (2006) A theory of network localization, *IEEE Transactions on Mobile Computing*, **5**(12), 1663–1678.

Astolfi, A. (1999) Exponential stabilization of a wheeled mobile robot via discontinuous control, *ASME Journal on Dynamic Systems, Measurement, and Control*, **121**(1), 121–127.

Aurenhammer, F. (1991) Voronoi diagrams: A survey of a fundamental geometric data structure, *ACM Computing Surveys*, **23**(3), 345–405.

Axelrod, R. (1997) The dissemination of culture: A model with local convergence and global polarization, *Journal of Conflict Resolution*, **41**(2), 203–226.

Baillieul, J. and Suri, A. (2003) Information patterns and hedging Brockett's theorem in controlling vehicle formations, in *IEEE Conference on Decision and Control*, pages 556–563, Maui, HI.

Balch, T. and Parker, L. E., (editors) (2002) *Robot Teams: From Diversity to Polymorphism*, A. K. Peters, ISBN 1568811551.

Barbašin, E. A. and Krasovskiĭ, N. N. (1952) On stability of motion in the large, *Doklady Akad. Nauk SSSR*, **86**, 453–456, (In Russian).

Barlow, G. W. (1974) Hexagonal territories, *Animal Behavior*, **22**, 876–878.

Barooah, P. (2007) *Estimation and Control with Relative Measurements: Algorithms and Scaling Laws*, Ph.D. thesis, University of California at Santa Barbara.

Barooah, P. and Hespanha, J. P. (2007) Estimation from relative measurements: Algorithms and scaling laws, *IEEE Control Systems Magazine*, **27**(4), 57–74.

Bartle, R. G. (1995) *The Elements of Integration and Lebesgue Measure*, Wiley-Interscience, ISBN 0471042226.

Batalin, M. A. and Sukhatme, G. S. (2004) Coverage, exploration and deployment by a mobile robot and communication network, *Telecommunication Systems Journal*, **26**(2), 181–196, Special Issue on Wireless Sensor Networks.

Bauso, D., Giarré, L., and Pesenti, R. (2006) Nonlinear protocols for optimal distributed consensus in networks of dynamic agents, *Systems & Control Letters*, **55**(11), 918–928.

Beard, R. W., McLain, T. W., Goodrich, M. A., and Anderson, E. P. (2002) Coordinated target assignment and intercept for unmanned air vehicles, *IEEE Transactions on Robotics and Automation*, **18**(6), 911–922.

Belta, C. and Kumar, V. (2004) Abstraction and control for groups of robots, *IEEE Transactions on Robotics*, **20**(5), 865–875.

Bertozzi, A. L., Kemp, M., and Marthaler, D. (2004) Determining environmental boundaries: Asynchronous communication and physical scales, in *Cooperative Control*, V. Kumar, N. E. Leonard, and A. S. Morse, editors, volume 309 of *Lecture Notes in Control and Information Sciences*, pages 25–42, Springer, ISBN 3540228616.

Bertsekas, D. P. and Castañón, D. A. (1991) Parallel synchronous and asynchronous implementations of the auction algorithm, *Parallel Computing*, **17**, 707–732.

— (1993) Parallel primal-dual methods for the minimum cost flow problem, *Computational Optimization and Applications*, **2**(4), 317–336.

Bertsekas, D. P. and Tsitsiklis, J. N. (1997) *Parallel and Distributed Computation: Numerical Methods*, Athena Scientific, ISBN 1886529019.

Biggs, N. (1994) *Algebraic Graph Theory*, second edition, Cambridge University Press, ISBN 0521458978.

Blondel, V. D., Hendrickx, J. M., Olshevsky, A., and Tsitsiklis, J. N. (2005) Convergence in multiagent coordination, consensus, and flocking, in *IEEE Conference on Decision and Control and European Control Conference*, pages 2996–3000, Seville, Spain.

Boinski, S. and Campbell, A. F. (1995) Use of trill vocalizations to coordinate troop movement among whitefaced capuchins – a 2nd field-test, *Behaviour*, **132**, 875–901.

Boissonnat, J.-D. and Cazals, F. (2002) Smooth surface reconstruction via natural neighbour interpolation of distance functions, *Computational Geometry: Theory and Applications*, **22**(1), 185–203.

Bollobás, B. (2001) *Random Graphs*, second edition, Cambridge University Press, ISBN 0521809207.

Bollobás, B. and Riordan, O. (2006) *Percolation*, Cambridge University Press, ISBN 0521872324.

Boltyanski, V., Martini, H., and Soltan, V. (1999) *Geometric methods and optimization problems*, volume 4 of *Combinatorial optimization*, Kluwer Academic Publishers, ISBN 0792354540.

Boyd, S. (2006) Convex optimization of graph Laplacian eigenvalues, in *Proceedings of the International Congress of Mathematicians*, pages 1311–1319, Madrid, Spain.

Bruckstein, A. M., Cohen, N., and Efrat, A. (1991) Ants, crickets, and frogs in cyclic pursuit, Technical Report CIS 9105, Center for Intelligent Systems, Technion, Haifa, Israel, available at `http://www.cs.technion.ac.il/tech-reports`.

Bullo, F., Cortés, J., and Piccoli, B. (2009) Special issue on control and optimization in cooperative networks, *SIAM Journal on Control and Optimization*, **48**(1), vii–vii.

Bullo, F. and Lewis, A. D. (2004) *Geometric Control of Mechanical Systems*, volume 49 of *Texts in Applied Mathematics*, Springer, ISBN 0387221956.

Bulusu, N., Heidemann, J., and Estrin, D. (2001) Adaptive beacon placement, in *International Conference on Distributed Computing Systems*, pages 489–498, Mesa, AZ.

Burgard, W., Moors, M., Stachniss, C., and Schneider, F. E. (2005) Coordinated multi-robot exploration, *IEEE Transactions on Robotics*, **21**(3), 376–386.

Caicedo-Nùñez, C. H. and Žefran, M. (2008) Performing coverage on nonconvex domains, in *IEEE Conference on Control Applications*, pages 1019–1024, San Antonio, TX.

Cao, M. (2007) *Multi-Agent Formations and Sensor Networks*, Ph.D. thesis, Yale University.

Cao, M., Morse, A. S., and Anderson, B. D. O. (2008) Reaching a consensus in a dynamically changing environment - convergence rates, measurement delays and asynchronous events, *SIAM Journal on Control and Optimization*, **47**(2), 601–623.

Cao, Y. U., Fukunaga, A. S., and Kahng, A. (1997) Cooperative mobile robotics: Antecedents and directions, *Autonomous Robots*, **4**(1), 7–27.

Carli, R. (2008) *Topics in the Average Consensus Problems*, Ph.D. thesis, Universitá di Padova, Italy.

Carli, R. and Bullo, F. (2009) Quantized coordination algorithms for rendezvous and deployment, *SIAM Journal on Control and Optimization*, (Submitted Dec. 2007) to appear.

Carli, R., Bullo, F., and Zampieri, S. (2009) Quantized average consensus via dynamic coding/decoding schemes, *International Journal on Robust and Nonlinear Control*, (Submitted May 2008) to appear.

Carli, R., Fagnani, F., Speranzon, A., and Zampieri, S. (2008) Communication constraints in the average consensus problem, *Automatica*, **44**(3), 671–684.

Casbeer, D. W., Kingston, D. B., Beard, R. W., Mclain, T. W., Li, S.-M., and Mehra, R. (2006) Cooperative forest fire surveillance using a team of small unmanned air vehicles, *International Journal of Systems Sciences*, **37**(6), 351–360.

Casbeer, D. W., Li, S.-M., Beard, R. W., Mehra, R. K., and McLain, T. W. (2005) Forest fire monitoring with multiple small UAVs, in *American Control Conference*, pages 3530–3535, Portland, OR.

Cassandras, C. G. and Lafortune, S. (2007) *Introduction to Discrete-Event Systems*, second edition, Springer, ISBN 0387333320.

Castañón, D. A. and Wu, C. (2003) Distributed algorithms for dynamic reassignment, in *IEEE Conference on Decision and Control*, pages 13–18, Maui, HI.

Chatterjee, S. and Seneta, E. (1977) Towards consensus: Some convergence theorems on repeated averaging, *Journal of Applied Probability*, **14**(1), 89–97.

Chavel, I. (1984) *Eigenvalues in Riemannian Geometry*, Academic Press, ISBN 0121706400.

Chen, C.-T. (1984) *Linear System Theory and Design*, Holt, Rinehart, and Winston, ISBN 0030602890.

Chopra, N. and Spong, M. W. (2009) On exponential synchronization of Kuramoto oscillators, *IEEE Transactions on Automatic Control*, **54**(2), 353–357.

Chorin, A. J. and Marsden, J. E. (1994) *A Mathematical Introduction to Fluid Mechanics*, volume 4 of *Texts in Applied Mathematics*, third edition, Springer, ISBN 0387979182.

Choset, H. (2001) Coverage for robotics – A survey of recent results, *Annals of Mathematics and Artificial Intelligence*, **31**(1-4), 113–126.

Chvátal, V. (1975) A combinatorial theorem in plane geometry, *Journal of Combinatorial Theory. Series B*, **18**, 39–41.

Clark, J. and Fierro, R. (2007) Mobile robotic sensors for perimeter detection and tracking, *ISA Transactions*, **46**(1), 3–13.

Clarke, F. H. (1983) *Optimization and Nonsmooth Analysis*, Canadian Mathematical Society Series of Monographs and Advanced Texts, John Wiley, ISBN 047187504X.

Cogburn, R. (1984) The ergodic theory of Markov chains in random environments, *Zeitschrift für Wahrscheinlichkeitstheorie und verwandte Gebiete*, **66**(1), 109–128.

Conradt, L. and Roper, T. J. (2003) Group decision-making in animals, *Nature*, **421**(6919), 155–158.

Cormen, T. H., Leiserson, C. E., Rivest, R. L., and Stein, C. (2001) *Introduction to Algorithms*, second edition, MIT Press, ISBN 0262032937.

Cortés, J. (2006) Finite-time convergent gradient flows with applications to network consensus, *Automatica*, **42**(11), 1993–2000.

— (2007) Distributed Kriged Kalman filter for spatial estimation, *IEEE Transactions on Automatic Control*, submitted.

— (2008a) Area-constrained coverage optimization by robotic sensor networks, in *IEEE Conference on Decision and Control*, pages 1018–1023, Cancún, México.

— (2008b) Discontinuous dynamical systems – a tutorial on solutions, nonsmooth analysis, and stability, *IEEE Control Systems Magazine*, **28**(3), 36–73.

— (2008c) Distributed algorithms for reaching consensus on general functions, *Automatica*, **44**(3), 726–737.

Cortés, J. and Bullo, F. (2005) Coordination and geometric optimization via distributed dynamical systems, *SIAM Journal on Control and Optimization*, **44**(5), 1543–1574.

Cortés, J., Martínez, S., and Bullo, F. (2005) Spatially-distributed coverage optimization and control with limited-range interactions, *ESAIM: Control, Optimisation & Calculus of Variations*, **11**, 691–719.

— (2006) Robust rendezvous for mobile autonomous agents via proximity graphs in arbitrary dimensions, *IEEE Transactions on Automatic Control*, **51**(8), 1289–1298.

Cortés, J., Martínez, S., Karatas, T., and Bullo, F. (2004) Coverage control for mobile sensing networks, *IEEE Transactions on Robotics and Automation*, **20**(2), 243–255.

Couzin, I. D., Krause, J., Franks, N. R., and Levin, S. A. (2005) Effective leadership and decision-making in animal groups on the move, *Nature*, **433**(7025), 513–516.

Cucker, F. and Smale, S. (2007) Emergent behavior in flocks, *IEEE Transactions on Automatic Control*, **52**(5), 852–862.

Cybenko, G. (1989) Dynamic load balancing for distributed memory multi-processors, *Journal of Parallel and Distributed Computing*, **7**(2), 279–301.

de Berg, M., van Kreveld, M., Overmars, M., and Schwarzkopf, O. (2000) *Computational Geometry: Algorithms and Applications*, second edition, Springer, ISBN 3540656200.

de Gennaro, M. C. and Jadbabaie, A. (2006) Decentralized control of connectivity for multi-agent systems, in *IEEE Conference on Decision and Control*, pages 3628–3633, San Diego, CA.

de Silva, V. and Ghrist, R. (2007) Coverage in sensor networks via persistent homology, *Algebraic & Geometric Topology*, **7**, 339–358.

Deffuant, G., Neau, D., Amblard, F., and Weisbuch, G. (2000) Mixing beliefs among interacting agents, *Advances in Complex Systems*, **3**(1/4), 87–98.

DeGroot, M. H. (1974) Reaching a consensus, *Journal of the American Statistical Association*, **69**(345), 118–121.

Dias, M. B., Zlot, R., Kalra, N., and Stentz, A. (2006) Market-based multirobot coordination: A survey and analysis, *Proceedings of the IEEE*, **94**(7), 1257–1270.

Diestel, R. (2005) *Graph Theory*, volume 173 of *Graduate Texts in Mathematics*, second edition, Springer, ISBN 978-3-540-26182-7.

Dimarogonas, D. V. and Kyriakopoulos, K. J. (2007) On the rendezvous problem for multiple nonholonomic agents, *IEEE Transactions on Automatic Control*, **52**(5), 916–922.

Drezner, Z., (editor) (1995) *Facility Location: A Survey of Applications and Methods*, Series in Operations Research, Springer, ISBN 0-387-94545-8.

Drezner, Z. and Hamacher, H. W., (editors) (2001) *Facility Location: Applications and Theory*, Springer, ISBN 3540421726.

Du, Q., Faber, V., and Gunzburger, M. (1999) Centroidal Voronoi tessellations: Applications and algorithms, *SIAM Review*, **41**(4), 637–676.

Dubins, L. E. (1957) On curves of minimal length with a constraint on average curvature and with prescribed initial and terminal positions and tangents, *American Journal of Mathematics*, **79**, 497–516.

Dullerud, G. E. and Paganini, F. (2000) *A Course in Robust Control Theory*, number 36 in Texts in Applied Mathematics, Springer, ISBN 978-0-387-98945-7.

Dunbar, W. B. and Murray, R. M. (2006) Distributed receding horizon control for multi-vehicle formation stabilization, *Automatica*, **42**(4), 549–558.

Durrett, R. (2006) *Random Graph Dynamics*, Series in Statistical and Probabilistic Mathematics, Cambridge University Press, ISBN 0521866561.

Dynia, M., Kutylowski, J., Meyer auf der Heide, F., and Schindelhauer, C. (2006) Smart robot teams exploring sparse trees, in *International Symposium of Mathematical Foundations of Computer Science*, Stará Lesná, Slovakia.

Efrat, A. and Har-Peled, S. (2006) Guarding galleries and terrains, *Information Processing Letters*, **100**(6), 238–245.

Eidenbenz, S., Stamm, C., and Widmayer, P. (2001) Inapproximability results for guarding polygons and terrains, *Algorithmica*, **31**(1), 79–113.

Emerson, A. E. (1994) Temporal and modal logic, in *Handbook of Theoretical Computer Science, Vol. B: Formal Models and Semantics*, J. van Leeuwen, editor, pages 997–1072, MIT Press, ISBN 0262720159.

Fagnani, F., Johansson, K. H., Speranzon, A., and Zampieri, S. (2004) On multi-vehicle rendezvous under quantized communication, in *Mathematical Theory of Networks and Systems*, Leuven, Belgium, Electronic Proceedings.

Fagnani, F. and Zampieri, S. (2009) Average consensus with packet drop communication, *SIAM Journal on Control and Optimization*, **48**(1), 102–133.

Fainekos, G. E., Kress-Gazit, H., and Pappas, G. J. (2005) Temporal logic motion planning for mobile robots, in *IEEE International Conference on Robotics and Automation*, pages 2032–2037, Barcelona, Spain.

Fang, L. and Antsaklis, P. J. (2008) Asynchronous consensus protocols using nonlinear paracontractions theory, *IEEE Transactions on Automatic Control*, **53**, 2351–2355.

Fekete, S. P., Mitchell, J. S. B., and Beurer, K. (2005) On the continuous Fermat–Weber problem, *Operations Research*, **53**(1), 61 – 76.

Ferrari-Trecate, G., Buffa, A., and Gati, M. (2006) Analysis of coordination in multi-agent systems through partial difference equations, *IEEE Transactions on Automatic Control*, **51**(6), 1058–1063.

Fiedler, M. (1986) *Special Matrices and their Applications in Numerical Mathematics*, Martinus Nijhoff Publishers, ISBN 9024729572.

Flocchini, P., Prencipe, G., Santoro, N., and Widmayer, P. (1999) Hard tasks for weak robots: The role of common knowledge in pattern formation by autonomous mobile robots, in *ISAAC 1999, 10th International*

Symposium on Algorithm and Computation (Chennai, India), A. Aggarwal and C. P. Rangan, editors, volume 1741 of *Lecture Notes in Computer Science*, pages 93–102, Springer, ISBN 3540669167.

— (2005) Gathering of asynchronous oblivious robots with limited visibility, *Theoretical Computer Science*, **337**(1-3), 147–168.

Fraigniaud, P., Gąsieniec, L., Kowalski, D. R., and Pelc, A. (2004) Collective tree exploration, in *LATIN 2004: Theoretical Informatics*, M. Farach-Colton, editor, volume 2976 of *Lecture Notes in Computer Science*, pages 141–151, Springer, ISBN 3540212582.

Francis, B. A. (2006) Distributed control of autonomous mobile robots, course Notes, Version 1.01, University of Toronto, Canada.

Frazzoli, E. and Bullo, F. (2004) Decentralized algorithms for vehicle routing in a stochastic time-varying environment, in *IEEE Conference on Decision and Control*, pages 3357–3363, Paradise Island, Bahamas.

Gallager, R. G. (1968) *Information Theory and Reliable Communication*, John Wiley, ISBN 0471290483.

Gallager, R. G., Humblet, P. A., and Spira, P. M. (1983) A distributed algorithm for minimum-weight spanning trees, *ACM Transactions on Programming Languages and Systems*, **5**(1), 66–77.

Ganguli, A., Cortés, J., and Bullo, F. (2007) Distributed coverage of nonconvex environments, in *Networked Sensing Information and Control (Proceedings of the NSF Workshop on Future Directions in Systems Research for Networked Sensing, May 2006, Boston, MA)*, V. Saligrama, editor, pages 289–305, Lecture Notes in Control and Information Sciences, Springer, ISBN 0387688439.

— (2009) Multirobot rendezvous with visibility sensors in nonconvex environments, *IEEE Transactions on Robotics*, **25**(2), 340–352.

Gao, C., Cortés, J., and Bullo, F. (2008) Notes on averaging over acyclic digraphs and discrete coverage control, *Automatica*, **44**(8), 2120–2127.

Gazi, V. and Passino, K. M. (2003) Stability analysis of swarms, *IEEE Transactions on Automatic Control*, **48**(4), 692–697.

Gerkey, B. P. and Mataric, M. J. (2004) A formal analysis and taxonomy of task allocation in multi-robot systems, *International Journal of Robotics Research*, **23**(9), 939–954.

Ghosh, S. K. (1987) Approximation algorithms for Art Gallery Problems, in *Proceedings of the Canadian Information Processing Society*, pages 429–434.

Godsil, C. D. and Royle, G. F. (2001) *Algebraic Graph Theory*, volume 207 of *Graduate Texts in Mathematics*, Springer, ISBN 0387952411.

Godwin, M. F., Spry, S., and Hedrick, J. K. (2006) Distributed collaboration with limited communication using mission state estimates, in *American Control Conference*, pages 2040–2046, Minneapolis, MN.

Goebel, R., Hespanha, J. P., Teel, A. R., Cai, C., and Sanfelice, R. G. (2004) Hybrid systems: generalized solutions and robust stability, in *IFAC Symposium on Nonlinear Control Systems*, pages 1–12, Stuttgart, Germany.

Goodman, J. E. and O'Rourke, J., (editors) (2004) *Handbook of Discrete and Computational Geometry*, second edition, CRC Press, ISBN 1584883014.

Graham, R. and Cortés, J. (2009) Asymptotic optimality of multicenter Voronoi configurations for random field estimation, *IEEE Transactions on Automatic Control*, **54**(1), 153–158.

Gray, R. M. and Neuhoff, D. L. (1998) Quantization, *IEEE Transactions on Information Theory*, **44**(6), 2325–2383, Commemorative Issue 1948-1998.

Gruber, P. M. (1983) Approximation of convex bodies, in *Convexity and its Applications*, P. M. Gruber and J. M. Willis, editors, pages 131–162, Birkhäuser, ISBN 3764313846.

Guckenheimer, J. and Holmes, P. (1990) *Nonlinear Oscillations, Dynamical Systems, and Bifurcations of Vector Fields*, volume 42 of *Applied Mathematical Sciences*, Springer, ISBN 0387908196.

Gueron, S. and Levin, S. A. (1993) Self-organization of front patterns in large wildebeest herds, *Journal of Theoretical Biology*, **165**, 541–552.

Gupta, P. and Kumar, P. R. (2000) The capacity of wireless networks, *IEEE Transactions on Information Theory*, **46**(2), 388–404.

Gupta, V., Chung, T. H., Hassibi, B., and Murray, R. M. (2006a) On a stochastic sensor selection algorithm with applications in sensor scheduling and sensor coverage, *Automatica*, **42**(2), 251–260.

Gupta, V., Langbort, C., and Murray, R. M. (2006b) On the robustness of distributed algorithms, in *IEEE Conference on Decision and Control*, pages 3473–3478, San Diego, CA.

Hatano, Y. and Mesbahi, M. (2005) Agreement over random networks, *IEEE Transactions on Automatic Control*, **50**(11), 1867–1872.

Hayes, J., McJunkin, M., and Kosecká, J. (2003) Communication enhanced navigation strategies for teams of mobile agents, in *IEEE/RSJ International Conference on Intelligent Robots & Systems*, pages 2285–2290, Las Vegas, NV.

Hegselmann, R. and Krause, U. (2002) Opinion dynamics and bounded confidence models, analysis, and simulations, *Journal of Artificial Societies and Social Simulation*, **5**(3).

Hendrickx, J. M. (2008) *Graphs and Networks for the Analysis of Autonomous Agent Systems*, Ph.D. thesis, Université Catholique de Louvain, Belgium.

Hendrickx, J. M., Anderson, B. D. O., Delvenne, J.-C., and Blondel, V. D. (2007) Directed graphs for the analysis of rigidity and persistence in autonomous agents systems, *International Journal on Robust and Nonlinear Control*, **17**(10), 960–981.

Hernández-Peñalver, G. (1994) Controlling guards, in *Canadian Conference on Computational Geometry*, pages 387–392, Saskatoon, Canada.

Hibbeler, R. C. (2006) *Engineering Mechanics: Statics & Dynamics*, 11th edition, Prentice Hall, ISBN 0132215098.

Horn, R. A. and Johnson, C. R. (1985) *Matrix Analysis*, Cambridge University Press, ISBN 0521386322.

Howard, A., Matarić, M. J., and Sukhatme, G. S. (2002) Mobile sensor network deployment using potential fields: A distributed scalable solution to the area coverage problem, in *International Conference on Distributed Autonomous Robotic Systems*, pages 299–308, Fukuoka, Japan.

Howard, A., Parker, L. E., and Sukhatme, G. S. (2006) Experiments with a large heterogeneous mobile robot team: Exploration, mapping, deployment, and detection, *International Journal of Robotics Research*, **25**(5-6), 431–447.

Hu, J., Prandini, M., and Tomlin, C. (2007) Conjugate points in formation constrained optimal multi-agent coordination: A case study, *SIAM Journal on Control and Optimization*, **45**(6), 2119–2137.

Hussein, I. I. and Stipanovìc, D. M. (2007) Effective coverage control for mobile sensor networks with guaranteed collision avoidance, *IEEE Transactions on Control Systems Technology*, **15**(4), 642–657.

Huygens, C. (1673) *Horologium Oscillatorium*, Paris, France.

Igarashi, Y., Hatanaka, T., Fujita, M., and Spong, M. W. (2007) Passivity-based 3D attitude coordination: Convergence and connectivity, in *IEEE Conference on Decision and Control*, pages 2558–2565, New Orleans, LA.

Jadbabaie, A., Lin, J., and Morse, A. S. (2003) Coordination of groups of mobile autonomous agents using nearest neighbor rules, *IEEE Transactions on Automatic Control*, **48**(6), 988–1001.

Jadbabaie, A., Motee, N., and Barahona, M. (2004) On the stability of the Kuramoto model of coupled nonlinear oscillators, in *American Control Conference*, pages 4296–4301, Boston, MA.

Jaromczyk, J. W. and Toussaint, G. T. (1992) Relative neighborhood graphs and their relatives, *Proceedings of the IEEE*, **80**(9), 1502–1517.

Ji, M. and Egerstedt, M. (2007) Distributed control of multiagent systems while preserving connectedness, *IEEE Transactions on Robotics*, **23**(4), 693–703.

Jiang, Z.-P. and Wang, Y. (2001) Input-to-state stability for discrete-time nonlinear systems, *Automatica*, **37**(6), 857–869.

Johansson, K. J., Egerstedt, M., Lygeros, J., and Sastry, S. S. (1999) On the regularization of Zeno hybrid automata, *Systems & Control Letters*, **38**(3), 141–150.

Justh, E. W. and Krishnaprasad, P. S. (2004) Equilibria and steering laws for planar formations, *Systems & Control Letters*, **52**(1), 25–38.

— (2006) Steering laws for motion camouflage, *Proceedings of the Royal Society A: Mathematical, Physical and Engineering Sciences*, **462**(2076), 3629–3643.

Kang, K., Yan, J., and Bitmead, R. R. (2006) Communication resources for disturbance rejection in coordinated vehicle control, in *IEEE Conference on Decision and Control and European Control Conference*, pages 5730–5735, Seville, Spain.

Kashyap, A., Başar, T., and Srikant, R. (2007) Quantized consensus, *Automatica*, **43**(7), 1192–1203.

Khalil, H. K. (2002) *Nonlinear Systems*, third edition, Prentice Hall, ISBN 0130673897.

Kim, Y. and Mesbahi, M. (2006) On maximizing the second smallest eigenvalue of a state-dependent graph Laplacian, *IEEE Transactions on Automatic Control*, **51**(1), 116–120.

Klavins, E. (2003) Communication complexity of multi-robot systems, in *Algorithmic Foundations of Robotics V*, J.-D. Boissonnat, J. W. Burdick, K. Goldberg, and S. Hutchinson, editors, volume 7 of *Tracts in Advanced Robotics*, Springer, Berlin Heidelberg, ISBN 3540404767.

Klavins, E., Ghrist, R., and Lipsky, D. (2006) A grammatical approach to self-organizing robotic systems, *IEEE Transactions on Automatic Control*, **51**(6), 949–962.

Klavins, E. and Murray, R. M. (2004) Distributed algorithms for cooperative control, *IEEE Pervasive Computing*, **3**(1), 56–65.

Korte, B. and Vygen, J. (2005) *Combinatorial Optimization: Theory and Algorithms*, volume 21 of *Algorithmics and Combinatorics*, third edition, Springer, ISBN 3540256849.

Krasovskiĭ, N. N. (1963) *Stability of motion. Applications of Lyapunov's second method to differential systems and equations with delay*, Stanford University Press, translated by J. L. Brenner.

Krick, L. (2007) *Application of Graph Rigidity in Formation Control of Multi-Robot Networks*, Master's thesis, University of Toronto, Canada.

Kuramoto, Y. (1975) Self-entrainment of a population of coupled nonlinear oscillators, in *International Symposium on Mathematical Problems in Theoretical Physics*, H. Araki, editor, volume 39 of *Lecture Notes in Physics*, pages 420–422, Springer, ISBN 978-3-540-07174-7.

— (1984) *Chemical oscillations, waves, and turbulence*, Springer, ISBN 0387133224.

Kwok, A. and Martínez, S. (2009) Deployment algorithms for a power-constrained mobile sensor network, *International Journal on Robust and Nonlinear Control*, to appear.

Lafferriere, G., Williams, A., Caughman, J., and Veerman, J. J. P. (2005) Decentralized control of vehicle formations, *Systems & Control Letters*, **54**(9), 899–910.

Landau, H. J. and Odlyzko, A. M. (1981) Bounds for eigenvalues of certain stochastic matrices, *Linear Algebra and its Applications*, **38**, 5–15.

Langbort, C. and Gupta, V. (2009) Minimal interconnection topology in distributed control, *SIAM Journal on Control and Optimization*, **48**(1), 397–413.

Langetepe, E. and Zachmann, G. (2006) *Geometric Data Structures for Computer Graphics*, A. K. Peters, ISBN 1568812353.

Lanthier, M., Nussbaum, D., and Wang, T.-J. (2005) Calculating the meeting point of scattered robots on weighted terrain surfaces, in *Computing: The Australasian Theory Symposium (CATS)*, volume 27, pages 107–118, Newcastle, Australia.

LaSalle, J. P. (1960) Some extensions of Liapunov's second method, *IRE Trans. Circuit Theory*, **CT-7**, 520–527.

— (1986) *The Stability and Control of Discrete Processes*, volume 62 of *Applied Mathematical Sciences*, Springer, ISBN 0387964118.

Laventall, K. and Cortés, J. (2009) Coverage control by robotic networks with limited-range anisotropic sensory, *International Journal of Control*, **82**, to appear.

Lee, D. and Spong, M. W. (2007) Stable flocking of multiple inertial agents on balanced graphs, *IEEE Transactions on Automatic Control*, **52**(8), 1469–1475.

Lee, D. T. and Lin, A. K. (1986) Computational complexity of art gallery problems, *IEEE Transactions on Information Theory*, **32**(2), 276–282.

Lekien, F. and Leonard, N. E. (2009) Non-uniform coverage and cartograms, *SIAM Journal on Control and Optimization*, **48**(1), 351–372.

Li, X.-Y. (2003) Algorithmic, geometric and graphs issues in wireless networks, *Wireless Communications and Mobile Computing*, **3**(2), 119–140.

Lin, J., Morse, A. S., and Anderson, B. D. O. (2007a) The multi-agent rendezvous problem. Part 1: The synchronous case, *SIAM Journal on Control and Optimization*, **46**(6), 2096–2119.

— (2007b) The multi-agent rendezvous problem. Part 2: The asynchronous case, *SIAM Journal on Control and Optimization*, **46**(6), 2120–2147.

Lin, Z. (2005) *Coupled Dynamic Systems: From Structure Towards Stability and Stabilizability*, Ph.D. thesis, University of Toronto, Canada.

Lin, Z., Broucke, M., and Francis, B. (2004) Local control strategies for groups of mobile autonomous agents, *IEEE Transactions on Automatic Control*, **49**(4), 622–629.

Lin, Z., Francis, B., and Maggiore, M. (2005) Necessary and sufficient graphical conditions for formation control of unicycles, *IEEE Transactions on Automatic Control*, **50**(1), 121–127.

— (2007c) State agreement for continuous-time coupled nonlinear systems, *SIAM Journal on Control and Optimization*, **46**(1), 288–307.

Lloyd, E. L., Liu, R., Marathe, M. V., Ramanathan, R., and Ravi, S. S. (2005) Algorithmic aspects of topology control problems for ad hoc networks, *Mobile Networks and Applications*, **10**(1-2), 19–34.

Lloyd, S. P. (1982) Least squares quantization in PCM, *IEEE Transactions on Information Theory*, **28**(2), 129–137, presented as Bell Laboratory Technical Memorandum at a 1957 Institute for Mathematical Statistics meeting.

Lorenz, J. (2007) *Repeated Averaging and Bounded Confidence – Modeling, Analysis and Simulation of Continuous Opinion Dynamics*, Ph.D. thesis, University of Bremen, Germany, available at `http://nbn-resolving.de/urn:nbn:de:gbv:46-diss000106688`.

Lorenz, J. and Lorenz, D. A. (2008) On conditions for convergence to consensus, available at `http://arxiv.org/abs/0803.2211`.

Lovász, L. (1993) Random walks on graphs: A survey, in *Combinatorics: Paul Erdös is Eighty*, T. S. D. Miklós, V. T. Sós, editor, volume 2, pages 353–398, János Bolyai Mathematical Society, ISBN 9638022744.

Lumelsky, V. J. and Harinarayan, K. R. (1997) Decentralized motion planning for multiple mobile robots: The cocktail party model, *Autonomous Robots*, **4**(1), 121–135.

Lygeros, J., Johansson, K. H., Simić, S. N., Zhang, J., and Sastry, S. S. (2003) Dynamical properties of hybrid automata, *IEEE Transactions on Automatic Control*, **48**(1), 2–17.

Lynch, N. A. (1997) *Distributed Algorithms*, Morgan Kaufmann, ISBN 1558603484.

Lynch, N. A., Segala, R., and Vaandrager, F. (2003) Hybrid I/O automata, *Information and Computation*, **185**(1), 105–157.

Marshall, J. A., Broucke, M. E., and Francis, B. A. (2004) Formations of vehicles in cyclic pursuit, *IEEE Transactions on Automatic Control*, **49**(11), 1963–1974.

Marthaler, D. and Bertozzi, A. L. (2003) Tracking environmental level sets with autonomous vehicles, in *Recent Developments in Cooperative Control and Optimization*, S. Butenko, R. Murphey, and P. M. Pardalos, editors, pages 317–330, Kluwer Academic Publishers, ISBN 1402076444.

Martínez, S. (2009a) Distributed interpolation schemes for field estimation by mobile sensor networks, *IEEE Transactions on Control Systems Technology*, to appear.

— (2009b) Practical multiagent rendezvous through modified circumcenter algorithms, *Automatica*, to appear.

Martínez, S. and Bullo, F. (2006) Optimal sensor placement and motion coordination for target tracking, *Automatica*, **42**(4), 661–668.

Martínez, S., Bullo, F., Cortés, J., and Frazzoli, E. (2007a) On synchronous robotic networks – Part I: Models, tasks and complexity, *IEEE Transactions on Automatic Control*, **52**(12), 2199–2213.

— (2007b) On synchronous robotic networks – Part II: Time complexity of rendezvous and deployment algorithms, *IEEE Transactions on Automatic Control*, **52**(12), 2214–2226.

Martínez, S., Cortés, J., and Bullo, F. (2007c) Motion coordination with distributed information, *IEEE Control Systems Magazine*, **27**(4), 75–88.

McLure, D. E. and Vitale, R. A. (1975) Polygonal approximation of plane convex bodies, *Journal of Mathematical Analysis and Applications*, **51**(2), 326–358.

Meester, R. and Roy, R. (2008) *Continuum Percolation*, Cambridge University Press, ISBN 0521062500.

Mehaute, A. L., Laurent, P. J., and Schumaker, L. L., (editors) (1993) *Curves and Surfaces in Geometric Design*, A. K. Peters, ISBN 1568810393.

Merris, R. (1994) Laplacian matrices of a graph: A survey, *Linear Algebra its Applications*, **197**, 143–176.

Mesbahi, M. (2005) On state-dependent dynamic graphs and their controllability properties, *IEEE Transactions on Automatic Control*, **50**(3), 387–392.

Meyer, C. D. (2001) *Matrix Analysis and Applied Linear Algebra*, SIAM, ISBN 0898714540.

Meyn, S. and Tweedie, R. (1999) *Markov Chains and Stochastic Stability*, Springer, ISBN 3540198326.

Miller, M. B. and Bassler, B. L. (2001) Quorum sensing in bacteria, *Annual Review of Microbiology*, **55**, 165–199.

Mirollo, R. E. and Strogatz, S. H. (1990) Synchronization of pulse-coupled biological oscillators, *SIAM Journal on Applied Mathematics*, **50**(6), 1645–1662.

Mitchell, J. S. B. (1997) Shortest paths and networks, in *Handbook of Discrete and Computational Geometry*, J. E. Goodman and J. O'Rourke, editors, chapter 24, pages 445–466, CRC Press, ISBN 0849385245.

Mohar, B. (1991) The Laplacian spectrum of graphs, in *Graph Theory, Combinatorics, and Applications*, Y. Alavi, G. Chartrand, O. R. Oellermann, and A. J. Schwenk, editors, volume 2, pages 871–898, John Wiley, ISBN 0471532452.

Moore, B. J. and Passino, K. M. (2007) Distributed task assignment for mobile agents, *IEEE Transactions on Automatic Control*, **52**(4), 749–753.

Moreau, L. (2003) Time-dependent unidirectional communication in multi-agent systems, available at `http://arxiv.org/abs/math/0306426`.

— (2004) Stability of continuous-time distributed consensus algorithms, available at `http://arxiv.org/abs/math/0409010`.

— (2005) Stability of multiagent systems with time-dependent communication links, *IEEE Transactions on Automatic Control*, **50**(2), 169–182.

Moses, Y. and Tennenholtz, M. (1995) Artificial social systems, *Computers and AI*, **14**(6), 533–562.

Moshtagh, N. and Jadbabaie, A. (2007) Distributed geodesic control laws for flocking of nonholonomic agents, *IEEE Transactions on Automatic Control*, **52**(4), 681–686.

Nijmeijer, H. (2001) A dynamical control view on synchronization, *Physica D*, **154**(3-4), 219–228.

Ögren, P., Fiorelli, E., and Leonard, N. E. (2004) Cooperative control of mobile sensor networks: Adaptive gradient climbing in a distributed environment, *IEEE Transactions on Automatic Control*, **49**(8), 1292–1302.

Oh, S., Schenato, L., Chen, P., and Sastry, S. S. (2007) Tracking and coordination of multiple agents using sensor networks: system design, algorithms and experiments, *Proceedings of the IEEE*, **95**(1), 163–187.

Okabe, A., Boots, B., Sugihara, K., and Chiu, S. N. (2000) *Spatial Tessellations: Concepts and Applications of Voronoi Diagrams*, second edition, Wiley Series in Probability and Statistics, John Wiley, ISBN 0471986356.

Okubo, A. (1986) Dynamical aspects of animal grouping: swarms, schools, flocks and herds, *Advances in Biophysics*, **22**, 1–94.

Olfati-Saber, R. (2005) Ultrafast consensus in small world networks, in *American Control Conference*, pages 2371–2378, Portland, OR.

— (2006) Flocking for multi-agent dynamic systems: Algorithms and theory, *IEEE Transactions on Automatic Control*, **51**(3), 401–420.

Olfati-Saber, R., Fax, J. A., and Murray, R. M. (2007) Consensus and cooperation in networked multi-agent systems, *Proceedings of the IEEE*, **95**(1), 215–233.

Olfati-Saber, R., Franco, E., Frazzoli, E., and Shamma, J. S. (2006) Belief consensus and distributed hypothesis testing in sensor networks, in *Network Embedded Sensing and Control. (Proceedings of NESC'05 Worskhop)*, P. Antsaklis and P. Tabuada, editors, volume 331 of *Lecture Notes in Control and Information Sciences*, pages 169–182, Springer, ISBN 3540327940.

Olfati-Saber, R. and Murray, R. M. (2002) Graph rigidity and distributed formation stabilization of multi-vehicle systems, in *IEEE Conference on Decision and Control*, pages 2965–2971, Las Vegas, NV.

— (2004) Consensus problems in networks of agents with switching topology and time-delays, *IEEE Transactions on Automatic Control*, **49**(9), 1520–1533.

Olshevsky, A. and Tsitsiklis, J. N. (2009) Convergence speed in distributed consensus and averaging, *SIAM Journal on Control and Optimization*, **48**(1), 33–55.

O'Rourke, J. (2000) *Computational Geometry in C*, Cambridge University Press, ISBN 0521649765.

Paley, D. A., Leonard, N. E., Sepulchre, R., Grunbaum, D., and Parrish, J. K. (2007) Oscillator models and collective motion, *IEEE Control Systems Magazine*, **27**(4), 89–105.

Pallottino, L., Scordio, V. G., Frazzoli, E., and Bicchi, A. (2007) Decentralized cooperative policy for conflict resolution in multi-vehicle systems, *IEEE Transactions on Robotics*, **23**(6), 1170–1183.

Papachristodoulou, A. and Jadbabaie, A. (2006) Synchronization in oscillator networks with heterogeneous delays, switching topologies and nonlinear dynamics, in *IEEE Conference on Decision and Control*, pages 4307–4312, San Diego, CA.

Parhami, B. (1999) *Introduction to Parallel Processing: Algorithms and Architectures*, Plenum Series in Computer Science, Springer, ISBN 0306459701.

Parrish, J. K., Viscido, S. V., and Grunbaum, D. (2002) Self-organized fish schools: an examination of emergent properties, *Biological Bulletin*, **202**, 296–305.

Passino, K. M. (2004) *Biomimicry for Optimization, Control, and Automation*, Springer, ISBN 1852338040.

Patterson, S., Bamieh, B., and Abbadi, A. E. (2007) Distributed average consensus with stochastic communication failures, in *IEEE Conference on Decision and Control*, pages 4215–4220, New Orleans, LA.

Pavone, M. and Frazzoli, E. (2007) Decentralized policies for geometric pattern formation and path coverage, *ASME Journal on Dynamic Systems, Measurement, and Control*, **129**(5), 633–643.

Pavone, M., Frazzoli, E., and Bullo, F. (2007) Decentralized algorithms for stochastic and dynamic vehicle routing with general target distribution, in *IEEE Conference on Decision and Control*, pages 4869–4874, New Orleans, LA.

— (2008) Distributed policies for equitable partitioning: Theory and applications, in *IEEE Conference on Decision and Control*, pages 4191–4197, Cancún, México.

Payton, D., Daily, M., Estowski, R., Howard, M., and Lee, C. (2001) Pheromone robotics, *Autonomous Robots*, **11**(3), 319–324.

Pearl, J. (1988) *Probabilistic Reasoning in Intelligent Systems: Networks of Plausible Inference*, Morgan Kaufmann, ISBN 1558604790.

Peleg, D. (2000) *Distributed Computing. A Locality-Sensitive Approach*, Monographs on Discrete Mathematics and Applications, SIAM, ISBN 0898714648.

Penrose, M. (2003) *Random Geometric Graphs*, Oxford Studies in Probability, Oxford University Press, ISBN 0198506260.

Picci, G. and Taylor, T. (2007) Almost sure convergence of random gossip algorithms, in *IEEE Conference on Decision and Control*, pages 282–287, New Orleans, LA.

Pimenta, L. C. A., Kumar, V., Mesquita, R. C., and Pereira, G. A. S. (2008) Sensing and coverage for a network of heterogeneous robots, in *IEEE Conference on Decision and Control*, pages 3947–3952, Cancún, México.

Pinciu, V. (2003) A coloring algorithm for finding connected guards in art galleries, in *Discrete Mathematical and Theoretical Computer Science*, volume 2731/2003 of *Lecture Notes in Computer Science*, pages 257–264, Springer.

Poduri, S. and Sukhatme, G. S. (2004) Constrained coverage for mobile sensor networks, in *IEEE International Conference on Robotics and Automation*, pages 165–172, New Orleans, LA.

Porfiri, M. and Stilwell, D. J. (2007) Consensus seeking over random weighted directed graphs, *IEEE Transactions on Automatic Control*, **52**(9), 1767–1773.

Preparata, F. P. and Shamos, M. I. (1993) *Computational Geometry: An Introduction*, Springer, ISBN 0387961313.

Radke, J. D. (1988) On the shape of a set of points, in *Computational morphology. A computational geometric approach to the analysis of form.*, G. T. Toussaint, editor, pages 105–136, North-Holland, ISBN 0-444-70467-1.

Rao, B. S. Y. and Durrant-Whyte, H. F. (1993) A decentralized Bayesian algorithm for identification of tracked targets, *IEEE Transactions on Systems, Man, & Cybernetics*, **23**(6), 1683–1698.

Rathinam, S., Sengupta, R., and Darbha, S. (2007) A resource allocation algorithm for multi-vehicle systems with non holonomic constraints, *IEEE Transactions on Automation Sciences and Engineering*, **4**(1), 98–104.

Reeds, J. A. and Shepp, L. A. (1990) Optimal paths for a car that goes both forwards and backwards, *Pacific Journal of Mathematics*, **145**(2), 367–393.

Rekleitis, I. M., Dudek, G., and Milios, E. (2001) Multi-robot collaboration for robust exploration, *Annals of Mathematics and Artificial Intelligence*, **31**(1-4), 7–40.

Ren, W. and Beard, R. W. (2005) Consensus seeking in multi-agent systems under dynamically changing interaction topologies, *IEEE Transactions on Automatic Control*, **50**(5), 655–661.

— (2008) *Distributed Consensus in Multi-vehicle Cooperative Control*, Communications and Control Engineering, Springer, ISBN 978-1-84800-014-8.

Ren, W., Beard, R. W., and Atkins, E. M. (2007) Information consensus in multivehicle cooperative control: Collective group behavior through local interaction, *IEEE Control Systems Magazine*, **27**(2), 71–82.

Robert, J.-M. and Toussaint, G. T. (1990) Computational geometry and facility location, in *International Conference on Operations Research and Management Science*, pages 1–19, Manila, The Philippines.

Roy, N. and Dudek, G. (2001) Collaborative exploration and rendezvous: Algorithms, performance bounds, and observations, *Autonomous Robots*, **11**(2), 117–136.

Sack, J. R. and Urrutia, J., (editors) (2000) *Handbook of Computational Geometry*, North-Holland, ISBN 0444825371.

Sanfelice, R. G., Goebel, R., and Teel, A. R. (2007) Invariance principles for hybrid systems with connections to detectability and asymptotic stability, *IEEE Transactions on Automatic Control*, **52**(12), 2282–2297.

Santi, P. (2005) *Topology Control in Wireless Ad Hoc and Sensor Networks*, John Wiley, ISBN 0470094532.

Santoro, N. (2001) Distributed computations by autonomous mobile robots, in *SOFSEM 2001: Conference on Current Trends in Theory and Practice of Informatics (Piestany, Slovak Republic)*, L. Pacholski and P. Ruzicka, editors, volume 2234 of *Lecture Notes in Computer Science*, pages 110–115, Springer, ISBN 3-540-42912-3.

Sarlette, A. (2009) *Geometry and Symmetries in Coordination Control*, Ph.D. thesis, University of Liège, Belgium.

Sarlette, A. and Sepulchre, R. (2009) Consensus optimization on manifolds, *SIAM Journal on Control and Optimization*, **48**(1), 56–76.

Savkin, A. (2004) Coordinated collective motion of groups of autonomous mobile robots: Analysis of Vicsek's model, *IEEE Transactions on Automatic Control*, **49**(6), 981–982.

Savla, K., Bullo, F., and Frazzoli, E. (2009a) Traveling Salesperson Problems for a double integrator, *IEEE Transactions on Automatic Control*, (Submitted Nov. 2006) to appear.

Savla, K., Frazzoli, E., and Bullo, F. (2008) Traveling Salesperson Problems for the Dubins vehicle, *IEEE Transactions on Automatic Control*, **53**(6), 1378–1391.

Savla, K., Notarstefano, G., and Bullo, F. (2009b) Maintaining limited-range connectivity among second-order agents, *SIAM Journal on Control and Optimization*, **48**(1), 187–205.

Scardovi, L., Sarlette, A., and Sepulchre, R. (2007) Synchronization and balancing on the N-torus, *Systems & Control Letters*, **56**(5), 335–341.

Schultz, A. C. and Parker, L. E., (editors) (2002) *Multi-Robot Systems: From Swarms to Intelligent Automata*, Kluwer Academic Publishers, ISBN 1402006799, Proceedings from the 2002 NRL Workshop on Multi-Robot Systems.

Schumacher, C., Chandler, P. R., Rasmussen, S. J., and Walker, D. (2003) Task allocation for wide area search munitions with variable path length, in *American Control Conference*, pages 3472–3477, Denver, CO.

Schuresko, M. and Cortés, J. (2007) Safe graph rearrangements for distributed connectivity of robotic networks, in *IEEE Conference on Decision and Control*, pages 4602–4607, New Orleans, LA.

Schwager, M., Bullo, F., Skelly, D., and Rus, D. (2008) A ladybug exploration strategy for distributed adaptive coverage control, in *IEEE International Conference on Robotics and Automation*, pages 2346–2353, Pasadena, CA.

Schwager, M., Rus, D., and Slotine, J. J. (2009) Decentralized, adaptive coverage control for networked robots, *International Journal of Robotics Research*, **28**(3), 357–375.

Seeley, T. D. and Buhrman, S. C. (1999) Group decision-making in swarms of honey bees, *Behavioral Ecology and Sociobiology*, **45**, 19–31.

Seneta, E. (1981) *Non-negative Matrices and Markov Chains*, second edition, Springer, ISBN 0387297650.

Sepulchre, R., Paley, D. A., and Leonard, N. E. (2007) Stabilization of planar collective motion: All-to-all communication, *IEEE Transactions on Automatic Control*, **52**(5), 811–824.

Sharma, V., Savchenko, M., Frazzoli, E., and Voulgaris, P. (2007) Transfer time complexity of conflict-free vehicle routing with no communications, *International Journal of Robotics Research*, **26**(3), 255–272.

Sibson, R. (1981) A brief description of natural neighbour interpolation, in *Interpreting Multivariate Data*, V. Barnett, editor, pages 21–36, John Wiley, ISBN 0471280399.

Simmons, R., Apfelbaum, D., Fox, D., Goldman, R., Haigh, K., Musliner, D., Pelican, M., and Thrun, S. (2000) Coordinated deployment of multiple heterogenous robots, in *IEEE/RSJ International Conference on Intelligent Robots & Systems*, pages 2254–2260, Takamatsu, Japan.

Sinclair, A. R. (1977) *The African Buffalo, A Study of Resource Limitation of Population*, The University of Chicago Press.

Sipser, M. (2005) *Introduction to the Theory of Computation*, second edition, Course Technology, ISBN 0534950973.

Skyum, S. (1991) A simple algorithm for computing the smallest enclosing circle, *Information Processing Letters*, **37**(3), 121–125.

Smith, R. S. and Hadaegh, F. Y. (2007) Closed-loop dynamics of cooperative vehicle formations with parallel estimators and communication, *IEEE Transactions on Automatic Control*, **52**(8), 1404–1414.

Smith, S. L., Broucke, M. E., and Francis, B. A. (2005) A hierarchical cyclic pursuit scheme for vehicle networks, *Automatica*, **41**(6), 1045–1053.

— (2007) Curve shortening and the rendezvous problem for mobile autonomous robots, *IEEE Transactions on Automatic Control*, **52**(6), 1154–1159.

Smith, S. L. and Bullo, F. (2009) Monotonic target assignment for robotic networks, *IEEE Transactions on Automatic Control*, **54**(10), (Submitted June 2007) to appear.

Sontag, E. D. (1998) *Mathematical Control Theory: Deterministic Finite Dimensional Systems*, volume 6 of *TAM*, second edition, Springer, ISBN 0387984895.

— (2008) Input to state stability: Basic concepts and results, in *Nonlinear and Optimal Control Theory*, P. Nistri and G. Stefani, editors, pages 163–220, Lecture Notes in Mathematics, Springer, ISBN 3540776443.

Spanos, D. P. and Murray, R. M. (2005) Motion planning with wireless network constraints, in *American Control Conference*, pages 87–92, Portland, OR.

Spanos, D. P., Olfati-Saber, R., and Murray, R. M. (2005) Approximate distributed Kalman filtering in sensor networks with quantifiable performance, in *Symposium on Information Processing of Sensor Networks*, pages 133–139, Los Angeles, CA.

Spong, M. W., Hutchinson, S., and Vidyasagar, M. (2006) *Robot Modeling and Control*, third edition, John Wiley, ISBN 0-471-64990-2.

Stewart, K. J. and Harcourt, A. H. (1994) Gorillas vocalizations during rest periods – signals of impending departure, *Behaviour*, **130**, 29–40.

Strogatz, S. H. (2000) From Kuramoto to Crawford: Exploring the onset of synchronization in populations of coupled oscillators, *Physica D*, **143**(1), 1–20.

— (2003) *SYNC: The emerging science of spontaneous order*, Hyperion, ISBN 0786868449.

Sundaram, S. and Hadjicostis, C. N. (2008) Distributed function calculation and consensus using linear iterative strategies, *IEEE Journal on Selected Areas in Communications*, **26**(4), 650–660.

Susca, S. (2007) *Distributed Boundary Estimation and Monitoring*, Ph.D. thesis, University of California at Santa Barbara, available at `http://ccdc.mee.ucsb.edu`.

Susca, S., Martínez, S., and Bullo, F. (2008) Monitoring environmental boundaries with a robotic sensor network, *IEEE Transactions on Control Systems Technology*, **16**(2), 288–296.

— (2009) Gradient algorithms for polygonal approximation of convex contours, *Automatica*, **45**(2), 510–516.

Suzuki, I. and Yamashita, M. (1999) Distributed anonymous mobile robots: Formation of geometric patterns, *SIAM Journal on Computing*, **28**(4), 1347–1363.

Tabuada, P., Pappas, G. J., and Lima, P. (2005) Motion feasibility of multi-agent formations, *IEEE Transactions on Robotics*, **21**(3), 387–392.

Tahbaz-Salehi, A. and Jadbabaie, A. (2006) A one-parameter family of distributed consensus algorithms with boundary: From shortest paths to mean hitting times, in *IEEE Conference on Decision and Control*, pages 4664–4669, San Diego, CA.

— (2007) Small world phenomenon, rapidly mixing markov chains, and average consensus algorithms, in *IEEE Conference on Decision and Control*, pages 276–281, New Orleans, LA.

— (2008) Consensus over random networks, *IEEE Transactions on Automatic Control*, **53**(3), 791–795.

Tang, Z. and Özgüner, Ü. (2005) Motion planning for multi-target surveillance with mobile sensor agents, *IEEE Transactions on Robotics*, **21**(5), 898–908.

Tanner, H. G., Jadbabaie, A., and Pappas, G. J. (2007) Flocking in fixed and switching networks, *IEEE Transactions on Automatic Control*, **52**(5), 863–868.

Tanner, H. G., Pappas, G. J., and Kumar, V. (2004) Leader-to-formation stability, *IEEE Transactions on Robotics and Automation*, **20**(3), 443–455.

Tel, G. (2001) *Introduction to Distributed Algorithms*, second edition, Cambridge University Press, ISBN 0521794838.

Toh, C.-K. (2001) *Ad Hoc Mobile Wireless Networks: Protocols and Systems*, Prentice Hall, ISBN 0130078174.

Triplett, B. I., Klein, D. J., and Morgansen, K. A. (2006) Discrete time Kuramoto models with delay, in *Network Embedded Sensing and Control. (Proceedings of NESC'05 Worskhop)*, P. J. Antsaklis and P. Tabuada, editors, volume 331 of *Lecture Notes in Control and Information Sciences*, pages 9–24, Springer, ISBN 3540327940.

Tse, D. and Viswanath, P. (2005) *Fundamentals of Wireless Communication*, Cambridge University Press, ISBN 0521845270.

Tsitsiklis, J. N. (1984) *Problems in Decentralized Decision Making and Computation*, Ph.D. thesis, Massachusetts Institute of Technology, available at `http://web.mit.edu/jnt/www/Papers/PhD-84-jnt.pdf`.

Tsitsiklis, J. N., Bertsekas, D. P., and Athans, M. (1986) Distributed asynchronous deterministic and stochastic gradient optimization algorithms, *IEEE Transactions on Automatic Control*, **31**(9), 803–812.

Tutuncu, R. H., Toh, K. C., and Todd, M. J. (2003) Solving semidefinite-quadratic-linear programs using SDPT3, *Mathematical Programming, Series B*, **95**, 189–217.

van der Schaft, A. J. and Schumacher, H. (2000) *An Introduction to Hybrid Dynamical Systems*, volume 251 of *Lecture Notes in Control and Information Sciences*, Springer, ISBN 1852332336.

Vazirani, V. V. (2001) *Approximation Algorithms*, Springer, ISBN 3540653678.

Vicsek, T., Czirók, A., Ben-Jacob, E., Cohen, I., and Shochet, O. (1995) Novel type of phase transition in a system of self-driven particles, *Physical Review Letters*, **75**(6-7), 1226–1229.

Wang, W. and Slotine, J.-J. E. (2006) A theoretical study of different leader roles in networks, *IEEE Transactions on Automatic Control*, **51**(7), 1156–1161.

Watton, A. and Kydon, D. W. (1969) Analytical aspects of the N-bug problem, *American Journal of Physics*, **37**(2), 220–221.

Whiteley, W. (1997) Rigidity and scene analysis, in *Handbook of Discrete and Computational Geometry*, J. E. Goodman and J. O'Rourke, editors, chapter 49, pages 893–916, CRC Press, ISBN 0849385245.

Wiener, N. (1958) *Nonlinear Problems in Random Theory*, MIT Press.

Winfree, A. T. (1980) *The Geometry of Biological Time*, Springer, ISBN 0387525289.

Wolfowitz, J. (1963) Product of indecomposable, aperiodic, stochastic matrices, *Proceedings of American Mathematical Society*, **14**(5), 733–737.

Wu, C. W. (2006) Synchronization and convergence of linear dynamics in random directed networks, *IEEE Transactions on Automatic Control*, **51**(7), 1207–1210.

Xiao, L., Boyd, S., and Lall, S. (2005) A scheme for robust distributed sensor fusion based on average consensus, in *Symposium on Information Processing of Sensor Networks*, pages 63–70, Los Angeles, CA.

Xue, F. and Kumar, P. R. (2004) The number of neighbors needed for connectivity of wireless networks, *Wireless Networks*, **10**(2), 169–181.

Yang, P., Freeman, R. A., and Lynch, K. M. (2008) Multi-agent coordination by decentralized estimation and control, *IEEE Transactions on Automatic Control*, **53**(11), 2480–2496.

Yu, C., Anderson, B. D. O., Dasgupta, S., and Fidan, B. (2009) Control of minimally persistent formations in the plane, *SIAM Journal on Control and Optimization*, **48**(1), 206–233.

Yu, J., LaValle, S. M., and Liberzon, D. (2008) Rendezvous without coordinates, in *IEEE Conference on Decision and Control*, pages 1803–1808, Cancún, México.

Zavlanos, M. M. and Pappas, G. J. (2005) Controlling connectivity of dynamic graphs, in *IEEE Conference on Decision and Control and European Control Conference*, pages 6388–6393, Seville, Spain.

— (2007a) Dynamic assignment in distributed motion planning with local information, in *American Control Conference*, pages 1173–1178, New York.

— (2007b) Potential fields for maintaining connectivity of mobile networks, *IEEE Transactions on Robotics*, **23**(4), 812–816.

Zhang, F. and Leonard, N. E. (2005) Generating contour plots using multiple sensor platforms, in *IEEE Swarm Intelligence Symposium*, pages 309–316, Pasadena, CA.

— (2007) Coordinated patterns of unit speed particles on a closed curve, *Systems & Control Letters*, **56**(6), 397–407.

Zheng, Z., Spry, S. C., and Girard, A. R. (2008) Leaderless formation control using dynamic extension and sliding control, in *IFAC World Congress*, pages 16027–16032, Seoul, Korea.

Zhong, M. and Cassandras, C. G. (2008) Distributed coverage control in sensor network environments with polygonal obstacles, in *IFAC World Congress*, pages 4162–4167, Seoul, Korea.

Zhu, M. and Martínez, S. (2008a) Dynamic average consensus on synchronous communication networks, in *American Control Conference*, pages 4382–4387, Seattle, WA.

— (2008b) On the convergence time of distributed quantized averaging algorithms, in *IEEE Conference on Decision and Control*, pages 3971–3976, Cancún, México.

Algorithm Index

Subject Index

Symbol Index

Symbol	: Description and page(s) when applicable
γ_{arc}	: arc-length parametrization, 5
$O(g)$	: big O Bachmann–Landau symbol, 3
$\Omega(g)$	: big Omega Bachmann–Landau symbol, 3
$\Theta(g)$	: big Theta Bachmann–Landau symbol, 3
∂S	: boundary of the set S, 1
$|S|$	: number of elements of the finite set S, 1
$S_1 \times S_2$	: Cartesian product of S_1 and S_2, 2
$\prod_{a \in A} S_a$	: Cartesian product of the collection of sets $\{S_a\}_{a \in A}$, 2
S^n	: Cartesian product of n copies of S, 2
S_δ	: δ-contraction of S, 97
δ^S	: symmetric difference, 252
$\phi : \mathbb{R}^d \to \mathbb{R}_{\geq 0}$	: density function on $\mathbb{R}^d$, 101
$\emptyset$	: the empty set, 1
$G \cap G'$	: intersection of graphs G and G', 21
$G \cup G'$	: union of graphs G and G', 21
$\mathbb{G}(S)$	: set of all undirected graphs whose vertex set is an element of $\mathbb{F}(S)$, 104
$H_{p,q}$	: closed halfspace defined by p and q, 96
$H_S(v)$	: internal tangent halfplane of v with respect to S, 98
$[a, b]$	: closed interval between the numbers a and b, 2
$]a, b[$	: open interval between the numbers a and b, 2
$f : S \to T$	: map f from set S to set T, 2
$f \circ g$	: composition of the maps f and g, 2
f^{-1}	: inverse map of a function f, 2
$f^{-1}(x)$	: level set of a function f corresponding to a value x, 2
T_f	: overapproximation map associated to a time-dependent evolution f, 20
$h : S \rightrightarrows T$	: set-valued map h from set S to set T, 2
$A > 0$	: a symmetric positive definite matrix A, 6

$w^{[i]}$: state of processor i, 38

$w_0^{[i]}$: initial state of processor i, 39

$W^{[i]}$: state set of processor i, 38

$W_0^{[i]}$: set of allowable initial values for processor i, 38

$r_{\exp}(A)$: exponential convergence factor of $A \in \mathbb{R}^{n \times n}$, 59

Σ^{b} : body reference frame, 151

Σ^{fixed} : fixed reference frame, 151

$[p, q]$: closed segment with extreme points p and q, 96

$]p, q[$: open segment with extreme points p and q, 96

rbt-sns : $\mathbb{R}^d \to \mathbb{A}_{\mathrm{rbt}}$:
  sensing function, 153

env-sns : $\mathbb{P}(\mathbb{R}^d) \to \mathbb{A}_{\mathrm{env}}$:
  environment sensing function, 153

$\{S_a\}_{a \in A}$: collection of sets indexed by the index set A, 2

$x \in S$: x is an element of the set S, 1

$R \subset S$: R is a subset of S, 1

$R \subsetneq S$: R is a strict subset of S, 1

$S_1 \cap S_2$: intersection of sets S_1 and S_2, 2

$\cap_{a \in A} S_a$: intersection product of the collection of sets $\{S_a\}_{a \in A}$, 2

$S_1 \cup S_2$: union of sets S_1 and S_2, 2

$\cup_{a \in A} S_a$: union of the collection of sets $\{S_a\}_{a \in A}$, 2

e_i : the vector in $\mathbb{R}^d$ whose entries are zero except for the ith entry, which is one, 2

$\mathbf{1}_d$: the vector in $\mathbb{R}^d$ whose entries are all equal to one, 2

$\mathbf{1}_{d-}$: shorthand for $(1, -1, 1, \ldots, (-1)^{d-2}, (-1)^{d-1}) \in \mathbb{R}^d$, 65

$\mathbf{0}_d$: the vector in $\mathbb{R}^d$ whose entries are all equal to zero, 2

$\mathcal{X}_{\mathrm{disk}}(p^{[i]}, p^{[j]})$: pairwise connectivity constraint set of agent at $p^{[i]}$ with respect to agent at $p^{[j]}$, 182

$\mathcal{X}_{\mathrm{disk}}(p^{[i]}, \mathcal{P})$: connectivity constraint set of agent at $p^{[i]}$ with respect to $\mathcal{P}$, 184

$\mathcal{X}_{\mathrm{disk},\mathcal{G}}(p^{[i]}, \mathcal{P})$: $\mathcal{G}$-connectivity constraint set of agent at $p^{[i]}$ with respect to $\mathcal{P}$, 186

$\mathcal{X}_{\mathrm{vis\text{-}disk}}(p^{[i]}, p^{[j]}; Q_\delta)$:
  line-of-sight connectivity constraint set in Q_δ of agent at $p^{[i]}$ with respect to agent at $p^{[j]}$, 189

$\mathcal{X}_{\mathrm{vis\text{-}disk}}(p^{[i]}, \mathcal{P}; Q_\delta)$:
  line-of-sight connectivity constraint set in Q_δ of agent at $p^{[i]}$ with respect to $\mathcal{P}$, 189

$\mathcal{X}_{\mathrm{lc\text{-}vis\text{-}disk}}(p^{[i]}, \mathcal{P}; Q_\delta)$:
  locally cliqueless line-of-sight connectivity constraint set in Q_δ of agent at $p^{[i]}$ with respect to $\mathcal{P}$, 190